DRIVE

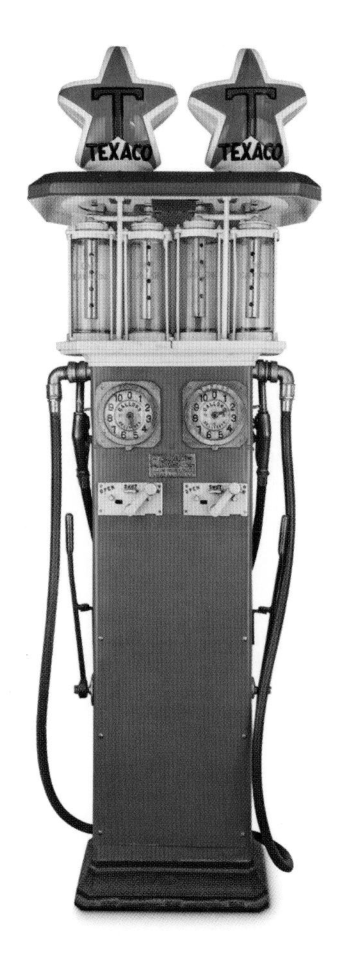

DRIVE

THE DEFINITIVE HISTORY OF MOTORING

Foreword Jodie Kidd
Editor-in-chief Giles Chapman
Contributors Andrew Noakes, Chris Rees, Martin Gurdon, Richard Truett, Sam Skelton, Richard Bremner, Peter Nunn, Simon Heptinstall, Alexandra Black

DK

DK LONDON
Senior Editors Chauney Dunford, Hugo Wilkinson, Andrew Szudek
Senior Art Editors Stephen Bere, Nicola Rodway, Anthony Limerick
Editors Katie John, Helen Ridge
Editorial Assistant Kate Taylor
Picture Researchers Sarah Smithies, Nic Dean
Managing Editor Gareth Jones
Senior Managing Art Editor Lee Griffiths
Producer, Pre-Production Andy Hilliard
Senior Producer Mandy Inness
Jacket Designer Surabhi Wadhwa
Design Development Manager Sophia M.T.T.
Jacket Editor Claire Gell
Associate Publishing Director Liz Wheeler
Art Director Karen Self
Publishing Director Jonathan Metcalf

DK DELHI
Project Art Editor Vikas Chauhan
Art Editor Jomin Johny
Assistant Art Editor Anukriti Arora
Project Editors Arani Sinha, Virien Chopra
Editor Madhurika Bhardwaj
Assistant Editor Devangana Ojha
Managing Editor Soma B. Chowdhury
Managing Art Editor Arunesh Talapatra
Senior Cartographer Subhashree Bharati
Cartographer Reetu Pandey
Cartography Manager Suresh Kumar
Picture Researchers Sumedha Chopra, Nishwan Rasool
Picture Research Manager Taiyaba Khatoon
DTP Designers Pawan Kumar, Vijay Kandwal, Nand Kishor Acharya
Production Manager Pankaj Sharma
Pre-production Manager Balwant Singh
Senior Jackets DTP Designer Harish Aggarwal
Jacket Designers Suhita Dharamjit, Tanya Mehrotra
Jackets Editorial Coordinator Priyanka Sharma
Managing Jackets Editor Saloni Singh

First published in Great Britain in 2018 by
Dorling Kindersley Limited
80 Strand, London, WC2R 0RL

10 9 8 7 6 5 4 3 2 1
001–308093–April/2018

A CIP catalogue record for this book is available from the British Library.

ISBN: 978-0-2413-1766-2

Printed and bound in China

A WORLD OF IDEAS:
SEE ALL THERE IS TO KNOW

www.dk.com

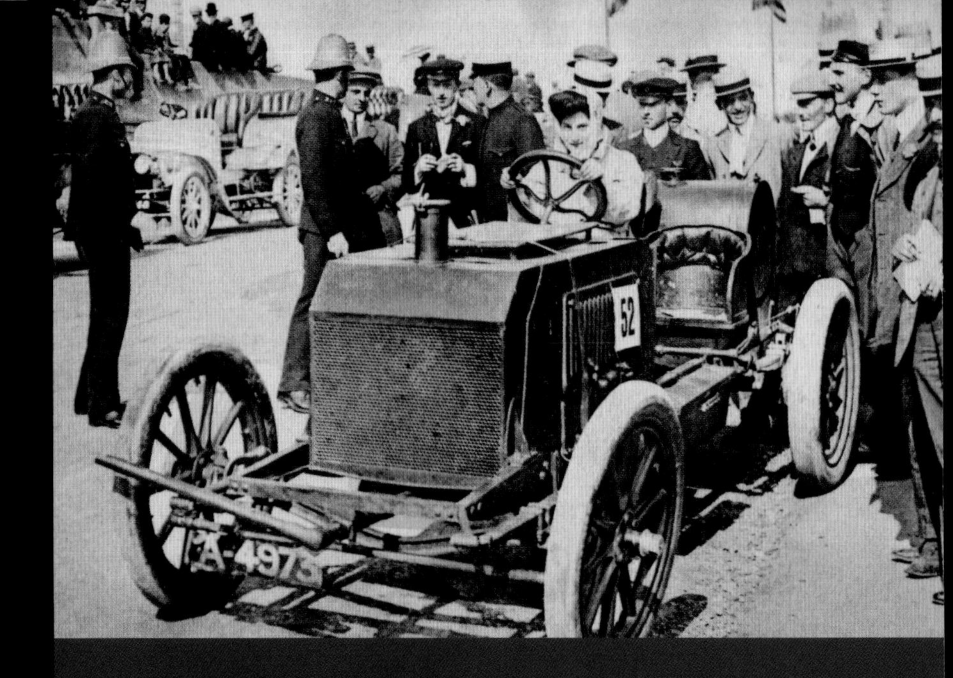

CONTENTS

EDITOR-IN CHIEF

Giles Chapman is an award-winning journalist and author of more than 40 books on a wide spectrum of car subjects. *Drive* is the third major title he has directed for DK after the internationally successful *The Car Book* (2011) and *The Classic Car Book* (2016), bringing his knowledge of car culture, history, and industry from 35 years in the motoring media. He was editor of *Classic & Sports Car*, the world's best-selling classic car magazine, and since 1994 he has contributed to dozens of major newspapers and magazines. Besides his own books – which include *My Dad had One Of Those*, *Chapman's Car Compendium*, and *Britain's Toy Car Wars* – he has worked as an advisor to many other authors and publishers. He founded the Royal Automobile Club's Motoring Book Of The Year Award, and appears regularly on TV and radio commenting on motoring and automotive industry issues.

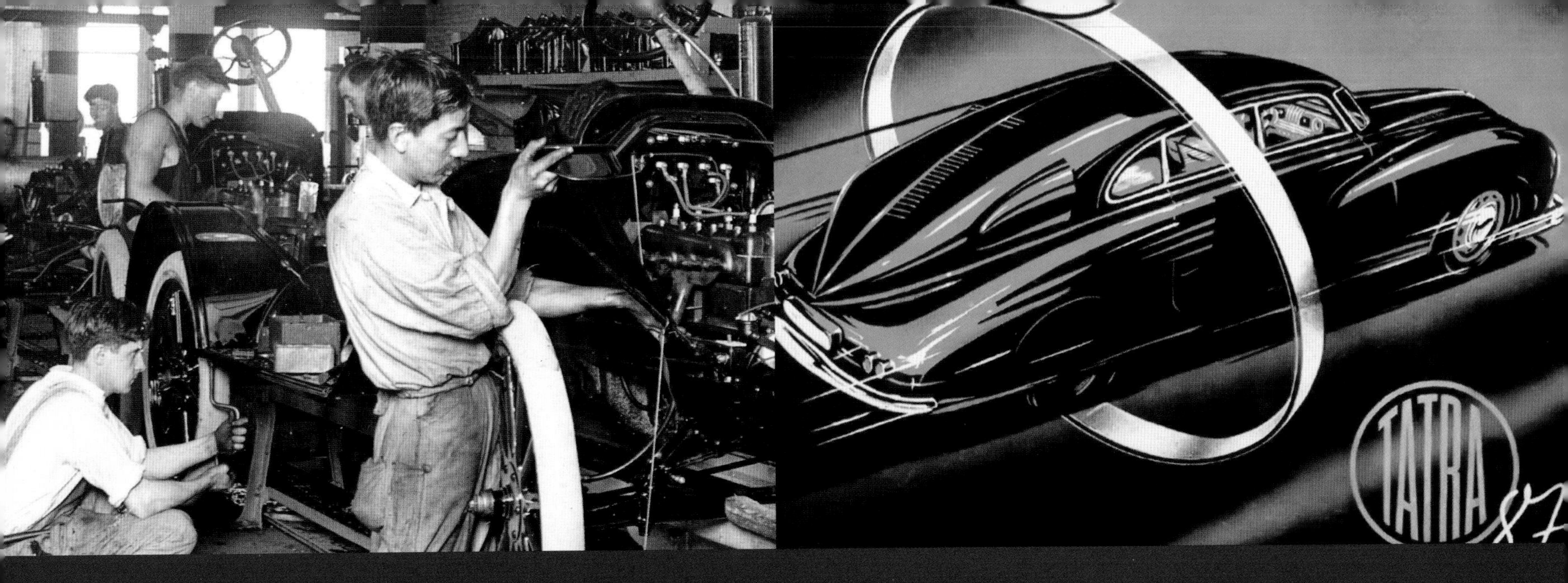

2

AN INDUSTRY IS BORN
1906–1925

3

SPEED, POWER, AND STYLE
1926–1935

4

THE CAR COMES OF AGE
1936–1945

5

REBUILDING THE WORLD
1946–1960

6

TECHNOLOGY AND SAFETY
1961–1980

7

THE CHANGING WORLD
1981–2000

8

DRIVING INTO THE FUTURE
2001–PRESENT

GREAT DRIVES

Foreword

From the race-track to the driveway, the car is a uniquely appealing, iconic, and useful machine. But where did it come from, and where is it going?

The story of the car begins with the pioneers of mechanized transport – far-sighted inventors who dreamed of replacing the horse-driven carriage with machines. It finds itself now on the cusp of a brave new world in which those machines might even do the driving for us.

But the story of motoring is not simply about how the car has evolved. It is about how it has come to define the world around us. Cars have become faster, safer, more reliable, more comfortable; they have opened up new ways of living and working; they have given us a freedom to move that we so often take for granted.

Cars are so much more than just a means of getting from A to B. At their best, they are both feats of engineering and works of art. A good car has the capacity to stir the emotions, delight the senses, and nourish the soul.

The story of the car, then, is the story of technology and industry, of romance and glamour, of full-throttle excitement and the thrill of the open road.

Sit back, buckle up, and enjoy the ride.

JODIE KIDD

DINERS
MOYNAT
MOYNAT

INVENTING THE CAR

1885–1905

Inventing the car

Glance between the pioneer automobiles of the late 19th century and the slick, targeted consumer products of today and it is almost impossible to believe that they are one and the same thing: the car. It is a very big leap indeed.

Nowadays, the built environment seems to be absolutely subordinated to serving the motor vehicle. However, in the beginning, townscapes and countrysides across the world were ill-prepared for the self-propelled, individually directed vehicles that would shortly – and unremittingly – be imposed on these communal spaces.

But for the advent of the reliable, rapid, and widely popular railway between 1820 and 1850, the transport revolution of motoring might have progressed a whole lot sooner. Steam power applied to the roads did not work anything like as well, and there was an interval of some 125 years between the earliest (in effect) boilers-on-wheels hissing at French citizens and the invention in Germany of the small internal combustion engine that could finally make a carriage "horseless".

An uncertain reception

Across wider society, there was less love and more loathing for the first automobiles. In the UK, they were restricted to virtually walking pace, and in France road racing was banned after a series of accidents in which out-of-control drivers, ignorant spectators, and even wandering cattle collided, with devastating results. Across the rural vastness of the US, there were precious few roads consisting of anything but dirt. So it was little wonder that, while Europe busied itself making playthings for the rich, American car designers concentrated on mechanical durability and the

PIONEERING MOTORISTS IN THE 19TH CENTURY

BATTERY-POWERED VEHICLES SHOW EARLY PROMISE

"Across the **rural vastness** of the US, there were **precious few roads**..."

ability of their products to conquer long distances without breaking into pieces. Indeed, ambitious individuals, partnerships, and companies everywhere faced tremendous problems in turning automobile designs into manufactured products. It was an era of hand tools, and of trial and error. Perhaps surprisingly, electricity was equal to petrol in these early years as a source of motive power – not for concerns about air pollution, but for sheer ease and dependability of use.

Reliability and refinements

Once the nascent mechanical technology had mostly been mastered, carmakers and their customers turned to creature comforts. The advent of the pneumatic tyre was a huge advance in improving the motoring experience. Further improvements came thick and fast – from attention to the layout of cars to make them safer and easier to handle, to solid improvements in lighting and weather protection. This all helped to broaden the appeal of the automobile from its place as a Sunday afternoon diversion into something more significant.

Mass car commuting, though, was something no-one had yet considered. For one thing, motor vehicles were simply not trustworthy enough. That was why so-called "reliability trials" rose to prominence much earlier than competitive racing as a way to break down widespread public scepticism. Slowly, however, the increasing versatility and performance of the internal combustion engine meant that the automobile crept into every sphere of modern life. Something unstoppable was clearly afoot.

RELIABILITY TRIALS BRING IMPROVEMENTS TO CAR DESIGN

MOTOR RACING SOON CAPTURES THE PUBLIC'S IMAGINATION

Before the car

The need to travel long distances over land has been a part of human history for millennia, from the time of the first roads to the birth of mechanized transport.

The earliest roads emerged as well-trodden routes through the untamed landscape towards sources of food and water. They also became links between settlements and, with the invention of the wheel around 7,000 years ago, they began to be used for trade and commerce.

The first roads

The first paved roads originated in the Indian subcontinent and Mesopotamia in around 4000 BCE. Later, the Romans pioneered multi-layer road surfaces and purpose-built, direct routes as they grew their empire. These roads were used mainly for moving troops, but also by two-wheeled carts and chariots, and four-wheeled wagons pulled by oxen – the first traffic. Roman roads were built with crushed stone bedrock, which drained water efficiently.

Centuries later, British surveyor Thomas Telford added a camber to road surfaces to drain standing water on either side. In the early 19th century, Scottish engineer John McAdam introduced uniform-sized stones to create a smooth, robust top surface. His process was called "macadamization" and spread to the US and Australia. However, the modern road was created in the early 20th century when tar was added to the mix as a binding agent. This became known as "tar-macadamization", or "tarmac".

Horses and pedestrians

Horse-drawn vehicles remained the main form of transport on roads until the late 19th century. However, working horses were also a threat to public health: in New York in 1898, there were 200,000 horses plodding the streets, each one producing around 24 lb (11 kg) of manure every day. It is little wonder the city staged the world's first urban planning conference that year, with sanitation issues among the main concerns. As well as horses, the road environment in villages, towns, and cities was the domain of the pedestrian. As late as the 1890s, horse-drawn

△ **Hornellsville Erie Pennsylvania Railway, US, 1874**
Steam-powered passenger transport on the road could not compete with the speed, reliability, and safety offered by the railways.

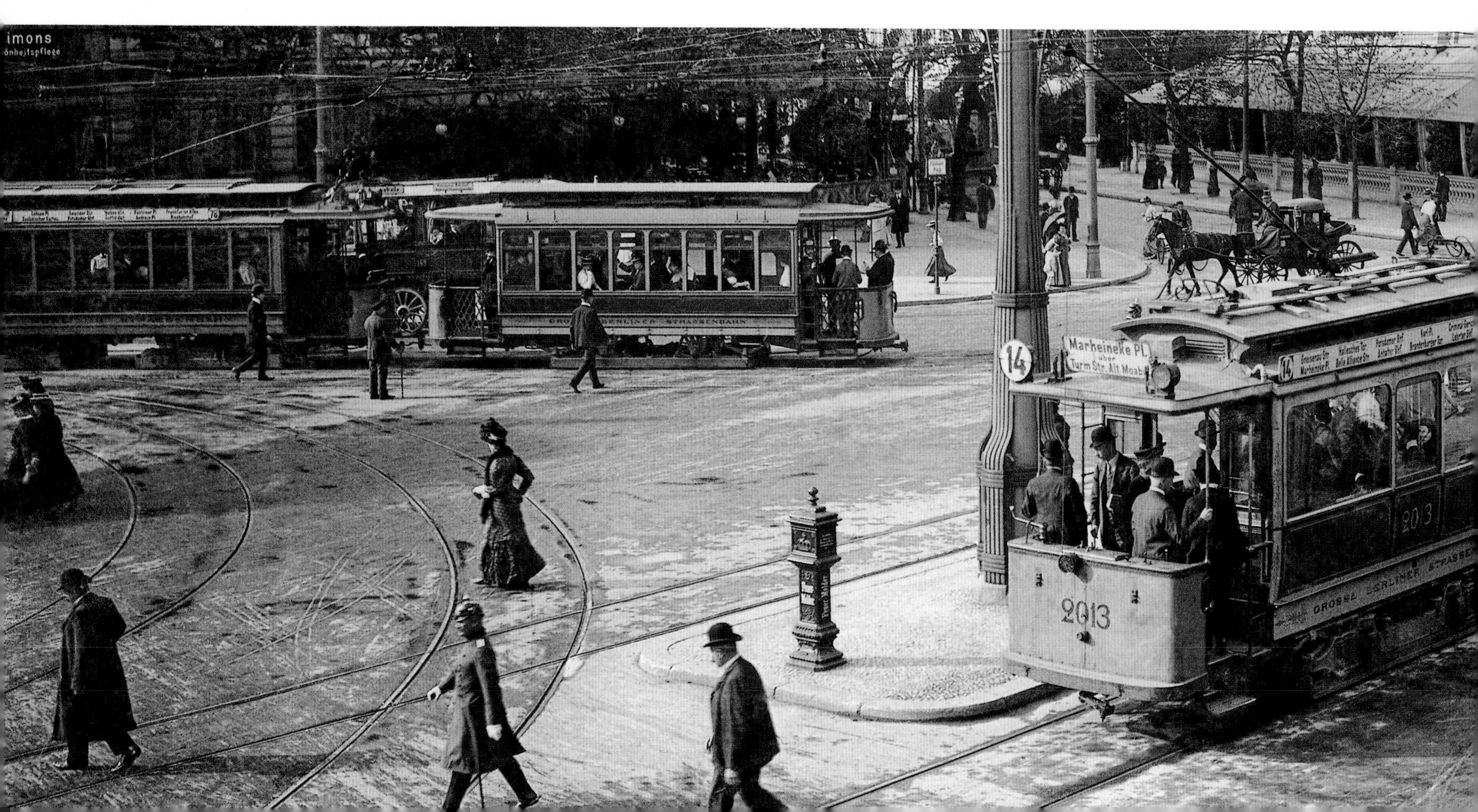

▽ **Potsdamer Platz in Berlin, Germany, 1908**
This scene was typical of city centres even after motor vehicles gained acceptance. People on foot had little to fear from the few large vehicles using the roads, and the horse was still the primary means of transport for traders.

vehicles on US streets had to avoid the people who thronged both roads and pavements. Street vendors used the thoroughfares as a marketplace, and children played on them without fear. It is perhaps not surprising that the earliest motor vehicles posed a threat to the bustle of street life. The US auto industry later coined the term "jaywalkers" for pedestrians obstructing traffic – a "jay" was a country bumpkin – as it sought to dominate the roads for its products.

The dawn of mechanization

The first steam-powered roadgoing passenger vehicles were lumbering, self-propelled coaches with industrial-sized boilers as powerplants. As well as being unwieldy, they faced various setbacks. In the UK, due to the threat to horse-drawn transport, the 1865 Locomotives on Highways Act (see pp.22–23) restricted the top speed to 4 mph (6.5 km/h), which is little faster than walking pace. The act also stipulated that three people must always be in attendance: one to steer, one to stoke, and another to walk 60 yards (55 m) ahead holding a red flag. There were also concerns for public safety: in the US, a steam carriage mowed down a mother and child in Cleveland, creating a public outrage. Rumours of steam carriages exploding also spread fear. Another setback to personal mechanized transport came in the form of railway technology, which enjoyed stellar growth in this period in Europe and the US. Fast, efficient, and affordable, trains quickly took business from steam-powered road vehicles, while tram services catered for the needs of city dwellers travelling short distances. As the 20th century drew nearer, it seemed that the horse, as well as the railway, was here to stay.

◁ **Hansom cab, London, late 19th century**
Small enough to be pulled by a single horse, the hansom cab was a popular form of transport in Europe and the US.

> "**People will travel**... by steam engines... almost **as fast as birds fly**, 15 or 20 miles **in an hour**."
>
> OLIVER EVANS, AMERICAN INVENTOR, 19TH CENTURY

KEY DEVELOPMENT

The bicycle craze

The bicycle had uncertain beginnings. German manufacturers originated the first early, impractical designs in 1817 with the scoot-along "Dandy Horse", leading to the crude, uncomfortable "bone-shaker", and the positively dangerous, large-wheeled "penny farthing". However, J.K. Starley from Coventry came up with the concept of frame-mounted pedals driving the rear wheel via a chain. This machine became known as the "safety bicycle" and was a vast improvement on the earlier versions. In 1888, Scotsman John Boyd Dunlop invented the first pneumatic tyres, which could be fitted to the bicycle, and cycling boomed. It caused a worldwide revolution in personal mobility – the concept of a personally-owned machine that could take its operator wherever he or she wanted to go in speedy comfort.

THE "SAFETY BICYCLE" **WAS ONE OF THE FIRST FORMS OF MECHANICAL PERSONAL TRANSPORT. THEY WERE VIVIDLY ADVERTISED.**

△ **A photograph from 1886** shows Gottlieb Daimler being driven by his son, Paul, in the world's first four-wheeled car.

Internal combustion

After a tentative start in the early 19th century, development of the internal combustion engine accelerated from the 1870s thanks to Karl Benz and Gottlieb Daimler. These men produced the first motorized, petrol-fuelled vehicles in 1885.

KEY EVENTS

- **1807** The de Rivaz engine is patented.
- **1860** The Lenoir engine enters production.
- **1872** Gottlieb Daimler joins Nikolaus Otto at Gasmotoren-Fabrik Deutz AG.
- **1876** Nikolaus Otto commercializes the compressed-charge four-stroke engine.
- **1880** Daimler is fired from Deutz, and sets up with designer Wilhelm Maybach.
- **1883** Daimler and Maybach patent Daimler's Dream – a high-speed, horizontally opposed, compressed-charge, four-stroke engine.
- **1885** Benz installs a two-stroke engine into a three-wheeled vehicle, creating the first car.
- **1890** Daimler-Motoren-Gesellschaft (DMG) is founded to make and sell engines.
- **1892** Daimler and Maybach are ousted from DMG (although Daimler is reinstated in 1895).
- **1894** The 161st Benz Motorwagen is sold.
- **1895** Daimler makes its 1,000th engine.

***GOTTLIEB DAIMLER*, PICTURED IN A HAND-COLOURED LITHOGRAPH OF 1910.**

The Franco-Swiss inventor Isaac de Rivaz may well have invented the internal combustion engine and been the first to use it in a vehicle. However, Belgian Étienne Lenoir patented the first functioning internal combustion engine in 1858. This engine was noisy and inefficient, and turned noisier still with extended use, but it prompted *Scientific American* magazine to announce, if prematurely, that the age of steam was over.

Germany's Nikolaus Otto built his first internal combustion engine in 1861. By 1876 this had evolved into a "four-stroke" that compressed, ignited, combusted, and exhausted the petrol-air mixture. Alphonse Beau de Rochas first patented the four-stroke engine in 1862, but only Otto successfully manufactured an engine, using his "Otto Cycle". It was quieter, more efficient, and more reliable than the Lenoir engine, and sold over 30,000 in 10 years – until Otto's patent was revoked in favour of de Rochas' in 1886.

Daimler and Benz

Engineer Gottlieb Daimler worked with Otto before setting up with designer Wilhelm Maybach in 1880. Only 60 miles (96 km) away, Karl Benz was also working with petrol engines; having patented his tiny two-stroke single-cylinder engine in 1879, he fitted it to a three-wheeler to create the first internal combustion car in 1885. His wife Bertha later took it, supposedly without his knowledge, and successfully drove it 65 miles (105 km) from Mannheim to Pforzheim.

Despite the achievement, the Benz Patent Motorwagen attracted widespread ridicule for moving at no more than walking pace. In the same year, Daimler fitted his superior engine to a bicycle. The two men never met, but their names would eventually be joined to create Daimler-Benz in 1926.

△ **Gottlieb Daimler's home workshop**
Daimler and Wilhelm Maybach developed the first petrol-fuelled internal combustion engine in 1883. Here, a later engine, dubbed the "Grandfather Clock", is installed in a bicycle in 1885.

▷ **The first internal combustion engine**
This engine was developed by Belgian Étienne Lenoir in 1860. Most of the 400–500 engines that were made were used as stationary power units for printing presses and machine tools, but a few appeared in roadgoing vehicles.

Motoring into business

Licences for new automobile technology began to change hands while enhancements to vehicles continued and communications across the globe developed. France became the first country to start a manufacturing industry as the car turned into a consumer product.

The horseless carriage was barely a credibly functioning machine before entrepreneurs spotted its profit potential. Only five years after Karl Benz had developed his internal-combustion-engine-propelled three-wheeler in 1885, a rival engine-maker, Gottlieb Daimler, had gained the backing of financiers to form Daimler-Motoren-Gesellschaft (DMG). The company's aim was to manufacture and sell these fundamentally superior internal combustion engines, and the first licence that allowed local manufacture went to France's Panhard et Levassor.

DMG went on to sell more licences, with one going to the Steinway piano company, and another to bicycle and pepper-grinder manufacturer Peugeot. More significantly, a commercial arrangement was reached with Hamburg-born Briton Frederick Simms.

△ **Pierre-Alexandre Darracq, 1901**
Automobiles Darracq France produced more than 10 per cent of all cars in France by 1904, helping the company to develop licensing partnerships in the UK.

Seizing control

An engineer, Simms met Daimler in 1889, and subsequently imported one of his cars to the UK. By 1893 Simms had established a British Daimler subsidiary to sell both Daimler and Panhard cars.

Soon Simms began producing cars in Coventry, UK, after having sold the Daimler rights to the calculating entrepreneur Harry Lawson and his company, the British Motor Syndicate, which retained Simms as a consultant.

The syndicate, founded in 1895, attracted carmaker Thomas Humber and other luminaries into the evolving industry, plus substantial funding too. Lawson's aim was to gain control over all of the car manufacturing in the UK, but he was convicted of fraud before he could do so. Simms detached himself from the company to lead a successful motor industry career, but Lawson's dishonest enterprise damaged the British motor industry's early development.

Car as commodity

Meanwhile, the German and French industries were thriving. While Germany made the motorcar concept feasible, it was the French who refined and popularized it. Armand Peugeot's 1896 factory made Peugeot France's largest carmaker by 1913, selling 10,000 cars a year.

A greater influence on car design was Émile Levassor, who licence-built Daimler engines with his friend René Panhard. By 1890 they were making

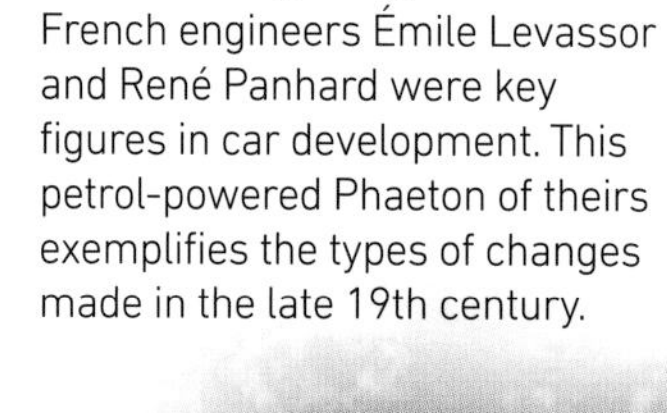

▷ **Innovating design**
French engineers Émile Levassor and René Panhard were key figures in car development. This petrol-powered Phaeton of theirs exemplifies the types of changes made in the late 19th century.

Flared back mudguards alongside polished black bodywork

Gear lever connected to steering column

Engine has twin vertical cylinders

◁ **Darracq 12 hp, 1904**
Shipped to London in 1905, this particular Darracq eventually rose to fame in the 1953 film *Genevieve*. In the story it took part in the London to Brighton Veteran Car Run, and so came to represent the generic veteran car.

◁ **Innovative design**
This inventive three-wheeled voiturette – created and patented by Léon Bollée Automobiles – seats the passenger, here Louis Paul, in the front, with the driver, Charles Rolls (partner of motorcar company Rolls-Royce) behind.

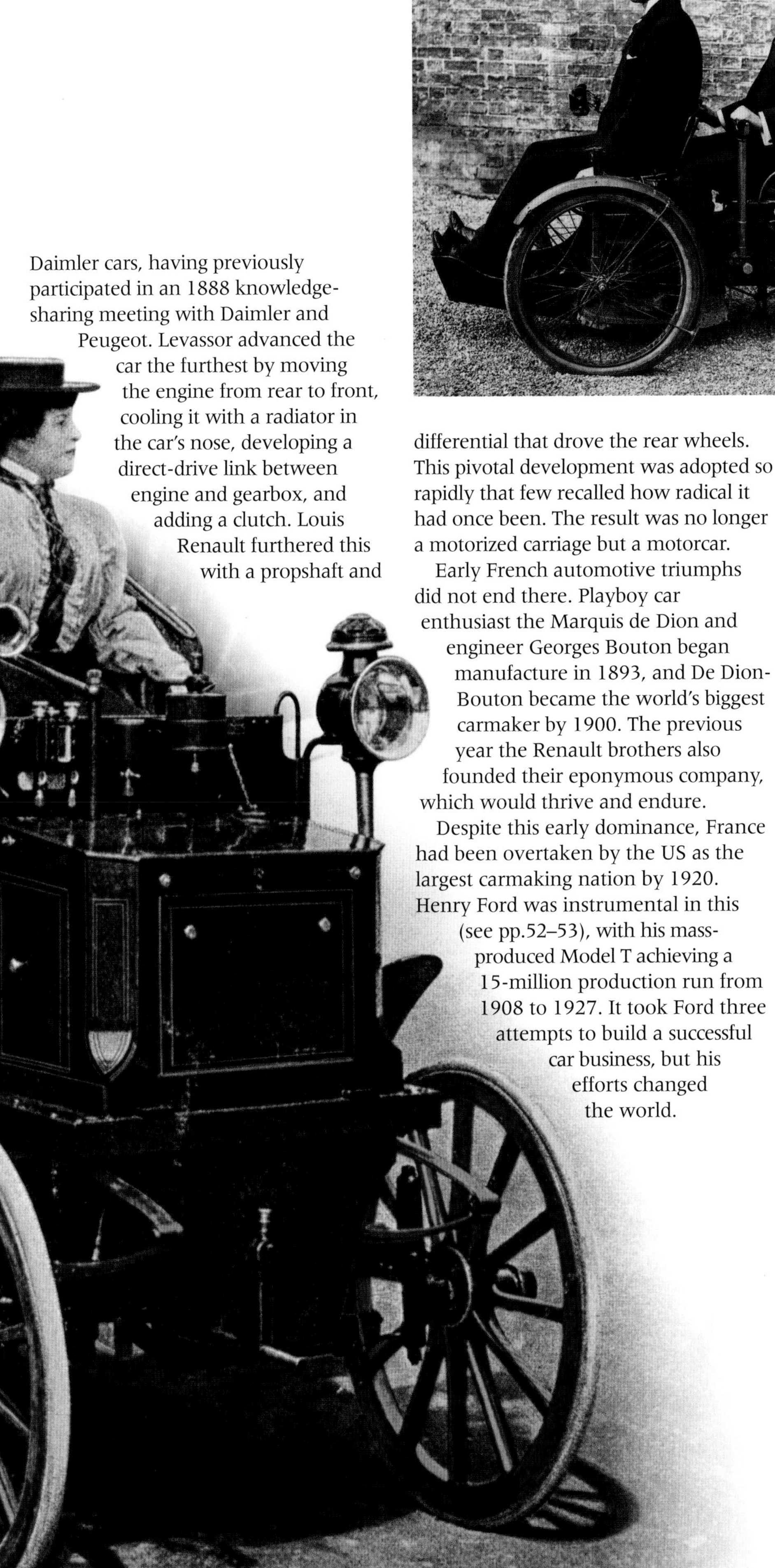

Daimler cars, having previously participated in an 1888 knowledge-sharing meeting with Daimler and Peugeot. Levassor advanced the car the furthest by moving the engine from rear to front, cooling it with a radiator in the car's nose, developing a direct-drive link between engine and gearbox, and adding a clutch. Louis Renault furthered this with a propshaft and differential that drove the rear wheels. This pivotal development was adopted so rapidly that few recalled how radical it had once been. The result was no longer a motorized carriage but a motorcar.

Early French automotive triumphs did not end there. Playboy car enthusiast the Marquis de Dion and engineer Georges Bouton began manufacture in 1893, and De Dion-Bouton became the world's biggest carmaker by 1900. The previous year the Renault brothers also founded their eponymous company, which would thrive and endure.

Despite this early dominance, France had been overtaken by the US as the largest carmaking nation by 1920. Henry Ford was instrumental in this (see pp.52–53), with his mass-produced Model T achieving a 15-million production run from 1908 to 1927. It took Ford three attempts to build a successful car business, but his efforts changed the world.

"Cars **don't** have a **homeland... like classic love**, they can **easily cross borders**."

ILYA EHRENBURG, *THE LIFE OF THE AUTOMOBILE*

KEY DEVELOPMENT

You say motorcar, I say automobile

In around 1895, "automobile", or "véhicule automobile" in full, was the term used by the French, after surpassing the popular word "locomobile". In English, however, "motorcar" or "autocar" were the words most frequently used. The "auto" element was derived from the Greek for "self", while "mobile" was French for "moving".

The etymology of "car" goes back much further, to around 1300, when a "carre" was an Anglo-French term for a wheeled vehicle, sometimes a chariot. In the US, "car" tended to refer to a railway carriage, but "automobile" was cemented when it was mentioned in an 1899 editorial in *The New York Times*.

THE WORLD'S FIRST AUTOMOBILE **WAS NICOLAS-JOSEPH CUGNOT'S FARDIER À VAPEUR, CREATED IN 1770. IT IS NOW HOUSED IN PARIS.**

▽ La Mouche ("The Fly"), 1900
This Art Nouveau car advertisement by artist Francisco Tamagno conveys the freedom and exhilaration of the road, while showing the practical modern features of the time, including a horn and headlamps.

ORETTE

A MOUCHE"

TAMAGNO

CONSTRUCTEURS

MORET & Cie. LYON-VAISE

The freedom of the road

For those who could afford them, the very first cars offered unheard-of freedom of movement. However, that freedom came at a price, and lawmakers were sometimes slow to keep up with quickly advancing vehicle technology.

Although true cars did not appear until the 1880s, powered road vehicles had existed for decades. They were often big and bulky: steam-driven carriages were a good example.

In the UK, an extensive but often poorly surfaced road network, accidents, and regular problems such as vehicles scaring horses led to three "Locomotive Acts" designed to control the use of mechanical vehicles (see box, opposite).

◁ **The right to motor**
Dorothy Elizabeth Levitt drives her Napier 80hp at the Brighton Motor (Speed) Trials on 21 July 1905.

An era of emancipation

When British motoring laws were relaxed in 1896, motorists celebrated. In honour of this brave new world, the Automobile Club arranged a drive from London to Brighton that it called the Emancipation Run. Thirty motorists took part in an event that is now known as the London to Brighton Run, which continues to this day.

Competitor Walter Arnold had the distinction of receiving the world's first speeding ticket under the old law, for reaching 8 mph (13 km/h) on Paddock Wood High Street. He was chased by a policeman on a bicycle, who fined him a shilling. Sadly, just weeks after the law was relaxed, Bridget Driscoll became the UK's first road fatality when a car struck her at London's Crystal Palace. The coroner at her inquest said he hoped this event "would never be repeated". Emancipation had its price.

Despite the hazards, the car was rapidly becoming established. In 1900, Claude Johnson organised the 1,000

Mile Trial, in which cars travelled all over the UK in a reliability test and demonstrated the capabilities of the "horseless carriage" to a wider public.

By 1903 cars were allowed to reach 20 mph (32 km/h), and drivers had to be over 17 and have licences (14-year-olds could ride motorcycles), although there was still no driving test. Cars needed registration numbers, brake tests, lights, and "an audible warning".

The motoring age

Cars were still the preserve of the rich because they were prohibitively expensive. In the US, however, mass-production techniques – first tried by Ransom Olds for his 1901 Oldsmobile Curved-Dash Runabout, but perfected by Henry Ford and his famous Model T from 1907 – soon enabled cars to be built and sold for a fraction of their former cost. As the 20th century progressed, thousands of people acquired their first cars, gaining instant personal mobility, and changing where they lived, worked, and spent their leisure time. By 1909, the UK had introduced a tax on petrol and on the vehicles themselves. To some this seemed the opposite of emancipation, but the money was spent on improving roads that were otherwise little more than dirt tracks on which cars created ruts and dust clouds that irritated pedestrians and motorists alike. This led to the arrival of smoother, cleaner, bitumen-coated roads.

◁ **The open road**
British Motoring Club secretary Charles McRobie Turrell (left) and Harry John Lawson, organiser of the London to Brighton run (right), enjoy their right to drive without a flag-waving chaperone in 1896.

Motoring heroine

Among the early competitive drivers was Dorothy Levitt, probably the first woman to become a racing driver. A secretary at the Napier Car Company, she soon proved to be a formidable competitor, winning her class in the 1903 Southport Speed Trials. She was reportedly the first woman to win a motor race outright on the Isle of Wight. Then, in 1905, driving from London to Liverpool and back in two days with her dog Dodo and a revolver, she achieved the longest continuous distance covered by a female driver. A keen exponent of "the woman's right to motor", Levitt was initially banned from driving at the Brooklands motor circuit when it opened in 1907 (female drivers were allowed on to its famous banked circuit the following year). Ultimately, though, she competed in a string of world-class motoring competitions, piloted power boats, and flew some of the first aircraft. For her, the car's power to emancipate was not in doubt.

◁ **Arrival in Brighton, 1896**
To celebrate the relaxation of motoring laws in 1896, pioneering car owners drove from London to Brighton, drawing huge crowds.

LIFE BEHIND THE WHEEL

Legal limitations

In 1865 the British Parliament introduced three "Locomotive Acts", limiting powered vehicle speeds to 4 mph (6.5 km/h) on the open road and 2 mph (3.2 km/h) in towns, and requiring cars to follow a man on foot with a red flag. The "Red Flag Act" became a subject of satire, and motor vehicle enthusiasts celebrated when it was repealed in 1896.

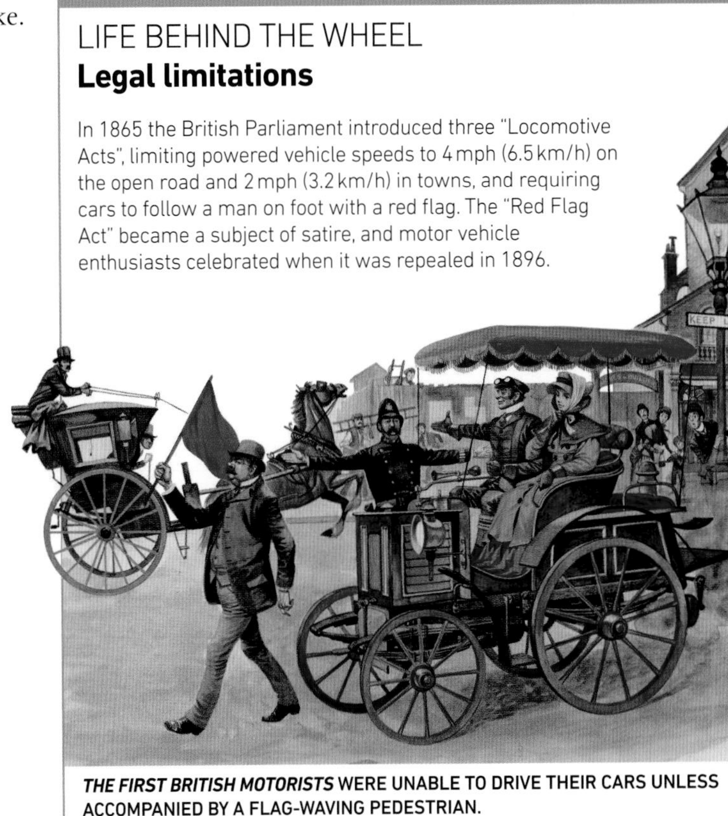

THE FIRST BRITISH MOTORISTS **WERE UNABLE TO DRIVE THEIR CARS UNLESS ACCOMPANIED BY A FLAG-WAVING PEDESTRIAN.**

The triumph of petrol

Some pioneering early cars were powered by steam or electricity, but in the first decades of the 20th century these were surpassed by the technically superior petrol-burning internal combustion engine.

When German engineer Karl Benz revealed the first petrol-powered car in 1885, steam road vehicles had already been around for decades.

△ **Electric cars**
This French poster from 1899 advertises fine coachwork for vehicles – including "autos electriques". Around this time, electric cars were just one of the options available for enterprising drivers.

Pioneering designs

Passenger-carrying steam-carriages were on the roads in the early 19th century, but they were killed off by railways for mass transport. By the late 1880s, however, there was a big industry in steam-powered cars, particularly in France, with firms such as Serpollet, De Dion-Bouton, and Peugeot. At one point the US boasted 125 steam car-makers, including Oldsmobile.

These steam vehicles were generally more reliable than their early petrol rivals, and easier to start (although they could take up to half an hour to fire up). In 1902 Americans actually bought more steam than petrol cars. By then electric cars were also gaining ground, again thanks to smooth performance and often greater reliability. In the 1910s they accounted for almost 40 per cent of US car sales. Electric and steam cars held the first world land speed records. The Jeantaud electric car got to 39 mph (63 km/h) in 1898; by 1900 it had reached 66 mph (106 km/h). In 1902 the Gardner-Serpollet "Easter Egg" steam car achieved 75 mph (121 km/h). This record was broken in 1906, when American Fred Marriott hit 127 mph (204 km/h) in his steam-driven Stanley Rocket. The following year he survived a near-150 mph (240 km/h) crash in a similar car.

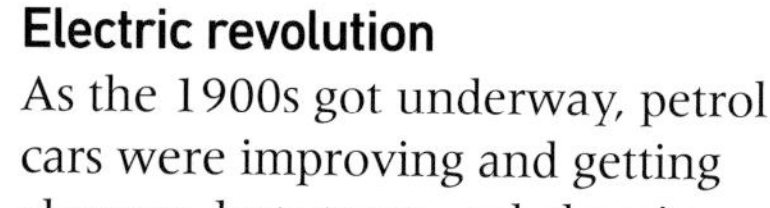

Electric revolution

As the 1900s got underway, petrol cars were improving and getting cheaper, but steam and electric cars were also carving their own niches in the market. The US was a centre for electric car production, with around 124 carmakers by 1912. Detroit Electric, which began in 1907, was making 13,000 cars a year at its peak. Electric vehicles tended to be popular in urban areas, where their limited range between charges – 80–100 miles

> "**Electricity** is the thing. There are no **whirring and grinding gears**"
>
> THOMAS EDISON

(130–160 km) – was less important because they travelled shorter distances. In cities such as San Francisco, doctors, particularly those who delivered babies, often preferred electric cars over their petrol counterparts because they were more likely to start and reach patients.

Electric cars were also popular with wealthy urban American women. Petrol cars had open roofs to prevent the build-up of fumes, but emission-free electrics could have closed cabins. This made them secure places to keep valuables. Some female electric car owners left their chauffeurs at home and drove themselves, collecting friends who welcomed the chance to socialize and have private conversations without the risk of chauffeurs reporting what they had heard to prying husbands.

These cars were beginning to supplant horses and carriages; some entrepreneurs bought redundant stables and turned them into electric car garages, where owners could park their cars overnight to recharge their batteries.

◁ **De Dion-Bouton steam car, 1894**
The French vehicle pioneer's second steam car took its inspiration from a horse and carriage. Later designs were quickly refined and became less cumbersome.

However, those lead/acid batteries were heavy. Thomas Edison, creator of the electric light bulb, invested in lighter nickel-iron batteries, but where these removed weight they added cost. A Detroit Electric with Edison's batteries cost $600 more than one without.

Electric and, increasingly, steam cars were already more expensive than their petrol rivals. The latter were becoming easier to drive and more reliable. Places to refuel and repair petrol cars were also becoming far more common.

Petrol pulls ahead

The arrival of a petrol Cadillac with an electric starter in 1912 added to the pressure on steam and electric cars. Electric starters were soon common; petrol cars, which could be fuelled quickly, with no waiting while water heated or batteries charged, were no longer at a disadvantage. They were also becoming cheaper in ways that electric and steam cars could not match, thanks to Henry Ford's mass-production system. Ford could make thousands of petrol Model Ts, which cost $850 each in 1909 and dropped to $260 by 1925. Despite efficiency gains, steam and electric cars could not compete on sales, and manufacturers began failing. When the price of commodities for batteries, such as lead, brass, and copper, doubled due to their use in armaments during World War I, many electric cars were priced out of existence, a fate that also befell steam cars.

◁ **Serpollet Easter Egg steam car, 1902**
This odd vehicle was an unlikely record breaker. In 1902, it became the world's fastest car, achieving 75 mph (121 km/h). However, this record was soon eclipsed by other steam vehicles.

DRIVING TECHNOLOGY

The Bersey taxi

The very first non-horse-drawn London taxi was powered by electricity. Named after its creator, Victorian entrepreneur Walter Bersey, the cab began work in 1897. Twelve of these 9-mph (14.5-km/h) vehicles were built, and became known as "hummingbirds" because of their yellow and black livery and distinctive sound.

However, the vehicles had their drawbacks. Shy passengers were not keen on being revealed by the electric interior lights, the cabs were slow, and after just two years cost and reliability issues killed them off.

BERSEY CABS **WERE IMMEDIATELY IDENTIFIABLE BY THEIR VIVID BLACK AND YELLOW COLOURING.**

Behind the wheel

Few of the earliest cars had interfaces that a modern driver would be familiar with. Pioneering motorists had to master a complex set of controls – but many left such technical matters to a professional.

Many of the earliest cars lacked steering wheels, but were steered using a tiller, which was little more than a lever at the top of the steering column. Other controls would be equally unfamiliar to modern eyes: brakes were usually operated by a long lever similar to that found on a horse-drawn carriage, and there might be two gear levers, one to engage drive and another to select the ratio required.

To warn other road users of the horseless carriage's approach, there was usually a horn that the driver sounded by squeezing a rubber bulb on the end of a bugle-style flared brass tube. Lighting was provided by carriage lamps, which burned acetylene gas, generated by dripping water from a reservoir at the top of the lamp on to chunks of calcium carbide loaded in at the bottom.

Drivers and passengers had to be hardy since few cars had fixed roofs, and if a folding top was provided it was often rudimentary. The dashboard was open to the elements, too, meaning that delicate instruments were vulnerable to water damage. As keeping an eye out for engine overheating was vital in early cars, this lent an element of unpredictability to even the shortest of journeys.

When it came to tuition, you were on your own, and car owners at this time often preferred to employ a professional chauffeur. Alternatively, sometimes a servant would drive, after being trained by the motor manufacturer. Mastering the pedal controls could be tricky, but when the three-pedal layout of accelerator, brake, and clutch became the industry-wide default, drivers eventually adapted to the "dance" involved.

◁ **Early Panhard et Levassor controls**
This 10-hp Panhard, pictured in 1903, belonged to automotive pioneer Charles Stewart Rolls. It features a remotely activated horn, an external brake lever, battery-powered headlights, and pedals.

▷ **Travelling in style**
A wealthy gentleman wearing driving gear poses for a photograph in his leather-upholstered car in the early 1900s. The comfortable seats were a typical feature of the luxury vehicles of the era.

△ **De Dion-Bouton 8hp Type O, 1902**
The leading French manufacturer followed prevailing design trends for this popular model, moving the engine to the front and fitting a steering wheel rather than a tiller.

△ **Cadillac Model A, 1903**
Henry Leland began selling these single-cylinder, four-seater cars in the thousands every year under his Cadillac brand. They were simply engineered and sturdy.

△ **Wolseley 6 hp, 1904**
This neat, single-cylinder car designed by Herbert Austin for Wolseley in Birmingham featured a modest 714 cc engine, yet was capable of reaching 25 mph (40 km/h).

Cars for the rich

Between 1901 and 1904, when the models pictured above were first introduced, a new car might have cost the equivalent of five years' earnings for the typical office worker. This put motoring well beyond the reach of anyone but the very rich. Even then, precious few car owners were everyday motorists, preferring to use their vehicles at the weekend as playthings for family outings. As most cars were open-top and exposed to the elements, exploring the surrounding countryside could be an invigorating pastime, and drivers and passengers alike had to dress appropriately for local weather conditions (see pp.32–33). However, snug and thickly padded leather seats emphasized the impression of lofty prosperity and luxurious exclusivity.

◁ **King Edward VII and his Daimler, 1900**
Wealthy landowners and aristocracts in the UK warmed to cars after the British monarchy bought a Coventry-built Daimler in 1900; this prestigious endorsement earned the car real respectability. The British royal family remained faithful Daimler customers throughout the 20th century, latterly thanks to the Queen Mother's patronage.

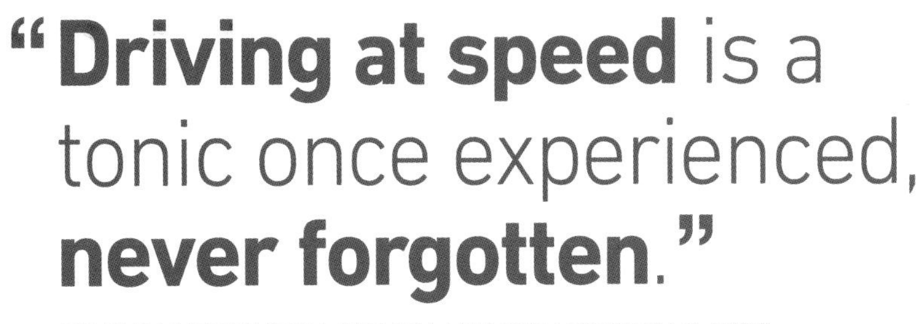

"**Driving at speed** is a tonic once experienced, **never forgotten.**"

DR F.W. HUTCHINSON, *HEALTH AND THE MOTOR CAR*, 1902

△ ***A Terrible Accident*, 1903**
Despite the initial enthusiasm for motor sport, a few early motor races ended in tragedy. The Paris–Madrid race of 1903 was perhaps the worst of these, with three spectators and five drivers killed – this print from French paper *Le Petit Journal* depicts one of the accidents. Among the fatalities in the race was Marcel Renault, one of the founders of the Renault car company. The French government responded by banning all open-road motor races.

Against the elements

Motoring was undoubtedly a thrill for pioneering drivers, but travelling any kind of distance in an early car could also be an endurance test for driver and passengers alike.

During the early days of the car, simply making progress along a road was an achievement in itself. If not thwarted by mechanical failure, a driver was likely to be undermined by poor-quality roads that regularly caused punctures and became wheel-swamping quagmires when it rained. Designed for horses, carts, and pedestrians, most roads of the time were barely adequate for motor vehicles, let alone cars capable of travelling at speed.

At the start of the 20th century, only France had a widespread network of metalled (gravel-surfaced) roads. In North America, less than 10 per cent of roads were surfaced in this way by 1903. Instead, across most of Europe and the US, motorists had to contend with rocks and roots – and, in the earliest days, hostile locals.

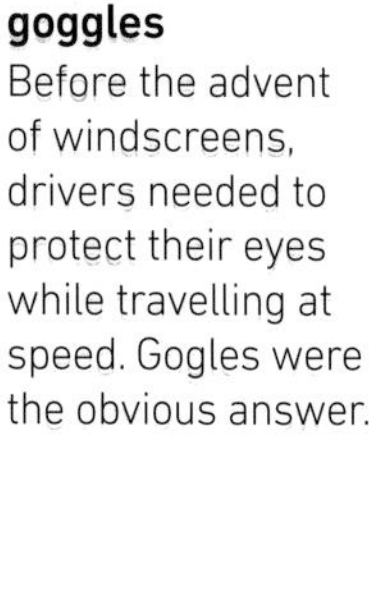

△ **Motoring goggles**
Before the advent of windscreens, drivers needed to protect their eyes while travelling at speed. Gogles were the obvious answer.

Early drivers frequently had to grapple with tyre changes caused by bad road surfaces. André Michelin developed a pneumatic car tyre as early as 1895, following Dunlop's pneumatic bicycle tyre of 1888, and their adoption by the motoring public was hastened by legislation outlawing solid tyres. Tubeless tyres were patented in 1903, and the demountable rim arrived a year later, enabling motorists to carry out roadside puncture repairs.

Comfort and protection

Since cars provided almost no protection from the elements, the weather could also be a challenge. Drivers and passengers had to wrap themselves up against cold, wind, rain, sun, and dust. Plenty of early vehicles had such marginal power-to-weight ratios that burdening them with the accoutrements of comfort risked turning them into immovable objects. However, as cars became more popular and more powerful, comfort and convenience came to the fore.

△ **Magazine illustration, 1904**
A chauffeur endures winter storms while his passenger sits in a snug cabin.

Driving after dark became increasingly necessary, not least because breakdowns and punctures often prolonged journeys beyond the hours of daylight. The candle lamps used by horse-drawn carriages were tried at first, but their range was insufficient for motoring, even while travelling at an uncertain 10 mph (16 km/h). Acetylene gas lamps proved more effective, their light provided by a self-contained chemical reaction.

KEY DEVELOPMENT

Large wheels and high ground clearance

One of the noticeable differences between early car designs and their modern counterparts is the prevalence of large wheels and high ground clearance. The cushioning effect of pneumatic rubber tyres was not enough to cope with early roads that had not been surfaced: many remote tracks were not graded, and even if they were, poor weather and traffic could soon destroy their surfaces. Cars rode high on big wheels to provide ground clearance over ruts, potholes, and stones, although these were also a test for the crude suspension systems of the time. However, even with larger wheels, carrying the tools to dig your car out was a necessity for the intrepid motorist.

MOTORING PIONEER C.S. ROLLS **DRIVES A PANHARD CAR IN 1903. LARGE WHEELS WERE COMMON AT THE TIME.**

The earliest electric lamps were provided on the 1898 Columbia Electric Car from Connecticut, US, while Peerless made electric headlights standard on their cars in 1908. The first set of headlights, sidelamps, and tail-lights was offered by the Pockley Automobile Electric Lighting Syndicate in 1908, and in 1912 Cadillac standardized a Delco system in tandem with electric ignition, effectively giving birth to the vehicle wiring loom. Dynamos generated the power for tungsten filaments, and similar systems appeared on European cars a year later. By 1912, electricity had replaced gas as standard.

◁ **Early protection, 1896**
C.S. Rolls drives his first car – an early Peugeot, which he bought at the age of 18. With its canopy and vertical windscreen, it offered some limited protection to its occupants despite its antiquated design.

Even in daylight, it was vital that drivers could see while the air rushed at their exposed faces. In the earliest days this was solved by wearing goggles (see far left), but it was not long before domestic glass was used in windbreakers. These were often two-piece constructions, with upper sections that could be folded back once they became dirty. Ford offered a windscreen, speedometer, and headlights as a $100 option on the $850 1908 Model T, while Oldsmobile was first to offer a windscreen along with a roof in 1915.

"**Be ready** to make your **tyre changes** quickly."

FIRESTONE TYRE ADVERTISEMENT, EARLY 1900s

▽ **Stuck in the mud, 1903**
Two US motorists attempt to use logs to lever their car out of a bog. In wet conditions, the US's dirt roads quickly became hazardous for drivers.

Built by hand

In the era before large-scale factories, early car companies produced everything by hand. Balancing technological innovation with creating a saleable product was tricky, but a few pioneering industrialists led the way.

△ **Poster advertising Hurtu cars and bicycles** Hurtu was a French bicycle firm, which, like many others, joined the car-making gold-rush in 1896 by producing licence-built and copied vehicles.

Before 1895, German and French engineers produced only a small number of vehicles each year – largely experimental, almost entirely handbuilt one-offs, each with small improvements on previous versions.

French company Panhard et Levassor became the first organised manufacturer in 1892, after buying an engine-making licence from Gottlieb Daimler. In its bustling Parisian workshops, forged metal components were pieced together in batches. By the end of 1894, it had made 90 cars. It also produced 350 engines, some of which were acquired under licence by the Peugeot family business of Montbéliard. The Peugeots had been making "consumer durables" – pepper and coffee grinders, chisels, bicycles, and more – for decades. Armand Peugeot was determined to build cars, making 29 of them in 1892, and 156 by 1898. With factory space and experience in mass production, Peugeot had a significant headstart; components were made in its foundries and its staff assembled cars individually. Meanwhile, in the US, Brothers Charles and Frank Duryea began American commercial car production in 1895, with 10 vehicles.

A slow start

After chassis frame assembly, or "setting out", cars left most early factories half-finished. Bodywork, seats, and fittings were added by outside companies called coachbuilders, each order being tailored to the customer, just as they were for horse-drawn carriages. Individual assembly, however, meant limited output. US manufacturer Ransom E. Olds realised this and patented the first production line to make his Oldsmobile Curved Dash, with parts delivered to assembly workers at fixed workstations, performing the same task repeatedly. This was the first truly mass-produced car: output soared from 425 vehicles in 1902 to 5,000 in 1905.

KEY EVENTS

- **1892** Former woodwork machine engineers René Panhard and Emile Levassor become the first manufacturers of series-produced cars, thus founding the motor industry.
- **1896** The UK's Daimler Company is established in a former mill in Coventry.
- **1899** Peugeot opens a dedicated factory at Audincourt and builds 300 cars – a quarter of France's entire annual output.
- **1899** Benz & Cie, Karl Benz's company, becomes the world's largest car-producer, making 572 vehicles in a year.
- **1899** Fiat is founded in Italy – Fabbrica Italiana Automobili Torino meaning "Italian Car Manufacturer of Turin".
- **1899** In the US, the Winton Motor Carriage Company delivers 100 cars, making it the country's best-selling carmaker.
- **1900** At the turn of the century, the number of makes of car totals 209 worldwide.
- **1901** Ransom Olds' factory in Lansing, Michigan, burns down; his new one is the first to use mass-production techniques.

THE FIRM FOUNDED BY KARL BENZ, **INVENTOR OF THE CAR, STILL LED MANUFACTURING IN 1899.**

◁ **Oldsmobile Curved Dash, 1901** A couple take their Curved Dash for a spin in the US. Their seat is directly above the centrally-mounted 5-bhp engine.

△ **Painters work** on a chassis at the Daimler factory in Untertürkheim near Stuttgart, Germany, in 1904.

Reliability trials

Long-distance reliability competitions were an early way of demonstrating that the new-fangled car was a viable replacement for the horse and carriage – and nothing to be afraid of.

The Paris newspaper *Le Petit Journal* organized the first competitive event for horseless carriages in the summer of 1894. More than 100 entries were received, with most of them powered by steam or petrol, although others were claimed to be driven by compressed air, hydraulics, and even gravity. In the end, only 26 cars took part in the qualifying events, which were tours of Parisian suburbs over the three days prior to the main trial. Each car had to complete one of the 30-mile (48-km) qualifying routes in less than three hours.

The 21 qualifiers then headed 78 miles (126 km) north to Rouen. Comte Jules-Albert de Dion, a French aristocrat and car manufacturer, was the first to finish in a little under seven hours in his steam-powered car, ahead of petrol-powered Peugeots and Panhards. But because de Dion's steamer needed a stoker to tend to its boiler, it was considered too difficult to use on a day-to-day basis, and the main prize was split between the leading petrol cars.

The far more ambitious 1,000 Mile Trial of 1900 marked the beginning of competitive motoring in the UK. It was organized by the Automobile Club, which a few years later would become the Royal Automobile Club (RAC). More than 60 horseless carriages left London on 23 April, St George's Day, and pounded the old coaching roads west to Bristol, then turned north and travelled as far as Edinburgh before venturing back south. At every town, the cars were scrutinized by fascinated locals, many of whom had never seen a self-propelled car before.

After three weeks of dust, mud, punctures, and peril – punctuated by champagne breakfasts and hearty dinners hosted by members of the aristocracy along the way – a remarkable 46 of the starters finally made it back to the capital. Charles Rolls – shortly to be of Rolls-Royce fame – drove a Panhard, which performed well in speed tests and hillclimbs in the competition, and was awarded a gold medal as the best-performing car.

▷ **1,000 Mile Trial**
The 12-hp Panhard, driven by Charles Rolls, outclassed its rivals in the 14-day 1,000 Mile Trial held in 1900. The race was the first public demonstration in the UK of the potential of the motorcar as a practical means of long-distance travel.

“**Poor quality** is remembered long after **low prices** are forgotten.”

ATTRIBUTED TO CHARLES ROLLS

Coast to coast

At the beginning of the 20th century, the car was still seen as a novelty. However, after the first successful coast-to-coast drive across the US in 1903, it started to be taken seriously as a mode of transport.

In 1903, the car was considered by many to be little more than a passing fad, a rich man's mechanized toy. There were few places a car could be driven – of the US's 2.6 million miles (4.2 million km) of roads, only 150 miles (240 km) were paved. There were no petrol stations, road maps, or road signs. If you owned one of the 8,000 cars on the road, the chances were you would have sent away for spare parts from the factory and fixed it yourself.

The idea that a motorcar could be a serious mode of transport began to take hold in the spring and summer of 1903. Two men, later joined by a goggle-wearing bulldog named Bud, slowly made their way across the US in an epic 64-day, 5,600-mile (9,000-km) journey from San Francisco to New York City, completing the first coast-to-coast drive by motorcar.

△ **Peugeot poster**
As seen in this advertisement from 1896, driving the open-topped cars of the era was an outdoors activity that required heavy-duty clothing, especially over long distances.

Jackson and Crocker

An adventurous 31-year-old Vermont physician named Dr Horatio Nelson Jackson overheard a group of wealthy men in a gentlemen's club disparage the automobile; Jackson bet them $50 that he could drive one across the country in 90 days or less. Four days later, on 23 May, Jackson and a 21-year old petrol engine mechanic named Sewall K. Crocker headed east down Market Street in San Francisco in a slightly used 1903 two-cylinder Winton Touring Car.

The car, which Jackson bought for $3,000 and named the *Vermont*, sported a two-cylinder, 20-bhp, twin-carburettor engine mounted under the driver's seat. It had a two-speed sliding gear transmission that drove the rear wheels via a chain. On a smooth road, the Winton had a top speed of 30 mph (50 km/h).

Blowouts, breakdowns, and bad luck dogged Jackson and Crocker almost from the start. The first flat tyre came

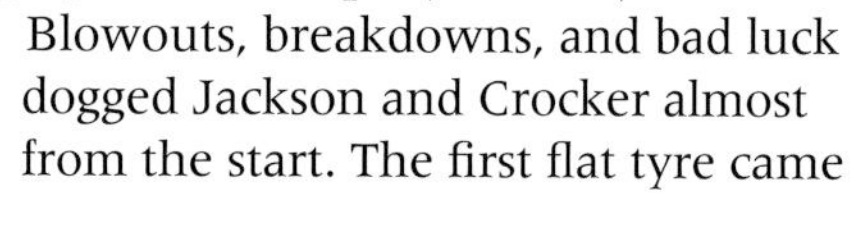

▷ **Jackson and Crocker**
Horatio Nelson Jackson and Sewall K. Crocker are seen here with their car – and Bud the bulldog – on the first stage of their Pacific-Atlantic ride, in Ohio, 1903.

just hours into the journey. In order to avoid the treacherous desert sands of Nevada, Jackson and Crocker went north to Wyoming, adding 1,000 miles (1,600 km) to the trip, but putting them on a course through the centre of the country, where they would closely follow railway lines.

Difficulties and dangers

On 23 June, the *Vermont* had travelled just 1,024 miles (1,647 km). As the two men crossed the rough terrain, supplies fell out of the car, parts broke, and life-threatening dangers were faced getting the car across streams. The men used a block and tackle to pull the car through, and had to manhandle boulders out of the way on narrow mountain trails. They purchased fuel and oil from general stores along the way. Tyres and other supplies for the car were sent by train. Blacksmiths – the men whose jobs would be rendered obsolete by the motorcar – made repairs to the *Vermont*'s suspension and wheels.

In Caldwell, Idaho, Jackson paid $15 for Bud the bulldog, who, after being fitted with goggles to keep the sand out of his eyes, took a spot at the front and stayed there for the rest of the trip. Word of Jackson's drive spread, and crowds gathered as the *Vermont* passed through towns along the way – many people had never seen a car.

At 4:30 a.m. on Sunday 26 July, the *Vermont* crossed into Manhattan and pulled into the forecourt of the Holland House Hotel. More than 5,600 miles (9,000 km) had passed under the *Vermont*'s wheels. Jackson had spent some $8,000 on the journey, including costs for fuel, tyres, parts, supplies, food, and the car itself.

Two other successful cross-country drives took place in 1903, and not long afterwards, the US government began investing in the nation's highways. Meanwhile, the car was becoming more dependable, and fuel, spares, and repairs were becoming easier to obtain. In 1909, Alice Huyler Ramsey became the first female driver to complete a coast-to-coast drive. To preserve the memory of his adventure for prosperity, Jackson donated the *Vermont*, Bud's goggles, and his scrapbook of clippings to the Smithsonian Institution in Washington.

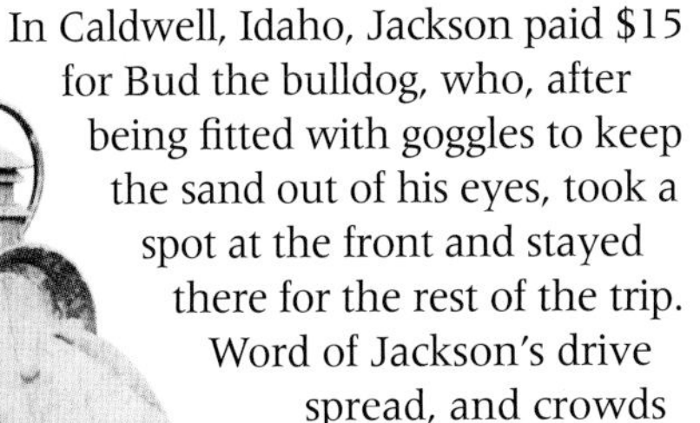

◁ **Early petrol station, 1900s**
Before the advent of filling stations, drivers had to buy petrol from shops or at crude roadside tanks.

> "He [Bud] was the **one member** of our trio who used **no profanity** on the entire trip."
>
> DR HORATIO NELSON JACKSON

LIFE BEHIND THE WHEEL

The demands of early motoring

Early cars like the *Vermont* were physically taxing to operate. Their large-diameter wheels made them strenuous to steer, and the driver had to astutely work a series of levers, pedals, switches, and knobs to control the engine, transmission, brakes, fuel, and battery. Drivers also needed good judgement, as the brakes were crude and took quite a distance to stop the vehicle. Early cars rarely had a windscreen or roof, so the driver was also at the mercy of the elements.

BEING ABLE TO CHANGE **A CAR WHEEL WAS AN ESSENTIAL SKILL FOR EARLY MOTORISTS AS ROAD SURFACES WERE OFTEN POOR QUALITY.**

DRIVING TECHNOLOGY
Cat's eyes

Cat's eyes – road fittings that reflect light from car headlamps – were invented in the UK in 1933. They proved particularly effective in the blackouts of World War II, when street lights were switched off and vehicles had to travel with reduced lighting to avoid detection by enemy planes. Inside each cat's eye is a pair of glass lenses with mirror-coated backs to reflect light towards an oncoming car. The lenses sit inside a rubber housing that squashes down when a car drives over it, at the same time wiping the lenses against a rubber blade to clean them. The latest innovation is to replace the glass lenses with solar-powered LEDs.

***CAT'S EYES* ARE AN EFFECTIVE WAY OF MARKING LANES AT NIGHT AND IN POOR WEATHER CONDITIONS.**

MILLER REAR OIL LAMP

ELECTRIC SIDELIGHT

ELECTRIC HEADLIGHT

SPARE ELECTRIC BULBS

SPARE BULB HOLDER

CANDLE LAMP

REAR AND SIDE OIL LAMP

Lighting the way

Pioneering motorists could see and be seen, to some extent, using oil or acetylene lamps. By the 1920s, electric lamps were in common use.

Candle-powered carriage lamps did little more than mark the corners of a vehicle. To light the way ahead, early cars used oil or acetylene lamps. The oil lamps were similar to the ones people used in their homes, and were fuelled by whale or olive oil. They needed regular attention, both to adjust the wick and to top up the oil, or the light would go out. They also had to be protected against bad weather. Acetylene lamps burned even when wet, but they had their own drawbacks. The acetylene gas was produced by dripping water on to calcium carbide inside the lamp, so supplies of carbide were kept onboard. However, these posed a fire risk – and made an unpleasant smell if they were not kept absolutely dry.

Although electric light bulbs were available from the early 1900s, the electrical systems in cars of the period were inadequate to power them. Instead, oil and gas lamps remained in common use until the 1920s, when more powerful electrical systems were developed. The familiar system of "main" and "dipped" beam also dates to this time.

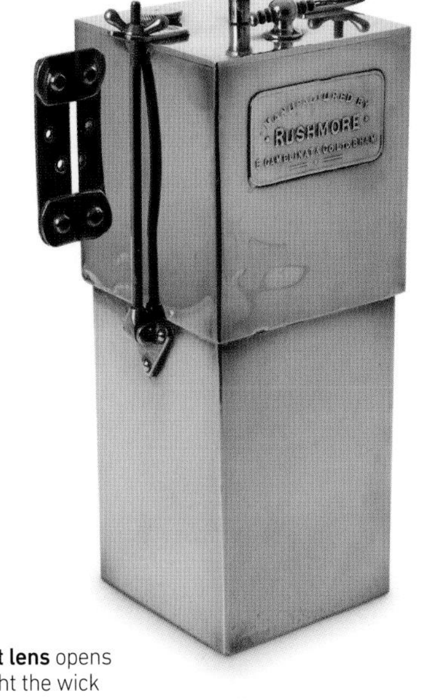

ACETYLENE LAMP AND GENERATOR

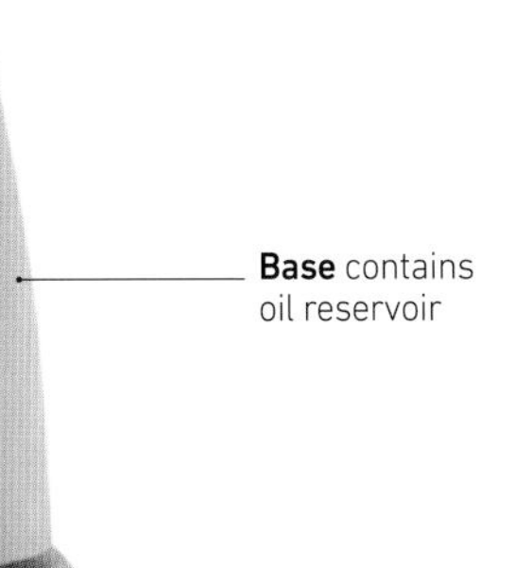

LUCAS OIL LAMP

LUCAS ACETYLENE REAR LAMP

ACETYLENE LAMP

The Gordon Bennett Cup

The first international car racing series was devised by the proprietor of the *New York Herald* newspaper. The epic city-to-city races were dominated by France, the world's leading car-making nation in the early years of the 20th century.

△ **1903 Cup winner**
This poster shows Belgian racing driver Camille Jenatzy in his Mercedes car, about to win the 1903 Gordon Bennett Cup in Ireland.

Millionaire newspaper owner James Gordon Bennett, Jr. put up a trophy for international motor racing in 1899, in consultation with the Automobile Club de France, as a way to promote his newspaper. The Gordon Bennett races were to become the pre-eminent racing series, and helped top-level motor racing spread throughout Europe.

▷ **Gordon Bennett**
James Gordon Bennett, Jr. (1841–1918), sponsor of the international race.

The first events

The Gordon Bennett Cup was a competition between national teams, each one entering up to three cars built entirely within its borders, with the winning nation hosting the following year's race. The series got off to a slow start, however, with only a low number of entries for the first event in 1900.

France, the biggest car-making nation in Europe, entered a team of three cars. There were two cars from the US, and lone entries from Belgium and Germany. Two cars withdrew before the race, leaving just five to tackle a 350-mile (565-km) route from Paris to Lyon. Only two finished: Fernand Charron's Panhard won at an average speed of 38.5 mph (60 km/h), with another Panhard driven by Léonce Girardot second, delayed after breaking a wheel. Public interest was limited, and only a small crowd greeted the winners. In 1901, Girardot was the only finisher of three entries, all French, a British challenger being forced to withdraw at the last moment. The third Cup race in 1902 saw a British winner, Selwyn Francis Edge's Napier, again the sole entrant to survive to the end of the race.

Growing popularity

The UK hosted the race the following year, in Ireland. A series of crashes in the Paris–Madrid road race a few weeks earlier had left eight people dead (see pp.30–31), so, to allay public concerns, a closed circuit was set up, centred on Ballyshannon. Three British Napiers faced three German Mercedes, two Panhards, and a Mors from France, and two Wintons and a Peerless from the US. Four cars finished, led by Mercedes driver Camille Jenatzy. Teams from Germany, France, the UK, Belgium, Austria, and Italy entered the 1904 Cup. The UK and

▷ **Selwin Edge triumphs in 1902 Cup**
Sporting dark green livery, Edge's Napier was one of the first cars painted in a colour that would later become known as "British Racing green". Napier was also the first British manufacturer to build cars specifically designed for racing.

DRIVING TECHNOLOGY
Mercedes 35 hp 1901

Early motorcars had much in common with horsedrawn carriages, but that changed in 1901 with the introduction of the 35-hp Mercedes designed by Wilhelm Maybach. It was the forerunner of the modern car. Instead of a wooden chassis used by rivals, Maybach designed a light but strong frame using pressed channel-section steel members. The engine adopted cam-operated valves, and there was a pressurized cooling system and a radiator using finned tubes. Wide, with a low centre of gravity, the Mercedes was a successful racing and road car.

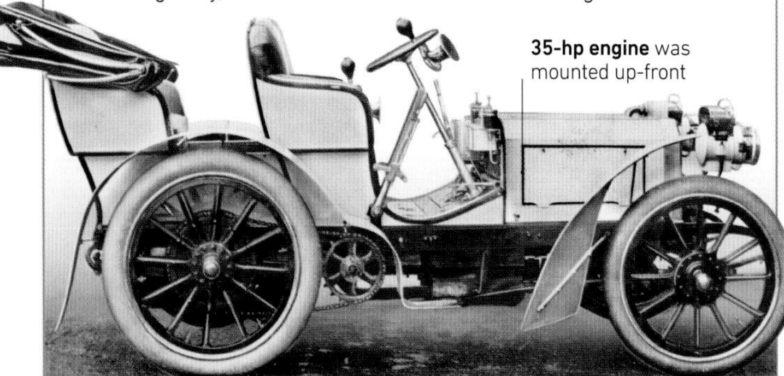

***THE FIRST MERCEDES* WAS CONSTRUCTED BY WILHELM MAYBACH, TO THE SPECIFICATIONS OF EMIL JELLINEK, DAIMLER'S SALES AGENT.**

France had so many contenders that qualifying heats had to be held to reduce their numbers to the three cars permitted. France won with a Richard-Brasier car driven by Leon Théry.

The 1905 Cup saw the largest ever field, of 18 cars. It was held on a circuit near the Michelin tyre company's headquarters at Clermont-Ferrand, where Théry and his Richard-Brasier triumphed again. However, the French became frustrated with the three-car limit on entries, so the fifth Gordon Bennett Cup was the last to be held. In its place, the international governing body of motor sport, the Alliance Internationale des Automobile-clubs Reconnus (AIACR), set up a new "Grand Prix" race for 1906.

◁ **Victory for France**
Camille Jenatzy won the 1903 Cup in Ireland, driving his 60-bhp Mercedes.

"... an **instant's inattention** would mean a **horrible death**."

SELWYN FRANCIS EDGE, COMPETITOR, DESCRIBING PART OF THE 1901 RACE

Travelling by forecar

In the frantic crossover period when pedalling gave way to petrol power, this curious hybrid offered an idiosyncratic way to hit the road.

At the end of the 19th century the world was gripped by the new craze for bicycles (see p.15). With advances in combustion engine technology being made at the same time, the motorcycle was a natural evolution of this, and another, more esoteric variant, the motorized tricycle, was especially favoured by tradesmen for light deliveries. Between the two front wheels, a cargo box or large basket could be installed to carry goods, with the rider or driver steering from behind it, using handlebars. Such vehicles became a common sight in towns and cities, where they were particularly suitable for negotiating narrow streets, and could also be used for selling goods such as ice-cream.

However, this vehicle also had applications for personal transport. With minimal changes, a single or double passenger seat could be installed instead of the cargo box. The resulting vehicle was known as a "forecar", because the passengers were carried ahead of the person in charge of the controls. Other popular appellations were "forecarriage" and "tri-car". Four-wheelers with better stability (but less agility) were also developed.

Forecars were never very fast, but the travelling experience could still be unnerving for those seated up front. There was absolutely no protection in the event of an accident, and they were fully exposed to the elements. These "cars" were shortlived: they mostly appealed to ex-cyclists who were resistant to the new-fangled automobile, and by 1905 they had mostly died out. However, a surprising number of famous car manufacturers began by making forecars, including Lagonda, Riley, A.C., and Singer.

▷ **Rexette Forecar in London, 1905**
This forecar was one of the more sophisticated models, featuring a steering wheel instead of handlebars and a decent suspension system, although the thin-spoked wheels were clearly based on bicycle design. In colder climates, passengers would have been well advised to wrap up warm, as seen here.

“For **signalling or scouting**, the forecar is to be **recommended**.”

ROYAL UNITED SERVICE JOURNAL VOL 48, 1904

Working vehicles

The arrival of the internal combustion engine in the 1890s sparked a petrol-driven commercial revolution. As utility vehicles multiplied, life in the city changed forever.

The needs of commerce and civic life helped to drive the adoption of motorized vehicles in the last few years of the 19th century. Petrol engines made working vehicles vastly more efficient and versatile than their horse-drawn equivalents, and operators everywhere were keen to exploit this.

The first small delivery van came into being in France in 1895, where Peugeot launched a model that could carry 1,000 lb (450 kg) at 9½ mph (15 km/h) or 650 lb (295 kg) at 12 mph (19 km/h). Drapers, department stores, and other retailers were quick to recognise the value of these vans for local deliveries. The concept of a metal box attached to the back of a car chassis had myriad applications, whether transporting milk and groceries to the ever-expanding reaches of suburbia, or carrying prisoners, or even moving dead bodies. Taxis based on modified car frames quickly replaced the horse-drawn cab. Even later, in World War I, the first "armoured cars" were regular cars fortified with heavy-gauge metal sheets.

△ **Sewing machine advert, 1905**
Companies quickly adopted motor vehicles to deliver their products, as shown in this advert.

The arrival of trucks

In 1897, the first two trucks entered commercial service in Germany, one with a haulage firm in Stuttgart and one with a Berlin brewery – both made by the Daimler Company of Stuttgart. The same year, British firm Thornycroft of London built two steam-powered dustcarts for a local authority. A year later, Thornycroft devised the concept of the articulated truck, with a separate tractor unit and detachable trailer.

Trucks provided a rolling platform for all kinds of other purposes, including buses for mass passenger transport and fire-fighting vehicles. At first, top speeds were limited to around 15 mph (24 km/h), but maintenance was vastly less than the unpredictable and messy business of stabling and caring for horses. Together, these vehicles laid the foundations for the commercial vehicle industry, and created an entirely new category of employment – the working driver.

KEY EVENTS

- **1896** The Daimler Company in Cannstatt, Germany, builds the first petrol-powered lorry.
- **1897** The British Post Office pioneers the use of motor vans for mail deliveries.
- **1898** The first motorized fire engine is demonstrated at the French Heavy Autocar Trials in Versailles.
- **1900** The first civilian motor ambulance enters service in Alençon, France, shortly after the country introduces the first military ambulance.
- **1901** The world's first motorized hearse officiates at a funeral in Coventry, establishing the custom of black paintwork.
- **1903** The first car to replace a horse in police service joins the Police Department at Boston, Massachusetts, US.

***DAIMLER LORRY* WITH BELT DRIVE AND COIL SPRINGS, BUILT IN 1896.**

◁ **Early fire engines, 1905**
Here the Birmingham fire service is pictured on two of their fire engines, which have solid tyres and hand-cranked starting mechanisms.

△ **A delivery truck is loaded** with goods using a ramp from the second storey of the Edison Lamp Works, New York, in the early 1900s.

1906–1925

AN INDUSTRY IS BORN

1906–1925

An industry is born

This period was dominated by Henry Ford and his Model T. The American engineer-entrepreneur set a blistering new pace of progress, harnessing automation and other mass-production techniques to churn out cars at extremely low prices. With his discipline and determination, the car became a commodity that was suddenly accessible to an enormous number of people.

Petrol stations, highways, car parks, repair workshops, and used car lots all boomed as the Model T and its General Motors compatriots cascaded across the US and, indeed, the globe.

Not that these were the most appropriate cars for everywhere in the world. In Europe, tighter streets and even tighter household budgets meant that manufacturers trod cautiously in Ford's wake. They devised smaller, more economical models that attracted lower taxes, and innovated in different ways – for example, by marketing to female buyers for the first time. In France, Germany, and the UK, the concept of a range of models arose, as manufacturers attempted to meet the needs of different kinds of drivers. This included the arrival of road-going sports cars that drew on experience and acclaim garnered at the racetrack.

Diversity of designs

Until 1914, the wealthy car buyer had dozens of large, impressive models to choose from, mostly to be driven by their chauffeurs. For manufacturers at the time, the emphasis was on creating models with near-silent mechanicals and elegant coachwork – the latter, finally, evolving away from the appearance of a "horseless carriage" and becoming a distinctive form of machine in its own right.

HENRY FORD AT THE WHEEL OF HIS FIRST MODEL A

PETROL PRODUCERS COMPETE TO ATTRACT CUSTOM

“Many **conscripted civilians** got their first **taste of driving** under dark **war clouds**.”

Very much at the other end of the economic scale was the merging of two- and four-wheelers in the form of the cyclecar. This vehicle was an early attempt to create motoring for everyman that would have a brief flowering before the advent of “proper” small cars.

Vehicle use spreads worldwide

The mechanized world, of which the car was an increasing part, was hijacked by World War I and its needs. The first tanks and armoured cars were built as hasty creations with urgent jobs to do. Many conscripted civilians got their first taste of driving under dark war clouds; the hitherto courtly mystique of conducting a motorcar became something mundane. Professional drivers might latterly expect to control anything from a delivery van to a fire engine as an everyday task. Once hostilities had ceased, European and US governments at last devoted serious thought to creating road networks that were fit for purpose – where cars, not horses, were the main users, and where systems to regulate motor traffic, such as traffic lights and roundabouts, could be introduced.

Much of the rest of the world was still a no-go area for car drivers, including most of Asia, Africa, and South America. However, colonial powers such as France and the UK pushed the boundaries in these regions. Clever technology made driving across deserts possible, while the aspirations of local leaders gave wealthy Indians a taste for Rolls-Royces that buoyed the demand for luxury cars in far-off places. Meanwhile, at elegant residences in Rome, Paris, and London, stables became motor houses as simple animals gave way to a rather different type of horsepower.

WORLD WAR I IS POWERED BY THE INTERNAL COMBUSTION ENGINE

DRIVING BECOMES THE NORM FOR MANY WHEN PEACE RETURNS

Ford's production line

The Model T was the car that put the US on wheels, bringing cheap motoring to the public thanks to the mass-production techniques perfected by Henry Ford. The man himself became a much admired, sometimes feared, public figure.

△ **A Model T leaves the London showroom, 1910s** Consumers flocked to buy this affordable car from dealerships across the US. It was later exported worldwide.

Henry Ford was a farmer's son who went on to become one of the great engineering and marketing geniuses of the modern world. He first showed engineering promise at the age of 13, when his father gave him a pocket watch; the young Ford duly took it apart and rebuilt it. While still a teenager he became a shipyard machinist, and in 1891 he joined the Edison Illuminating Company as an engineer. Two years later he was its chief engineer and began working on petrol-engined car designs, building his first vehicle, the Quadricycle, in 1896. After two attempts at starting car companies, he established the Ford Motor Co. in 1903.

The car that changed history

Ford's 1908 Model T automobile used strong, lightweight steel in its construction, and was robust, simple to make, and easy to fix. Demand quickly outstripped supply. In 1913, inspired by Chicago meat-packaging plants and advice from colleagues, Ford introduced moving assembly lines; workers remained at fixed stations, while parts were moved in lines to them for each stage of assembly.

This process revolutionized manufacturing, and by 1918 half of the cars in the US were Model Ts. Eventually, 15 million of them were made and prices fell from $850 to $260. For many years Model Ts were only available with black paint, because it dried more quickly than other colours – one of the many ways in which Ford was ruthless in his pursuit of efficiencies.

Henry Ford was a man of many contradictions. A far-sighted engineer who built the Model T until 1927, by which time it was outmoded; an employer who paid good wages to keep his staff, but who used a private army to break up strikes and union activity; and a peace campaigner who expressed anti-Semitic views. He nevertheless changed the industrialized world.

KEY EVENTS

- **July 30 1863** Henry Ford is born on his parents' family farm in Michigan, US.
- **1876** Aged 13, Ford is given a pocket watch, which he dismantles and successfully rebuilds.
- **1879** Ford becomes a shipyard worker.
- **1891** He joins Edison Illuminating Company and is promoted to chief engineer in 1893.
- **1896** Ford builds his first car, the Quadricycle.
- **1903** The Ford Motor Company is created.
- **1908** The Model T is launched, costing $850.
- **1914** Moving assembly lines are introduced.
- **1916** The Model T price is reduced to $345.
- **1927** The last (15 millionth) Model T is made.
- **April 7 1947** Henry Ford dies.

PROUD HENRY FORD **WITH HIS FIRST PRODUCTION CAR, THE 1903 MODEL A.**

◁ **Model T Ford** Known as the "Tin Lizzie", the Model T featured a monobloc engine and high ground clearance to allow for poor-quality roads.

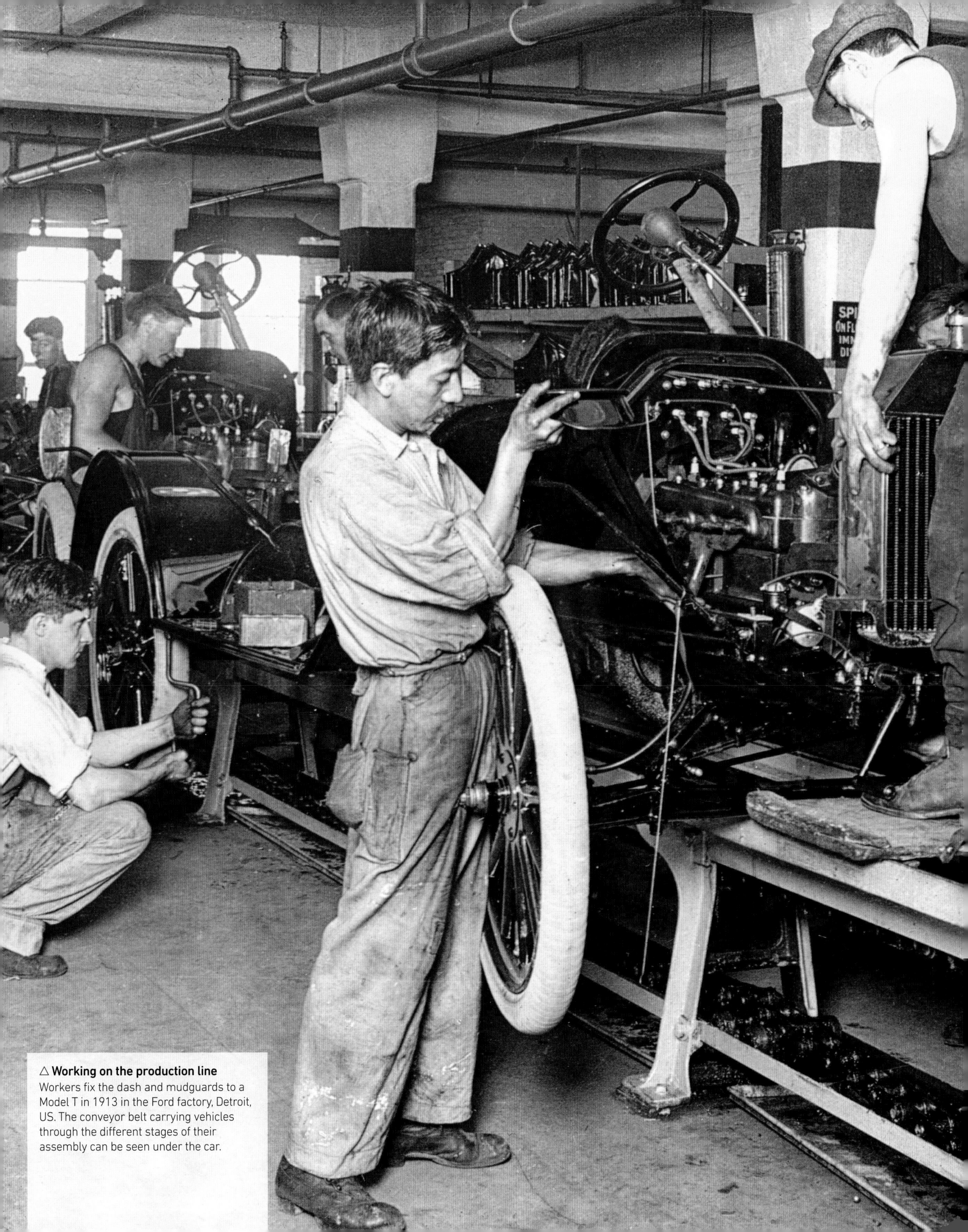

△ **Working on the production line**
Workers fix the dash and mudguards to a Model T in 1913 in the Ford factory, Detroit, US. The conveyor belt carrying vehicles through the different stages of their assembly can be seen under the car.

▷ **Shell advertising**
Much of the oil and petrol advertising of the era promoted a cheerful, dynamic image fuelled by the optimism of a new era of mobility.

The story of petrol

The widespread availability and unmatchable energy density of petrol soon made it the dominant propellant of the car, triggering the growth of massive networks of boldly branded filling stations.

KEY EVENTS

- **1745** The first oil well and refinery are built in Russia, to fuel lamps.
- **1848** The first modern oil well is drilled in Asia, on the Apsheron peninsula, near Baku (in modern Azerbaijan).
- **1851** James Young and partners build the first commercial oil refinery in Bathgate, UK.
- **1858** The Oil Springs well in Ontario, Canada, begins the first major oil boom.
- **1879** Standard Oil, owned by US businessman John D. Rockefeller, controls 90 per cent of America's oil refining capacity.
- **1905** The Baku oil fields are set alight during the Russian Revolution of 1905.
- **1907** The Royal Dutch Shell Group oil company, one of the global "supermajor" companies, is formed.
- **1910** The first petrol pumps are installed for public use in the US.
- **1914** World War I – the first global, industrialized war – makes petrol a matter of military and national importance.
- **1934** The first floating oil drilling rig is reported in use on the Caspian Sea.
- **1940** The UK's Royal Air Force launches its "oil campaign" during World War II. Nazi Germany's oil-producing facilities are specifically taregetted.

A RAIL STRIKE IN 1919 **LEADS TO CANS OF PETROL BEING DUMPED IN A CAR PARK.**

When Bertha Benz took her husband Karl's Motorwagen on its landmark drive in 1888, she fuelled it with petrol bought from chemists. Petroleum spirit was not widely available, and it had limited uses – it was a byproduct of kerosene, which was used for lighting before the advent of electricity in the late 19th century. Petrol itself, refined from oil, was not new – it had been used for medicinal purposes in ancient Persia, and in 7th-century China and Japan, where it was known as "burning water". Modern petroleum was patented by James Young in 1850, after he discovered the fluid bubbling up in his coalmine in Derbyshire, UK.

The real oil boom began in 1858 at the Oil Springs well in Ontario, Canada. This led to bigger finds in Pennsylvania, Texas, Oklahoma, and California. Global oil production grew consistently from four million barrels in 1859 to 57 million by 1899, but doubled by 1906 to 126 million, driven by demand from motorists: from around 8,000 cars on the road by 1900, the numbers exploded to 23 million by 1920, almost all petrol-fuelled.

Neverthless, supply was unreliable. Before World War I, most drivers carried supplies in two-gallon cans, usually mounted on the car's running boards and which were purchased from blacksmiths, chemists, or general stores. The US got its first pumps in 1910, and by 1921 it had 12,000 petrol stations, using large underground reservoirs to store the petrol. The US Bowser company installed the first pumps in the UK in 1915, and the first dedicated filling station appeared four years later. By 1929 there were 55,000 filling stations in the UK. Branding became increasingly important, and oil companies erected big logo landmarks all over the world.

Although the availability of fuel improved massively during the first part of the 20th century, the technology for transferring the precious liquid from filling station to fuel tank remained simple – hand-pumped by a paid attendant – for many decades.

△ **Pratts petrol can**
Before the arrival of petrol pumps, drivers bought fuel in two-gallon cans such as this.

▷ **The need for speed**
A racing driver hastily fills his car up with petrol at Brooklands racetrack in Surrey in 1907. Motor sports helped to increase the popularity of cars and motoring to the general public, leading to more demand for petrol.

Toy cars

Toy cars and model vehicles are almost as old as the car itself. As models became more sophisticated and realistic, they gave millions of young petrolheads a taste of motoring in miniature.

The first toy cars were crude models cast in lead, iron, or brass, and usually represented generic car types rather than specific vehicles. They were followed by more realistic models made from sheets of tinplate steel that were printed with colour and detail, before being shaped to form car bodies. Racing and record-breaking cars were popular toys, alongside everyday cars and lorries. Some toy cars were fitted with simple clockwork mechanisms that could be wound with a key to propel the cars along, adding to the fun. High-quality diecast models appeared in the late 1930s; household names included the UK's Dinky, Matchbox, and Corgi, France's Solido, and the US Hot Wheels®. After a heyday from the 1950s to the '70s, model cars as toys went out of fashion in the '80s, although they remain highly collectable among enthusiasts.

TYPE 35 BUGATTI, DIECAST, LESNEY, 1961

SINGER ROADSTER, DINKY, 1958

VAUXHALL TOWN COUPÉ, TRI-ANG MINIC, 1930s

COOPER-BRISTOL RACING CAR, DINKY, 1950s

KEY DEVELOPMENT
PEDAL CARS

In the early days of motoring, wealthy drivers often provided cars for their children – miniature machines with metal bodywork and pedals to give power. Over the years the best pedal cars became incredibly realistic and sophisticated, with chain drive, real rubber tyres, lights, suspension, and working brakes. Plastic-bodied pedal cars, which were lighter and easier to propel, started to appear in the 1950s. One of the most successful and fondly remembered pedal car was the Austin J40, a scaled-down version of the full-size 1948 Austin A40 Devon. They were made at Bargoed in South Wales between 1949–71 by disabled ex-miners, using metal off-cuts from Austin cars, and were exported worldwide.

CHILDREN LINE UP **TO COMPETE IN THEIR AUSTIN J40S. SIMILAR RACES ARE STILL RUN TODAY AT THE GOODWOOD REVIVAL.**

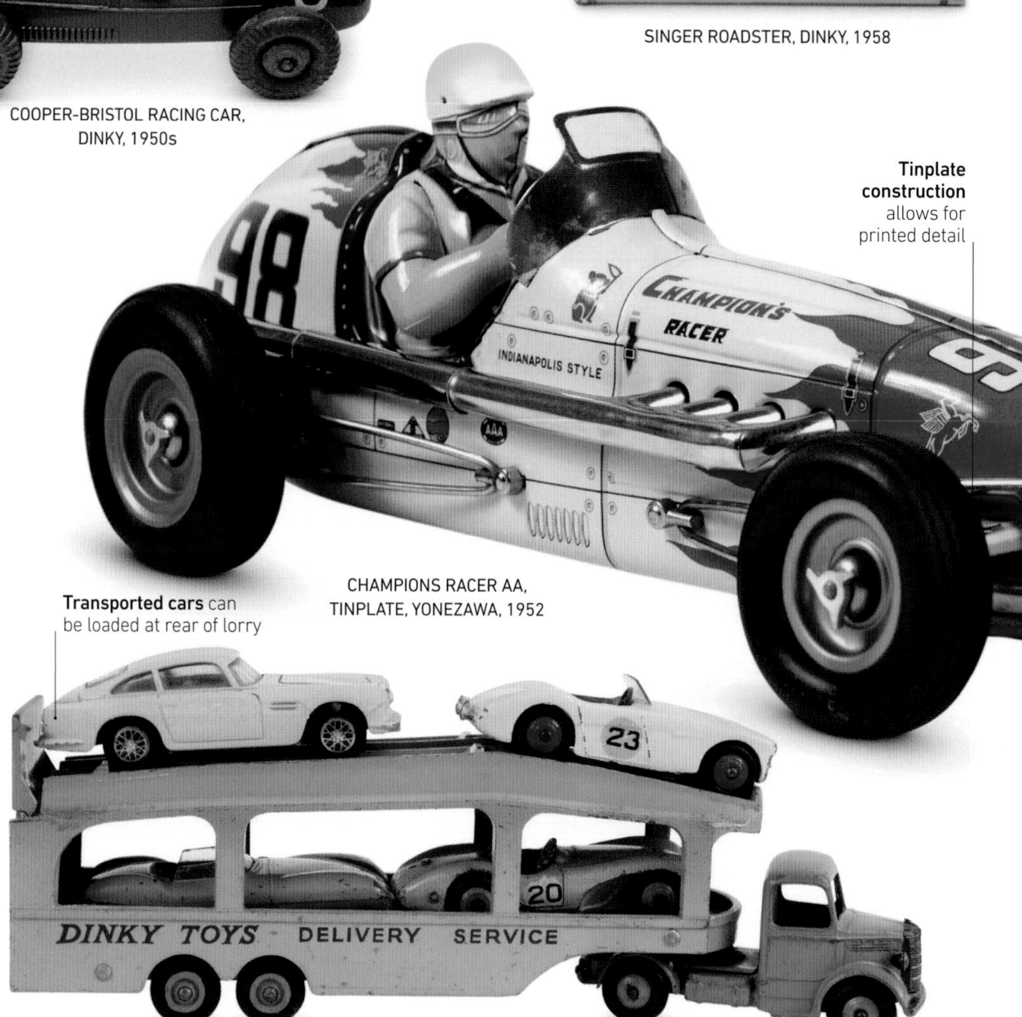

Tinplate construction allows for printed detail

CHAMPIONS RACER AA, TINPLATE, YONEZAWA, 1952

Transported cars can be loaded at rear of lorry

BEDFORD CAR TRANSPORTER, DINKY, 1950s

HEINKEL TROJAN, DIECAST, CORGI, 1962

Ejector seat fires passenger figure through roof hatch

Machine guns pop out of the front when button is pressed

JAMES BOND'S ASTON MARTIN DB5, CORGI, 1965

Rocket tubes raise and fire plastic projectiles

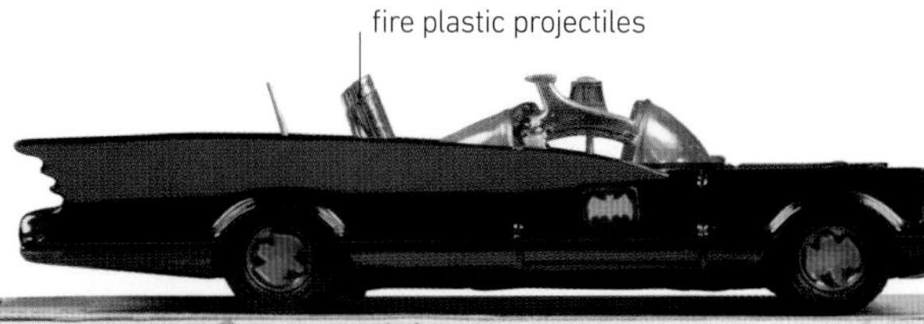

THE BATMOBILE. DIECAST, CORGI, 1966

Original packaging is as important as the model for collectors

Radio transmitter can control direction as well as speed

RADIO-CONTROLLED BMW M3, TAMIYA, 1980s

DODGE DEORA, DIECAST, HOT WHEELS®, MATTEL, 1968

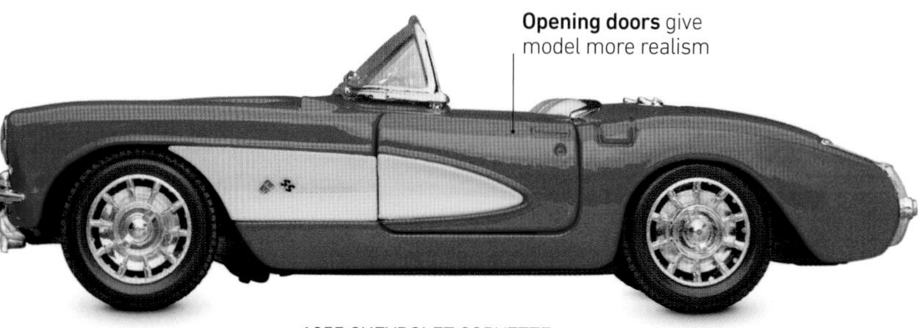

Opening doors give model more realism

1957 CHEVROLET CORVETTE, DIECAST KIT, MAISTO, 1990

Ferrari 312 T3 based on car driven by F1 champion Jody Scheckter

FERRARI F1 CAR, POLISTIL, 1978

FORD FOCUS SLOT CAR, SCALEXTRIC, 2013

Gridlock on the streets

As cars began to muscle their way into city centres, they shared space with pedestrians and horses in chaotic and often dangerous ways. Something had to be done to improve safety – and the results changed urban landscapes forever.

△ **Pollution concerns, 1910**
In a glimpse of future concerns over excess traffic, this German advertisement promotes the use of cars with reduced pollutant emissions.

In the first decade of the 20th century, roads were in a state of anarchy – they lacked stop signs, warning signs, traffic lights, traffic police, or even separated lanes. Cars barged onto roads already teeming with trams, buses, cycles, pedestrians, and horses; the latter were terrified by the noisy and often dangerously driven newcomers.

The number of cars on the roads grew exponentially. In 1909, there were just 200,000 motorized vehicles in the US; a mere seven years later, that figure had leapt to over two million. In San Francisco, the number of cars surpassed horse-drawn vehicles for the first time in 1914.

Accident statistics were shocking. In 1917 in Detroit, there were only 65,000 cars on the road, but they caused 7,171 accidents, including 168 fatalities. Congestion was also becoming severe; in 1920, Chicago's trams were travelling at half the speed they had in 1910.

Traffic management

The need for action was urgent. New York's solution was to hire ageing policemen to direct traffic; however, their signalling systems were confusing to motorists. The congestion was so severe that manpower alone could not cope. By 1920, one in four policemen in Detroit was being used for traffic duties. An obvious move was to separate traffic into lanes. The first lane marker appeared in Michigan in 1911, the same year that one-way streets were trialled in Detroit. Stop signs and traffic lights (see pp.66–67) followed in the same decade. Although roundabouts predated cars, some countries favoured them as a solution. The world's first modern roundabout, the Brautwiesenplatz, opened in Germany in 1899, while the first British roundabout went into use in Letchworth Garden City in 1909.

◁ **Early crashes**
Accidents in the early years of motoring were all too common: this pair of Ford Model Ts collided in 1910. Junctions were free-for-alls and separate traffic lanes did not exist.

KEY EVENTS

- **1868** The world's first "traffic lights" are installed outside the Houses of Parliament in London. These gas-lit lanterns last around four weeks before exploding.
- **1909** Nine European governments agree on pictorial traffic signs to warn drivers about road bumps, corners, intersections, and railway crossings.
- **1912** The first red-green electric traffic lights are developed in Salt Lake City, US.
- **1919** The first four-way, three-colour traffic lights are installed in Detroit, US.
- **1920** Los Angeles installs the first "Acme" automated semaphore traffic signals, as immortalized in Looney Tunes cartoons.
- **1926** The UK's first electric traffic lights arrive in Piccadilly Circus, London.

AN AMERICAN POLICEMAN **OPERATES TRAFFIC SIGNALS IN THE EARLY 1920S.**

▽ **The streets of Chicago, 1909**
The chaos and congestion of early urban streets can be readily seen in this image of Dearborn Street in Chicago, US, in 1909. Gridlock became a paralyzing scourge of city life.

▽ The Citroën 5CV Type C, 1922
Citroën's small, four-cylinder model took the market by storm. Thanks to its electric starting system, the driver no longer needed to manually crank the engine. For the first time, a manufacturer's marketing campaign targeted men and women by promoting easy-to-use features.

OLET"

par magnéto

923

Track racing

Manufacturers raced cars to prove their new designs under the greatest stress. Races provided challenges for the adventurers of the motor age, and spectators revelled in the contests at the new circuits.

△ **Indianapolis advertising poster, 1909**
When it first opened, the Indianapolis Motor Speedway hosted a variety of motor races before the Indy 500 was established.

The first motor races were city-to-city events, often covering epic mileages on narrow, rough, and unsurfaced roads. In dry weather the cars threw up vast clouds of dust that blinded anyone behind them, and made goggles and gauntlets essential for drivers. In wet conditions the cars sank to their axles in mud or slid off the road entirely. Onlookers unused to such high-speed vehicles were a hazard for the drivers; frightened animals were another, and car tyres were frequently punctured by stones and horseshoe nails. Crashes were commonplace, and the flimsy construction of the cars led to dire consequences. Carnage on the roads during these events began to turn public opinion against motor sport. Increasingly, racing drivers came to be seen not as heroic daredevils providing an exciting new spectacle, but as fools threatening public safety. The turning point was the Paris to Madrid race of 1903 (see pp.30–31), when a series of accidents on the first day left eight people dead and scores injured. Among the five competitors killed was Marcel Renault, one of the brothers who founded the Renault car company. The French government halted the race at Bordeaux, and Fernand Gabriel was declared the winner.

Brooklands and Indianapolis

To reduce the dangers, subsequent races were run on loops of roads that were closed to the public for the duration. The first purpose-built race track was opened in 1907. Hugh and Ethel Locke King built the banked track on their private estate at Brooklands, 20 miles (32 km) southwest of London, not just to provide a venue for racing but also to give the British motor industry a facility for high-speed testing. The track was

▽ **Brooklands race track, 1908**
Three cars line up at the start of a race at Brooklands.

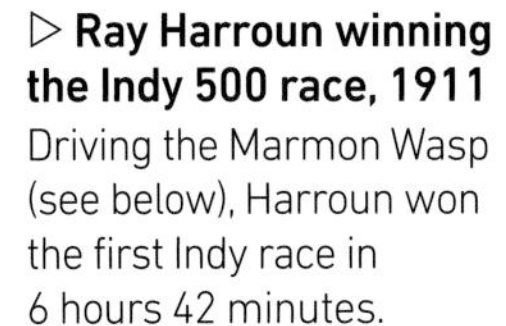

▷ **Ray Harroun winning the Indy 500 race, 1911**
Driving the Marmon Wasp (see below), Harroun won the first Indy race in 6 hours 42 minutes.

23/4 miles (4.5 km) long and 100 ft (30 m) wide, and the bends had concrete bankings up to 30ft (10 m) high to increase cornering speeds. Spectators in the vast infield area – which could hold a crowd of 250,000 or more – could see the whole track, and were immersed in the noise and spectacle. Race meetings were likened to a "motor Ascot" and were run in a similar way to the famous horse race – the drivers even wore jockeys' coloured silks for identification at first, with numbers coming later. Brooklands quickly became a mecca for motor sport, and remained so up to the start of World War II.

The first purpose-built track in the US was at Indianapolis, where motoring entrepreneur Carl G. Fisher was the central figure in the establishment of the Motor Speedway in 1909. At first the track had a surface of compacted stone, but this proved to be unsuitable. During the first car race meeting it quickly started to break up, with ruts and potholes forming, and there were numerous accidents. A few weeks later, the speedway was resurfaced with more than three million Indiana bricks, giving rise to the track's "Brickyard" nickname, which survives to this day. The "Indy 500" Memorial Day 500-mile (805-km) race at Indianapolis was established in 1911 and, except for the world war years, has been held annually ever since, becoming one of the world's most popular sporting events.

> "It will give them a **soft landing!**"
>
> ETHEL LOCKE KING ON THE SEWAGE FARM NEXT TO THE BROOKLANDS RACE TRACK

KEY DEVELOPMENT

Racing experience benefits road cars

Racing developments soon found their way into road cars. Multi-cylinder engines raised power outputs, shaft drive improved reliability, and detachable wheel rims dealt with the inevitable punctures. Some innovations were simple but brilliant, such as Ray Harroun's use of a rear-view mirror in the first Indianapolis 500 so he could save the weight of a "spotter" – a passenger who kept an eye on the cars behind. Harroun won the race, although his mirror actually vibrated so much that he could not see anything in it. Even so, rear-view mirrors were fitted to Marmon road cars, and soon became an essential accessory.

***THE MARMON WASP* CAR, DRIVEN TO VICTORY BY RAY HARROUN IN 1911, NAMED AFTER ITS BLACK-AND-YELLOW PAINT.**

The first sports cars

Sports cars were designed to make the open road a thrill for rich, speed-hungry drivers – but if they wanted even more excitement, they could fold down the windscreen, strap on some goggles, and actually go motor racing. Even sports car drivers who never ventured on to a race track could brag that their car was capable of racing performance. The UK's Vauxhall Prince Henry is commonly regarded as the first genuine sports car. It was soon followed by the tough Bentley 3-litre, and the Ferdinand Porsche-designed Austro-Daimler 27/80; the Spanish Hispano-Suiza Alfonso XIII; Bugattis and Delages from France; and Alfa Romeos from Italy. Meanwhile, in the US, decades before the Corvette and the Thunderbird, the Mercer Raceabout and Stutz Bearcat were the sports car pioneers.

△ **Bentley competing in a hillclimb in 1922**
The first Bentley production cars were delivered in 1921, and they soon found their way into motor sport. Here, Frank Clement competes in the Kop Hill Climb in Buckinghamshire.

△ **Vauxhall Prince Henry, 1913**
The Prince Henry was the first genuine sports car, designed with dual use on road and track in mind. It was produced from 1911–14, and was then replaced by the equally famous 30-98.

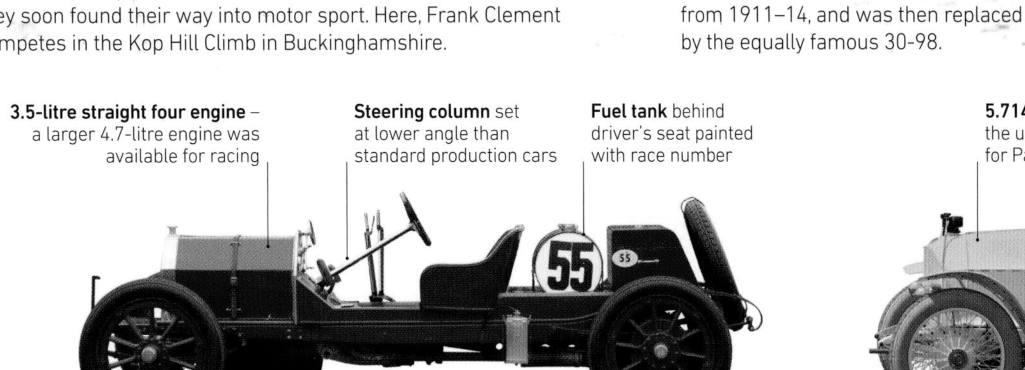

△ **Lancia Tipo 55 Corsa, 1910**
Vincenzo Lancia's 20-hp Tipo 55, also known as the Lancia Gamma, was a versatile car available in touring and sports forms. The frame could be configured in various ways to suit different uses; for sporting purposes the car was lower and had minimal, lightweight bodywork.

5.714-cc engine related to the unit Porsche designed for Parsifal airship

"Tulip" body shape narrower at the bottom than the top

△ **Austro-Daimler 27/80 Prince Henry, 1910**
Daimler's Austrian subsidiary produced this large, fast car, designed by Ferdinand Porsche and named after the German "Prince Henry" alpine trial. Competition versions could reach 90 mph (145 km/h) – but, as with all cars of the era, it only had brakes on the rear wheels.

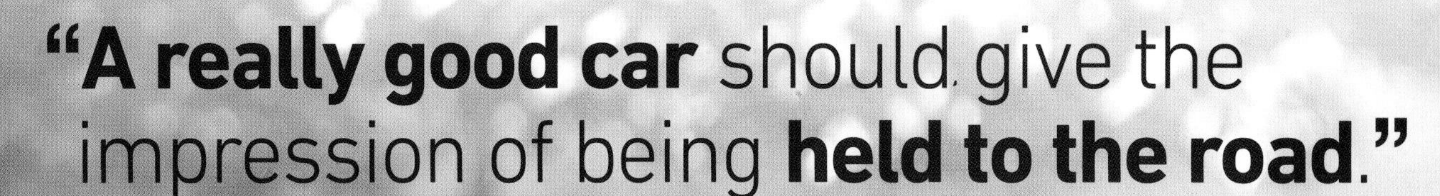

△ **Bugatti Type 13, 1910**
Bugatti's first production car was the 1910 Type 13, a lightweight sports car with a 1.4-litre engine delivering 30 hp, which was capable of 95 mph (153 km/h). The Type 15 was a long-wheelbase version with revised rear suspension, and was built until 1913.

△ **Mercer 35R Raceabout, 1910**
With a 5-litre, four-cylinder engine, this unusually low-slung sports car offered accomplished handling, and had a top speed of 90 mph (145 km/h). The addition of a four-speed gearbox in 1911 made it a more flexible performer and a successful competition car.

Taming the traffic

The story of how traffic lights became a crucial feature of the urban landscape does not start well, but it does begin with a bang.

The very first traffic signal, installed in London's Parliament Square in 1868, exploded after around four weeks, badly burning a policeman in the process. Its inventor, J.P. Knight, was a railway engineer, and his design resembled a railway crossing signal with flashing red and green lights and semaphore arms. Lit by gas at night, the traffic signal had to be manually operated by a policeman. Despite this limitation, the apparatus was considered a breakthrough in helping to control the increasingly heavy flow of horse-drawn carriages and allow pedestrians to cross safely.

However, the unfortunate accident, which had been caused by a leaky gas main, deterred all further plans for mechanical traffic control until 1914, when the first electric traffic lights were erected in Cleveland, Ohio, US. The idea had been hatched two years earlier by a safety-conscious policeman, Lester Wire, but James Hoge is credited with the invention of the electric traffic light, filing a patent for it in 1913. His system featured the alternating illuminated words "stop" and "move", and the rhythm of the lights could be altered by police or fire services to control traffic in emergencies.

In 1917, William Ghiglieri came up with the first red and green traffic lights in San Francisco; three years later Detroit policeman William Potts added an amber warning light to create a signal resembling the one used today. The three-colour traffic light system migrated across the Atlantic to London in 1925, when the first green-amber-red signals were put up at the junction between St James's Street and Piccadilly. These formative traffic lights still had to be manually operated by policemen, but the following year the first automatic signals made their debut at Princess Square in Wolverhampton, set to operate at timed intervals. In 1932, signals triggered by vehicle movement were trialled in London, forming the basis of modern-day computer-controlled traffic lights. But in a strange twist of fate they, too, were destroyed in a gas explosion.

▷ **Traffic lights in Atlanta in the 1920s**
Heavy traffic at Five Points intersection in Atlanta, Georgia, is regulated by an early set of traffic lights. The intersection is located where five major roads converge at what was then regarded as the city centre.

ENGINEERING LEADERSHIP
in every price class
WILLYS–OVERLAND — Fine Motor Cars
Richards-Wilcox
FRED THOMSON
DEVIL OVERADO
E & CO.

Refined design

Until mass production became widely established, car ownership was an obvious sign of wealth, and many manufacturers competed to meet the demands of the upper class motorist with truly luxurious models.

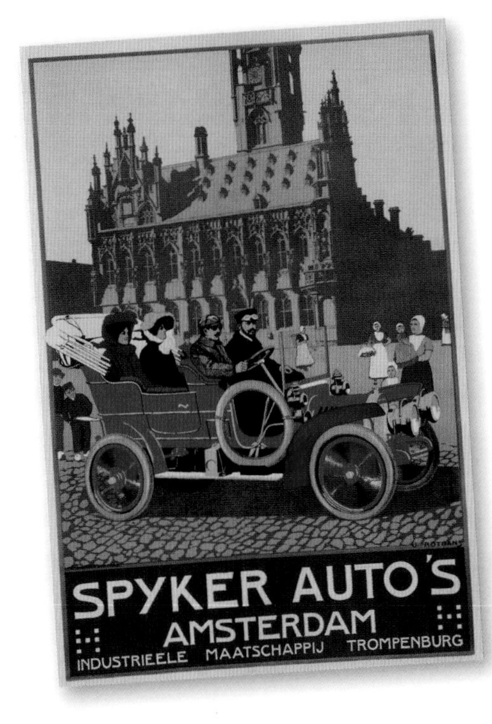

▷ **Poster for Spyker, 1910**
Founded in Holland in 1880, Spyker produced its first car in 1889, and quickly developed a reputation for quality.

The Roaring Twenties saw a surge in the production of high-end cars, leading to a huge choice for those who could afford them. Cheap labour and a booming economy (for the wealthy, at least) saw cars become increasingly refined, with sculpted coachwork, plush interiors, and powerful engines. It was an era of no limits for the affluent motorist. Manufacturers, including Isotta Fraschini, Hispano-Suiza, Delahaye, Cord, Voisins, Minerva, Maybach, Pierce-Arrow, Packard, Auburn, and Duesenberg (all now defunct), competed to satisfy the demands of their well-heeled clientele by producing increasingly luxurious and expensive machines.

International appeal

The fashion for flamboyance was most prevalent in Europe and the US, but also extended to India, where rich maharajahs commissioned large numbers of bespoke motors to reflect their status (see pp.80–81). For many of their owners, these cars were not so much about long-distance journeying in supreme comfort, as showing off their wealth and good taste at *concours d'elegance* events in Biarritz and Deauville in France, or Pebble Beach in California.

Most manufacturers of these cars have since floundered, although a number have survived and prospered, including Bentley, Alfa Romeo – very much a blue-blooded, sporting marque before World War II – Rolls-Royce, Cadillac, and Mercedes-Benz. While many of the cars these companies produced were works of art, the best came from Ettore Bugatti, an Italian whose factory was based at Molsheim in what was then Germany and is now France.

Bugatti came from a family of artists and considered himself one too, but he was also an engineer. The company went on to produce some of the most beautiful cars of the early 20th century, including the vast 1927 Bugatti Royale – an eight-cylinder, 13-litre limousine that drove like the racing cars for which Bugatti was famous. This model was later followed in 1936 by the sculptural Type 57 Atlantic coupé and the Type 57C Atalante Cabriolet.

Car couture

Bugatti's Atalante Cabriolet was the work of French coachbuilder Figoni et Falaschi, whose flowing, teardrop shapes were partly inspired by aircraft design. Fashion also played a role, however, with the coachbuilders

DRIVING TECHNOLOGY

Braking system advances

Engineers had largely perfected the mechanical smoothness and silence of a six- or eight-cylinder power unit by the 1920s. "Go" was one thing, but the '20s saw a close attention to "stop", as braking systems were vastly improved. Brakes on all four wheels were first offered by Scotland's Arrol-Johnston in 1909, but in the early 1920s this was still a rarity. Italy's Isotta Fraschini cars, however, featured them as standard from 1910. Hispano-Suiza was first to devise a servo-assisted system in 1919, and a year later the US's Duesenbergs included the first hydraulic set-up. The result of all these advances was to make braking a quicker-reacting and more predictable action. This meant a big, powerful car could be driven much more smoothly and with less lurching around on the move. At speed, of course, safety was hugely boosted too.

THE FIRST MARQUE **TO OFFER FOUR-WHEEL BRAKING, ARROL-JOHNSTON SOLD ITSELF ON TECHNONOLOGY.**

Retractable roof is an option according to the bodywork specified

Multiple vents cool the powerful 7.3-litre, straight-eight engine

◁ **Isotta Fraschini Tipo A8, 1924**
Supplied only as a rolling chassis, the car had its bodywork designed and fitted by a coachbuilder.

often displaying their cars adorned with models wearing the latest fashions to complement the colours and lines of the cars. Ovidio Falaschi, who ran the business while Giuseppe Figoni designed, described themselves as, "true couturiers of automotive coachwork, dressing and undressing a chassis one, two, three times, and even more before arriving at the definitive line that we wanted to give to a specific chassis-coachwork ensemble".

Advances in manufacturing

Manufacturers were motivated by artistry, imagination, money, and the desire to outdo their peers to create these incredible cars, but it was advances in technology and materials that made this possible. In particular, perfectionist engineers developed the high precision necessary to achieve both an exacting finish and the mechanical calibre to underpin these cars' styling. Improvements in the durability of metals and the precision with which they could be machined enabled these cars – mostly – to perform as well as they looked. Manufacturers strived for mechanical perfection, such as Bugatti's insistence on its car engine bays being free of unsightly cabling. Eventually, many of these pioneering new technologies would cascade their way down to the more affordable machinery, and benefit the engineering of all future motorcars.

◁ **Sir Henry Royce in his Silver Ghost asking for directions on the French Riviera, 1922**
Produced from 1906–26, the Silver Ghost (originally named 40/50 hp) was replaced by Rolls-Royce in 1925, as models from competing manufacturers began to rival its reputation for quality and reliability.

Vehicles of war

When war broke out in Europe in August 1914, engineering and manufacturing facilities were drafted to meet the demands of conflict in the trenches – not least in the fledgling automotive industry.

As the heavy artillery rolled out of munitions factories to support the foot soldiers mobilized in World War I, military strategists had to find a way to move it. Transportation became a priority, pushing the new automobile industry into overdrive.

△ **Medical Corps poster**
Proudly proclaiming, "The American ambulance in Russia", this poster solicits funds for the US ambulance service working on the Russian Front in World War I.

Armoured cars

Belgium was the first nation to respond to the challenge, developing an armoured vehicle based on the domestic Minerva car; this was ready for use in August 1914, just one month after war was declared. Meanwhile, the UK's Royal Naval Air Squadron was in France, using open-topped Rolls-Royce Silver Ghosts with machine-guns on board for spotting the German advance. Naval servicemen swiftly adapted the cars by welding iron sheets on to the sides to deflect enemy bullets.

Rolls-Royce was soon producing a purpose-built armoured vehicle based on the Silver Ghost – the Armoured Fighting Vehicle (AFV), an active contributor to the war effort. With armour cladding and a rotating machine-gun turret on some models, the Rolls-Royce AFV was most famously employed in the Middle East, where the Duke of Westminster led a squadron of the cars across the desert to rescue British prisoners in Egypt. (Rolls-Royce aero engines also became the powering force inside more than half of Allied aircraft.)

With AFV manufacture in full swing, the next challenge was to find enough experienced drivers, at a time when only the wealthy owned cars. Drivers were quickly trained, although mechanics were harder to find, so a recruitment drive began to source them from around the British Empire for both Navy and Army postings.

Medical transport

While combat vehicles were prioritized, provision for medical transport lagged behind. A fundraising campaign in *The Times* newspaper rallied support for the cause, and by January 1915 more than 1,000 ambulances and motorcars were in service, many donated by private owners. The Red Cross soon developed a specification for ambulance bodies that could be fitted to standard touring cars. Numerous manufacturers began producing these ambulance bodies, including Daimler, Morris, Sunbeam, Rover, Renault, Buick, and Ford,

△ **World War I tank helmet and mask**
Armoured vehicle crews wore masks to protect from "spalling" – flying shards of metal from vehicles' interiors caused by enemy fire.

▷ **Armoured convoy in World War I**
Belgian Minerva armoured cars travel in convoy to the Western Front in France, 1918. The Minerva entered service in 1914, and featured a four-cylinder 40-hp engine and an 8-mm Hotchkiss machine-gun.

some of whose Model T cars were converted into mobile field ambulances for the Red Cross.

The rise of the tank

When combat moved to the Western Front, armoured cars were no longer enough. In previous wars, cavalry was used to punch through enemy lines, but this was futile in the face of heavy machine-gun fire. To replace horsemen, tanks came in.

The British Mark I was the first tank to see combat, at the Battle of the Somme in September 1916. Soon, Renault in France revolutionized tank design with its lightweight FT, which was more sophisticated than the early French Schneider tanks. Seeing battle for the first time in May 1918, the FT proved its worth, breaking up a German advance east of Chaudun, and was then adopted by US forces on the Western Front. Germany would eventually develop its own tanks in 1917. Even with tanks at the vanguard of combat, the infantry that followed had to battle enemy fire, which made progress slow and also left tanks isolated and vulnerable in enemy territory. The solution was an armoured vehicle for troop transport, the Mark IX, but the war ended before the first of these could be deployed.

KEY DEVELOPMENT

Tank Mark IV

Although the Mark I tank had proved its value at the Western Front, it was notorious for breaking down in the field. After developing two further prototypes, British engineers William Tritton and Major Walter Gordon Wilson unveiled the improved Mark IV in 1917. Instead of the gravity fuel feed of the Mark I, a vacuum system was used to prevent the tank stalling when at a steep angle. The gun sponson was retractable, so that it did not have to be detached and then reassembled for transport.

6-pounder gun in sponson

MARK IV "MALE" **WITH TWO 6-POUNDER GUNS (AS HERE), OR "FEMALE", WITH FIVE MACHINE-GUNS.**

" ... towards us came **three huge mechanical monsters.**"

BERT CHANEY, SIGNAL OFFICER IN WORLD WAR I

Bygone marques

For every car manufacturer still in production, the early history of motoring is littered with many more that fell by the wayside.

BERLIET, FRANCE, 1899–1939

Sometimes carmakers were overtaken by technological change. The Stanley brothers built successful steam cars early in the 20th century, but the development of practical petrol cars left the company unable to compete on price or performance, and it closed in 1924. Changing economic conditions were another factor responsible for the demise of carmakers. Dozens of brands, even very successful names, were killed off by the slowdown in vehicle sales resulting from the 1929 Wall Street Crash. E.L. Cord's Auburn-Cord-Duesenberg group was a high-profile casualty, although it limped on until 1937.

The result is that many trademarks that were once instantly familiar to drivers and enthusiasts on the prow of bonnets are now merely obsolete emblems, occasionally seen at car shows, and collected as intriguing motoring curiosities.

Straight-eight badge was fitted to Duesenberg Model A

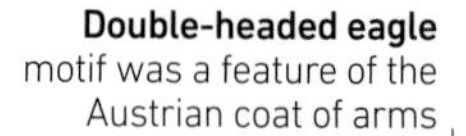

Double-headed eagle motif was a feature of the Austrian coat of arms

AUSTRO-DAIMLER, AUSTRIA, 1899–1934

SWIFT MOTOR COMPANY, UK, 1900–31

WOLSELEY MOTORS, UK, 1901–27

HAYNES, US, 1905–24

IMPERIA, BELGIUM, 1906–34

TALBOT, FRANCE, 1903–38

CROSSLEY MOTORS, UK, 1906–38

Radiator emblem appeared on early Hupmobile models, K and N

HUPMOBILE, US, 1909–40

Enamelled badge sits on the front of Stutz's model Vertical Eight

STUTZ MOTOR COMPANY, US, 1911–35

BEAN CARS, UK, 1919–29

ANSALDO, ITALY, 1921–31

DUESENBERG, US, 1921–37

Propeller motif represents the company's origins in aero engines

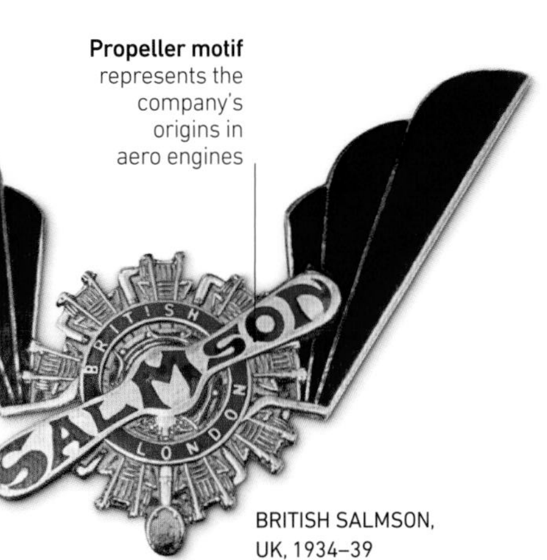

BRITISH SALMSON, UK, 1934–39

ITALA, ITALY, 1914-16

KEY DEVELOPMENT

Maybach revivals

Gottlieb Daimler's technical partner Wilhelm Maybach went into business for himself in 1909, building engines for Zeppelin aircraft and a series of luxury cars – until World War II put an end to car production. Daimler bought Maybach in 1960, and the company revived in 1997, when it produced super-luxury cars based on the Mercedes-Benz S-Class. However, next to new products from Rolls-Royce and Bentley, Maybach sales were poor, and Daimler ended production in 2012. In 2014, the company was revived again as a luxury sub-brand of Mercedes-Benz, producing the Mercedes-Maybach S600 V12 and S500 V8 models.

THE MAYBACH MASCOT **FEATURES ENTWINED LETTER "M"s, FOR MAYBACH MOTORENBAU (MAYBACH ENGINE).**

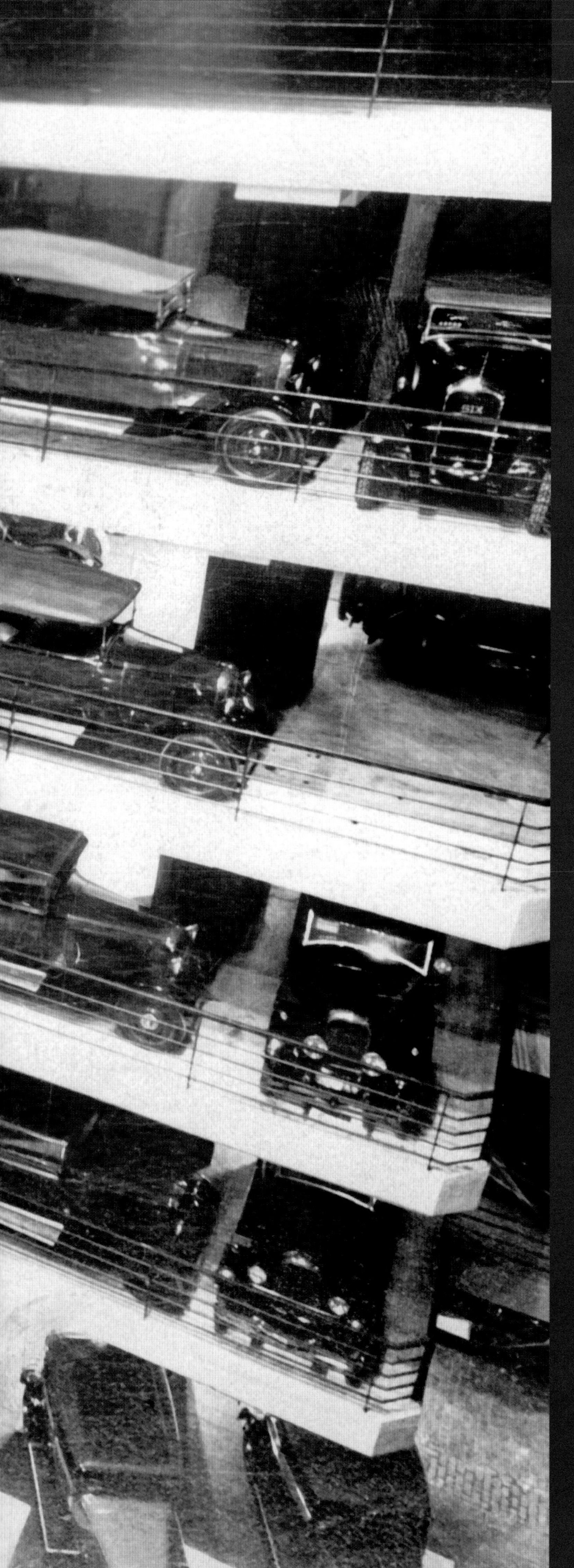

Carparks in the sky

Cars became popular so quickly that dedicated parking places soon became imperative. Often ingenious, some of these buildings were also aesthetically pleasing.

The world's first multi-storey carpark is thought to be the one built by the City & Suburban Electric Company on Denman Street in Soho, London, in 1901. With seven floors, it had room for 100 vehicles, and an electric lift to move them up and down. City & Suburban converted a second building in Westminster the following year, creating space for 230 cars. As the company's name implies, its carparks were filled with electric cars, which were cleaned and serviced, insured, and even delivered to, and collected from, their owners. One of the oldest carparks still in operation is on Carrington Street in London's Mayfair, where it was opened in 1907 as the Electromobile Garage.

The very first American multi-storey carpark is believed to have been built to serve the Hotel La Salle in downtown Chicago in 1918. Located several blocks away, it outlived the hotel by nearly 30 years, but was demolished in 2005.

Early carparks were often far more stylish than their humdrum purpose would imply, with architects incorporating Arts & Crafts, Art Nouveau, and, later, Art Deco elements into their designs. A notable example was The Kent Automatic Garage in New York, an Art Deco landmark that was later converted into apartments. Such automatic parking systems, in which cars were lifted and stacked using hydraulic mechanisms, were pioneered in France in 1905, but grew in popularity in the US in the 1920s. The "paternoster" automated parking system, which operates something like a Ferris wheel, was developed in the US, also in the 1920s, and is very popular in 21st-century Japan.

The 1920s and '30s saw an explosion in car-park building around the globe, with space created in the basements of many city apartments, while developers and local authorities bought up land and erected purpose-built carparks.

◁ **Form and function on a grand scale**
This striking Parisian multi-storey carpark – actually a garage built for Citroën – had eight floors and could hold up to 500 cars, making it the largest of its kind in the world at the time. Its angular architecture shows a distinct Art Deco influence.

Economy cyclecars

Cyclecars were a low-cost attempt to democratize car ownership. Although small, basic, and often unreliable, they offered a taste of motoring without the expense of having to buy a full-size car.

△ **Jappic cyclecar at Brooklands, 1925**
In this tiny car, the driver's jacket needed an elbow pad to prevent it from rubbing against the rear wheel.

Despite its minimalist crudeness, the cyclecar bridged the gap between the motorcycle and the motorcar, bringing motoring to those who could not afford a car. It evolved from the very first breed of small car, known as the *voiturette*, which appeared in France in the early 1900s. By 1912, there were around a dozen cyclecar makers apiece in the UK and France; two years later, the number had increased to more than 100 in each country. Manufacturers sprang up in the US, Germany, Austria, and elsewhere in Europe too.

Varied designs

The mechanical make-up of cyclecars varied considerably. Some had three wheels, although many had four, and some, such as the stark Buckboard, had five. The types of engine were equally varied, as were the methods of conveying power from the crankshaft to the wheels. There were single-, twin-, and occasionally four-cylinder engines, some of which were cooled by air, others by water. Tricky friction clutches could also be used in tandem with chain or drive-belts to transmit power from the gearbox to the axle, thus avoiding the weight (and cost) of a differential, which enabled the wheels of a conventional car to turn at slightly different speeds around a corner. There was plenty of choice of bodywork too, ranging from the eccentrically ill-proportioned to elegant miniatures of full-size cars. Nevertheless, all had the same large-diameter, spindly wheels that gave most cyclecars an appearance of fragility – one that was usually borne out in practice, despite promises of long-distance reliability.

Cyclecars were simple enough that half-competent engineers could create new models quite easily, using off-the-shelf engines, gearboxes, and wheels. This, coupled with a natural desire to develop new designs, quite often led to some bizarre-looking contraptions. With so many engineers working on so many designs in so many places, the evolution of these vehicles was inevitable. An advertisement for the £100 1921 Carden, for instance, boasted: "No belts! No chains! No frictions!"

△ **Griffon Tricar poster, 1900s**
This poster for one of the earliest cyclecars, by Griffon of France, shows the vehicle recklessly overtaking both a bicycle and a car.

A brief life

The UK had its own national magazine devoted to these vehicles – *The Cyclecar*, which was launched as a weekly in late 1912. The magazine was still selling a decade later as *The Light Car and Cyclecar*, but the change of name was ominous. During this time, full-size cars were evolving too, not just in the ways of functionality and reliability, but also in terms of cost. Major manufacturers such as Ford in the US, Peugeot in France, and Austin, Morris, and Singer in the

◁ **Mauser Monotrace car**
Mauser built this two-wheeled car in Germany after World War I. It had a 510-cc four-stroke engine and a pair of stabiliser wheels that could be raised or lowered by pulling a lever.

△ **Bedelia cyclecar**
Introduced in 1910, and built in Paris, France, the tandem-seat Bedelia was steered from the rear seat, but the front passenger had to help with changing gear.

UK, were harnessing mass-production methods to lower prices and improve quality, taking their offerings ever deeper into cyclecar territory.

The British 1922 Austin Seven in particular did much to kill the cyclecar, despite being diminutive itself. It came with a four-cylinder water-cooled engine, and early improvements included an electric starter, a cooling fan, and a speedometer – refinements that accelerated the Seven's mushrooming popularity and the cyclecar's demise.

Interest in small, economical cars was eventually rekindled after World War II, leading to the production of the "bubble cars" of the 1950s and '60s.

LIFE BEHIND THE WHEEL

Motorized buckboard car

This cyclecar's fifth wheel was a startling novelty, and had an integral engine providing propulsion from the rear. It was a low-cost attempt to overcome the need for a differential – an expensive gear system that enabled a car's driven wheels to rotate at different speeds during a turn to aid stability. The fifth wheel eliminated this need, its central position at the rear providing equal distribution of force. An even more economical arrangement was the single rear wheel, as employed by the Morgan Runabout.

Engine provides power to the rearmost wheel

△ **Briggs & Stratton Flyer, 1919**
Produced by lawnmower engine maker Briggs & Stratton, this buckboard car sold for $125 until 1925. The motorized fifth wheel lowered to drive the flexible wooden chassis.

Middle-class motoring

Early cars were for the well-heeled only, but smaller, cheaper models opened up car ownership to middle-class motorists for little more than the price of a motorcycle and sidecar – and sales boomed.

△ **Morris poster, 1920s**
The Morris Cowley Saloon was ideal for getting out into the spring sunshine, as depicted in this poster, and cost just £195.

In 1919, Fiat gave the Italian middle classes exactly what they hankered for: a neat, responsive, robust family car. This car, the 501, hugely expanded motoring across the country during the 1920s, with its four-cylinder engine and four-speed gearbox. It was not a "people's car" as such, but had great appeal to those in prosperous businesses or professional employment. Over 65,000 501-type cars were sold. Perhaps the 501's aspirational pull owed something to the presence of the firm's 41-year-old technical director, Carlo Cavalli, who had trained as a lawyer before becoming an automotive engineer.

The UK, France, and Germany

Parallel strategies were emerging all over Europe. France, still one of the major players in automotive vehicle development, had the Citroën Type C, Renault 6CV, and Peugeot 201. In Germany there was the Opel 4PS *Laubfrosch* (tree frog). In the UK, William Morris kept his production costs down by buying components from outside suppliers, which were used to assemble the "bullnose" Morris Oxford. For the later Morris Cowley, he used American engines and axles, as they proved to be even cheaper.

Austin steps in

Rival manufacturer Austin had concentrated on larger cars, but soon realised it was missing out on an unmet demand for cars that were light and cheap. Herbert Austin himself laid out the design for the Austin Seven on the billiard table of his home near the Longbridge factory, with the help of 18-year-old draughtsman Stanley Edge.

The Austin Seven was launched in 1922. Though small and inexpensive, the new car was built to the same standard as the larger Austins. It had a light steel chassis and mostly fabric bodywork. The engine was a tiny four-cylinder unit with "spit and hope" splash lubrication, but the Seven had technical innovation where it counted: it was one of the first production cars with front-wheel brakes as standard.

Austin sold tens of thousands of Sevens in various guises, from formal saloons to two-seater sports cars, but the influence of the Seven extended much further than that. The car was built under licence in France and Germany, while in Japan the first Nissan cars were laid out along very similar lines.

◁ **Austin Seven on the road**
Two women in the 1920s pose on the road with an Austin Seven. Its small size and affordability made it a hit.

KEY EVENTS

- **1911** The first Ford factory outside the US opens at Trafford Park, Manchester, UK.
- **1913** Morris uses bought-in components to build the Oxford.
- **1919** Citroën introduces the Type A light car.
- **1921** Morris cuts prices and its annual sales leap to 30,000 cars.
- **1922** Fiat's vast Lingotto plant in Turin starts mass-produing family cars.
- **1922** Ford becomes the first carmaker to build a million cars in one year.
- **1922** Austin introduces the tiny Seven which goes on to be built under licence in France, Germany, and the US.
- **1924** The 4PS is the first Opel to be built on a moving production line.

***UPHOLSTERY DEPARTMENT*, MORRIS FACTORY, IN THE MID-1920s, TRYING TO MEET DEMAND.**

△ **Motorists** refill their Fiat 501 with petrol in Italy, 1925.

Cars of the maharajas

Before World War II, British manufacturer Rolls-Royce's single most important group of customers were thousands of miles away.

India's regional rulers – its maharajas – nursed a passion for Rolls-Royces far beyond that of their colonial British occupiers. The finely engineered luxury cars with their distinctive Palladian radiator grilles were seen as the ultimate in transport, in which the heads of the dynasties could parade in regal magnificence before their subjects.

This display of automotive wealth was fully in line with the excess many maharajas enjoyed in their domestic lives. Indeed, these cars were miniature palaces on wheels, and the bespoke customization to make them stand out had never been seen before. No other carmaker could quite match them.

Their patronage was highly valued by Rolls-Royce back in the UK, at Derby. Their orders accounted for almost 10 per cent of the firm's entire output between 1908 and 1939, split between some 230 maharajas. His Highness the Nizam of Hyderabad, for example, bought his first Rolls-Royce Silver Ghost in 1913 as a yellow-painted, brocade-upholstered, gold-adorned mobile throne, and would eventually own 50 examples of the marque (he also reputedly had 12,000 servants). Meanwhile, the Maharaja of Mysore bought his cars in batches of seven, such an important order soon being referred to as "doing a Mysore" at the factory.

Rolls-Royce arranged for all manner of features to be built into the cars before they were shipped to India. These could range from colours matched to specific clothes or shoes, to powerful spotlights for tiger-shooting at night. The bodywork could be open so the owners could be admired, or closed-in with thick curtains so female members of the client's family need not be gawped at by ordinary men. Sometimes, exteriors were encrusted with jewels, necessitating guards to make sure none of the glittering splendour was stolen. India banned the export of these national treasures, so up to a quarter of the cars survive today in museums and private collections across the country.

▷ **Rolls-Royce Silver Ghost, 1920**
This heavily customized Rolls-Royce Silver Ghost was built for the Raja of Munger, Sir Raghunandan Prasad Singh, and was one of the most elaborate vehicles ever to grace India's roads.

Right or left?

There is no uniform code of conduct for the roads, and this extends to the side of the road that motorists drive on: in 163 countries they drive on the right, while in 76 they drive on the left.

Traditionally, traffic occupied the left-hand side of the road – for the simple reason that most people are right-handed. If you were right-handed, it was easier to mount your horse from the left, and to fight on horseback if your opponent was on your right. Being on the left also kept your scabbard away from the enemy, and so reduced your chances of being disarmed. And if you were simply walking your horse, you could hold the reins with your right hand and walk at the side of the road. As such, left-hand riding was natural, convenient, and almost universal, as it had been since medieval times, and perhaps even back to the days of ancient Rome, Greece, and Egypt.

The French and the Americans challenged all this in the 1700s. In the US, as larger freight wagons driven by multiple horses became popular, it was customary for the driver to sit near the back-left horse so that he could control all the horses with his right hand. To keep the whips clear of oncoming wheels, it made sense for these wagons to pass each other on the left, and so drivers began driving on the right-hand side.

The Napoleonic Wars had a similar effect in Europe. After his victory, Napoleon sought to impose his will on every aspect of France, including the

△ **Ford Model T, 1908**
The first cars had centrally located driver's seats, but by the time of the Model T it was clear that the driver needed a good view of the centre of the road. In the US, this was on the left-hand side.

▽ **Third Street, Minneapolis, US, 1915**
The first cars shared the road with trams and horse-drawn wagons. Use of the right-hand lane became common in the 18th century.

roads. Many believe that his decision to move from left to right was a protest against the old aristocratic practice of forcing peasants across to the right-hand side of the road – into oncoming traffic – so as not to impede your progress. Whatever the reason, the left was seen as the "old way", and as such it had to be scrapped. As France's empire expanded, right-hand traffic was imposed on vast swathes of Europe and Africa, as well as the West Indies and French Indochina.

Amherst Daily

AMHERST, NOVA SCOTIA, SATURDAY, APRIL 14, 1923

Drive to the Right After Two O'clock Sunday Morning on all Roads in Province Nova Scotia

One vehicle in passing another travelling in same direction will pass on left—Preparations being made for informing public of change—'Drive to Right' signs for autos—

Halifax, N. S. April 14—At two o'clock tomorrow morning and henceforth, highway traffic in Nova Scotia of wheeled vehicles, motor, electrically beast or man-propelled, and traffic of animals led, driven or ridden, will, when meeting and passing on the public highways, keep to the right hand side of the centre of the road, according to provisions of bill twenty four introduced in the Nova Scotia Legislature...

WILL EXTEND TIME OF CANADA HIGHWAYS ACT TO PROVINCES

(Canadian Press)

Montreal, April 12—S. L. Squires, chairman of the executive of the Canadian Good Roads Association speaking in regard to the attitude of the government on Federal aid to highways as expressed by Premier King in the House of Commons last Tuesday declared that "the Domin-

◁ **Road change reminder**
The front page of the *Amherst Daily News* reminds readers that from the following day, April 15 1923, drivers in Nova Scotia must use the right-hand lane.

Mixed traditions

In modern times, much of the world adapted to driving on the right as a result of the popularity of American cars such as the Ford Model T. Before it achieved independence, the US had been a patchwork of practices – its British, Dutch, Spanish, and Portuguese territories following the left-hand tradition; its French zones favouring the right. However, since it had opted for the right-hand lane, the US made cars with left-hand drive optimized for right-hand traffic. These were the first reliable, mass-produced cars, and they had markets all over the world. As they were exported, the countries they went to naturally shifted to right-hand traffic to match the vehicles they were buying.

The UK was an exception to all this. In 1773, the British government introduced the General Highways Act, which encouraged the tradition of driving on the left. The Highway Act of 1835 reinforced this, not just in the UK but in her colonies too. Japan also maintained the tradition, partly because they had always done so, but also because British engineers built their railways to be driven on the left, and their highways followed suit.

In the 1960s, the UK briefly considered switching to right-hand traffic to align itself with mainland Europe. However, the conservative nation concluded that the change would require too many resources. The UK remains one of just four European countries to drive on the left, alongside Ireland, Cyprus, and Malta.

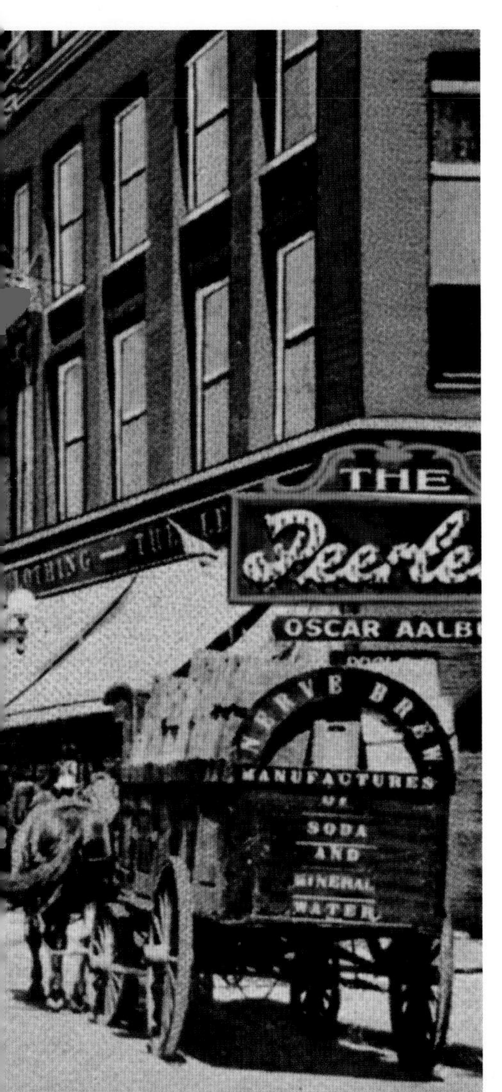

KEY DEVELOPMENT

Dagen H

Dagen H, or H-Day, was the day Sweden changed from driving on the left-hand side of the road to the right – "H" standing for "Högertrafik", meaning "right traffic" in Swedish. The day came on 3 September 1967, and its purpose was to align Swedish traffic laws with those of its neighbours, and to address the fact that 90 per cent of Swedes drove left-hand-drive vehicles. On the day, all non-essential traffic was banned from the roads between 1am and 6am. Then, at 4:50am, all traffic had to stop and carefully change to the right-hand side of the road. As a consequence of the change, many tram services had to be discontinued and buses were retrofitted to have their drivers' cabins and passenger doors switched to the correct side.

CARS CHANGE LANES **ON DAGEN H, THE DAY SWEDEN SWITCHED FROM DRIVING ON THE LEFT-HAND SIDE OF THE ROAD TO THE RIGHT.**

"[Sweden has] a **brief** but **monumental** traffic jam."

TIME MAGAZINE ON SWEDEN'S DAGEN H

Crossing the desert

To prove the resilience of his cars – and the colonial might of France – industrialist André Citroën first took on the Sahara Desert. To prevail, a new "go-anywhere" vehicle was needed.

A normal car would never have coped with the shifting sands and hostile terrain, but Citroën's *autochenilles* (motor-caterpillars), introduced in 1921, were perfect for this arduous challenge.

The idea for these "half-track" vehicles, with powered tracks replacing conventional rear wheels, came from engineer Adolphe Kégresse, who had found success with the concept in Russia, where he ran Tsar Nicholas II's garage. Unlike the heavy steel of World War I tank tracks, Kégresse's caterpillar tracks were made from rubber and canvas.

Citroën appointed his manager, Georges-Marie Haardt, to organize the first expedition. Ten men in five vehicles crossed the Sahara from Touggourt in Algeria to Timbuktu, Mali, and back again between 17 December 1922 and 7 March 1923. They reached Timbuktu in just 20 days, linking north and west Africa for the first time ever by motor vehicle. France needed a road link between the homeland and its colonial outposts, and here it was.

Emboldened by this success, Citroën embarked on a 12,430-mile (20,000-km) overland trip to the isolated French territory of Madagascar. He called the adventure the *Croisiere Noire* ("Black Cruise"), intending to conquer the whole of Africa, from Colomb-Béchar in Algeria to Cape Town in South Africa.

The convoy set off on 28 October 1924. The 16 men in eight vehicles were a hardy yet disciplined crew, ordered by Haardt to be immaculate and clean-shaven at all times. On 26 June 1925, the party arrived at Antananarivo on Madagascar. The survival of the cars was particularly impressive because the rubber tracks wore away quickly and needed almost constant repairs. A year later, a 70-minute film of the expedition was released to huge acclaim.

▷ **Setting up camp**
While Georges-Marie Haardt's team slept in makeshift camps when crossing the Sahara desert from Touggourt to Timbuktu in 1922–23, it was envisioned that the route would eventually become a major trade and tourist trail, dotted with luxury hotels and restaurants for wealthy patrons.

“[It] is intended to **transport as quickly as possible** the traveller wishing to go, **through the desert**, to the Nigerian region for **sport or business**.”

ANDRÉ CITROËN ON THE ESTABLISHMENT OF A MOTORIZED ROUTE ACROSS AFRICA

1926–1935

SPEED, POWER, AND STYLE

1926–1935

Speed, power, and style

Speed and style, as agents of success, attached themselves to the basic transportation role of cars in the late 1920s – and they stuck fast. The pace of competition in both Grands Prix and endurance racing events increased hugely as laps were achieved in ever-shorter times. New technology, such as the power-boosting supercharger, was devised for the racetrack and then crossed over to road-going cars, beginning a trend that continues to this day.

Industries abound

In the US and UK, a near obsession with breaking the world land speed record saw daredevil drivers punch through the 300 mph (482 km/h) barrier. However, their thunderous machines did not share much with road cars; gigantic aeroplane engines and aviation-style aerodynamics were a world away from the enjoyable and affordable sports cars that marques, such as Fiat and MG, were bringing to the showrooms.

Aerodynamics – or "streamlining" – did have an influence on popular car design, but it was not particularly scientific, and it was frequently bound up with the same Art Deco trends that were popular in architecture and household product design. The motor industry became a competitive place, with each country hungry for more attention-grabbing gimmicks. However, behind the scenes, more fundamental changes were being introduced by engineers, such as fully chassis-free car structures that reduced vibration and improved roadholding.

In both Italy and Germany planners created the first roads specifically for long-distance driving – smooth, uncluttered, multi-lane

MOTOR RACING TECHNOLOGY IS ADOPTED BY ROAD-GOING CARS

STREAMLINED DESIGN FOCUSES ON STYLE NOT AERODYNAMICS

"...with **ever more cars** joining the roads, **discipline** had to be **imposed**."

highways that seemed to shrink the landscape as if by magic. Entirely new businesses, such as motels, carwashes, and 24-hour rescue services, sprang up to sere this new generation of motorist. However, with ever more cars joining the roads, discipline had to be imposed. The early 1930s was a time when legislators examined issues such as speed limits, driving licences, and insurance with a new intensity. Popular small cars, such as the Austin Seven, gave rise to huge numbers of new drivers – even more so when the same vehicles became cheaply attainable secondhand, "pre-loved", cars.

Bust and boom

In the midst of this tumultuous period came the economic turmoil of the 1929 Wall Street Crash, and the consequent Great Depression – the lengthy hangover from the riotous party of the late 1920s "flapper" era. Manufacturers behind some of the finest cars on sale – including Stutz in the US, Bentley and Sunbeam in the UK, and Darracq in France – struggled to stay viable as customers deserted them, and were taken over, or in some cases vanished for good.

By the middle of the 1930s, optimism was returning. Newly-built suburban homes now routinely came with a garage. Motoring holidays, sometimes with caravans in tow, were a novel experience for the middle classes. Moreover, enterprising new marques, such as SS Cars, brought style to the masses at an affordable price – SS Cars itself being the foundation stone of Jaguar. Meanwhile, on the other side of the world, Japan was just waking up to the idea that it too could start a motor industry.

MANY LUXURY MARQUES FACE BANKRUPTCY IN THE 1930s, BUT NOT SS

THE MOTORCAR BECOMES ESSENTIAL FOR MIDDLE-CLASS LIFE

Mainstream motor racing

International Grand Prix racing, and races at Indianapolis and Le Mans, fuelled public interest in cars and speed. At the same time, the sport became a propaganda tool for a technologically resurgent Germany.

The first truly international, top-flight race to carry the title "Grand Prix" had taken place in France in 1906. Five more French Grands Prix had been held before World War I. By the 1920s, Grand Prix races were set up across Europe – first in Italy, Belgium, Spain, and the UK, then later in Germany, Monaco, and Switzerland. The Indianapolis 500, by now established as the premier race in the US, was also considered one of these "grandes épreuves", or great trials, from 1923 until a change in its regulations in 1930. The increasingly fast and specialized racing cars were outlawed at the American event in an attempt to make it more relevant to road car development.

Power and politics

Pure racing cars still ruled in European events, and regulations were adapted to allow pit crews to change tyres and service cars so a riding mechanic was no longer needed. In 1922 the practice of starting the cars one by one with a set time interval in between was replaced by a mass start with grid positions allocated at random. The inaugural Monaco Grand Prix in 1933 introduced starting grid positions determined using the drivers' practice times.

The impressive but unwieldy racing cars built at the turn of the century, prioritizing power over all else and generating it using ever larger engines, were replaced by smaller, lower cars with much better handling and braking. Smaller engines were fitted with superchargers, which pumped additional air into the cylinders so more power could be developed.

German aero-engineers were the masters of the art of supercharging, and their expertise was harnessed by Adolf Hitler so that he could use motor racing as a Nazi propaganda tool. The Third Reich bankrolled Auto Union and Mercedes-Benz to develop some of the most powerful and sophisticated racing cars the world had ever seen, and the silver-painted German cars came to dominate Grand Prix racing throughout the 1930s.

As racing cars developed and became more specialized, so motor sport branched into different forms. City-to-city racing on public roads had been abandoned as too dangerous, but a similar concept survived in the form of the Monte Carlo Rally, which dated back to 1911 and remained popular.

Rally on the roads

Competitors could make their way from starting points across Europe to "rally", or meet, at Monaco. They drove at a relatively sedate pace; the condition of the car on arrival was more important than outright speed, since the winners were chosen by a panel of judges who assessed design and passenger comfort. The rally really established itself from 1924, and became one of the best-known motoring events in the world.

Another of the great events was the 24 Hours of Le Mans race in France, which began in 1923. As well as its marathon length, the Le Mans race

△ **Poster for the Monaco Grand Prix, 1931**
Designed by Robert Falcucci, this poster advertised the third Monaco Grand Prix, which was won by Louis Chiron driving a Bugatti.

▽ **Ready to rally**
British competitor, Major J.A. Driscoll (left), prepares to leave London with his crew and Ford car for the 1934 Monte Carlo Rally. He did not win.

△ **Starting grid, Nice Grand Prix, 1933**
With his place on the grid determined by his previous practice times, Italian driver Tazio Nuvolari, finds himself in pole position at the front centre. He went on to the win the event, driving his Maserati 8CM.

was notable for its 3.7-mile (6-km) Mulsanne Straight, which formed part of the circuit, and favoured cars with a very high top speed. The lessons of aerodynamic "streamlining" learned from aeroplane design and applied to everything from steam locomotives to sidecars were soon applied to specialist Le Mans racing cars, which could then achieve ever faster speeds.

KEY DEVELOPMENT
Moonshine motor sport

In the southern states of the US, some people used illegal stills to make "moonshine" whiskey to avoid taxes on alcohol. They transported their product in "stock" cars, which looked like standard factory-made cars from the outside, but were fitted with heavy-duty suspension and highly-tuned engines so they could outrun the cars of law enforcement officers. The drivers began racing each other to prove who had the fastest car and the greatest skill, and a new motor sport was born. Rules were drawn up so that the competition was fair, and the first races sanctioned by the new National Association for Stock Car Auto Racing (NASCAR) were held early in 1948.

A 1930s PROHIBITION CAMPAIGN **IN WHICH A POLICEMAN SHOWS A CAR THAT WAS DAMAGED IN AN ACCIDENT CAUSED BY DRINK-DRIVING.**

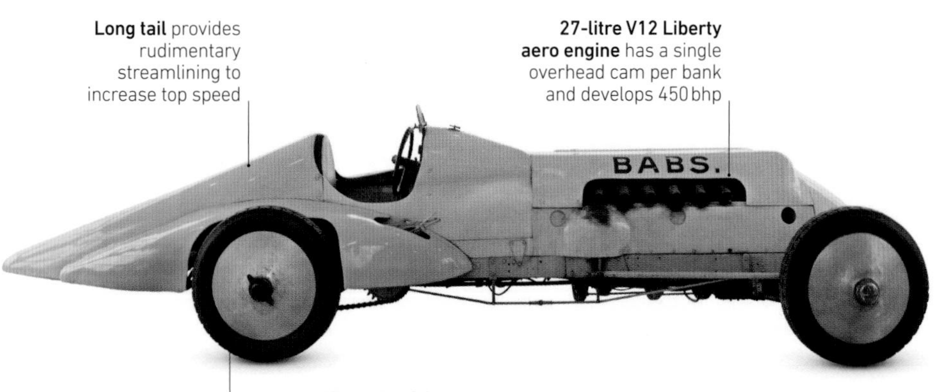

△ **"Babs", 1926**
Built by Polish racer Count Louis Zborowski, then rebuilt by Welshman John Parry-Thomas, "Babs" took the 1926 record at 171.02 mph (275.23 km/h). During another attempt in 1927 it crashed, killing Parry-Thomas.

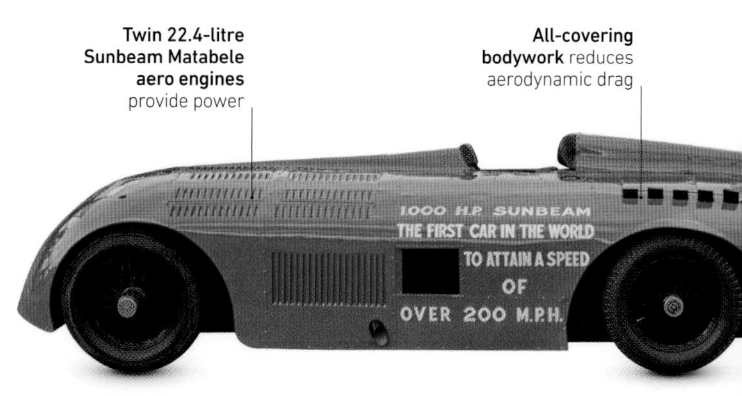

△ **Sunbeam 1,000 hp, 1927**
The Sunbeam, designed by Captain Jack Irving, was one of the first cars built specifically to break records. Henry Segrave drove it to a new record of 203.79 mph (327.97 km/h) at Daytona Beach in March 1927.

Record breakers

▽ **Blue Bird on the beach**
The Campbell-Napier-Railton Blue Bird on Daytona Beach, Florida, in 1931, having reclaimed the world land speed record with a speed of 246 mph (396 km/h).

Record-breaking feats – on land, on water, and in the air – were all the rage in the 1920s and '30s. The land speed record stood at 146 mph (235 km/h) early in 1925, but it was subsequently broken time and time again. The two most revered figures in this period were two British drivers, Sir Henry Segrave and Sir Malcolm Campbell. Segrave achieved 231 mph (372 km/h) at Daytona Beach in the Golden Arrow in 1929, then died attempting to raise his own record on water to the magic 100 mph (161 km/h). Campbell became the first to exceed 250 mph (402 km/h), then 275 mph (442 km/h), and finally topped 300 mph (483 km/h) on the Bonneville Salt Flats in the US in 1935. He then retired, but record-breaking continued. Even before the onset of World War II, machines built by John Cobb and George Eyston (see p.145) would go even faster.

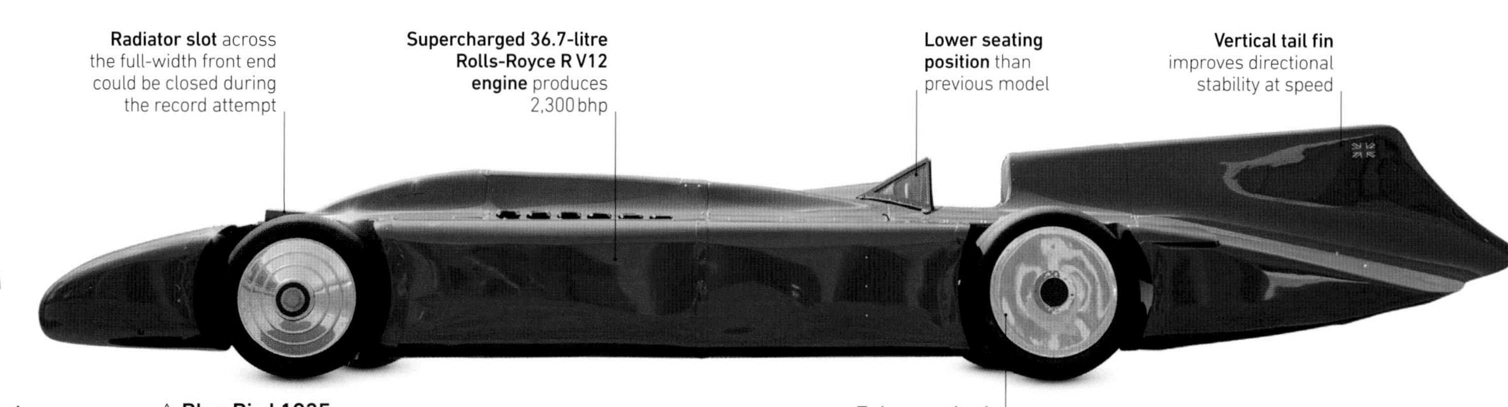

△ **Blue Bird 1935**
Sir Malcolm Campbell raised the record in stages from under 175 mph (282 km/h) in 1927 to over 300 mph (483 km/h) in 1935. This was his final record car, built at Brooklands in the UK by Thomson & Taylor.

"A tiny Delft-blue dot **grows to a machine** at full speed and **crashes past**."

THE AUTOCAR MAGAZINE DESCRIBES BLUE BIRD, 1931

▷ **Golden Arrow, Castrol poster, 1929**
This poster celebrates Sir Henry Segrave achieving a new land speed record.

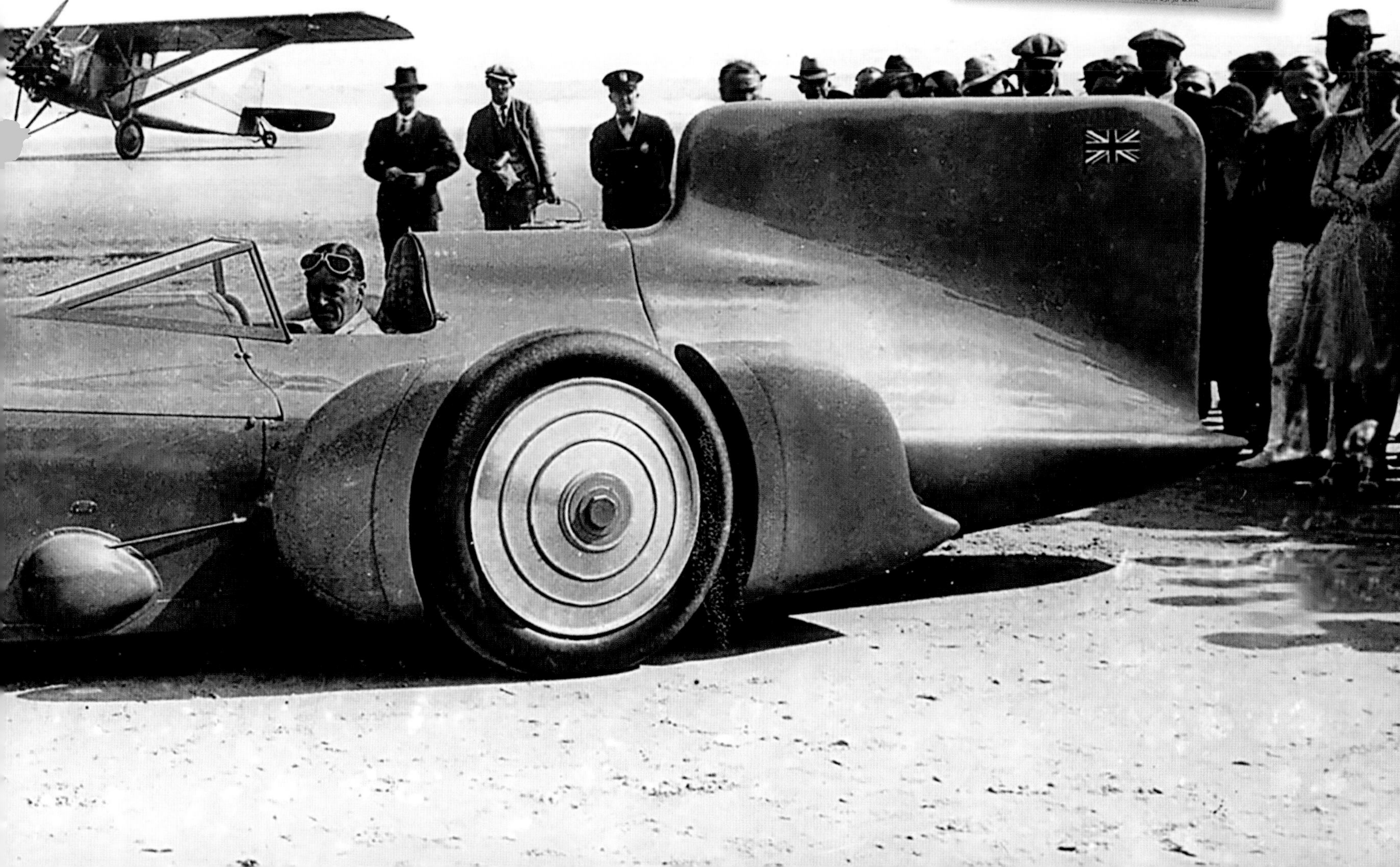

BP,
UK, 1920s

SHELL CROWN,
HOLLAND, 1920s

THOMPSON GARVIE,
UK/US, 1926

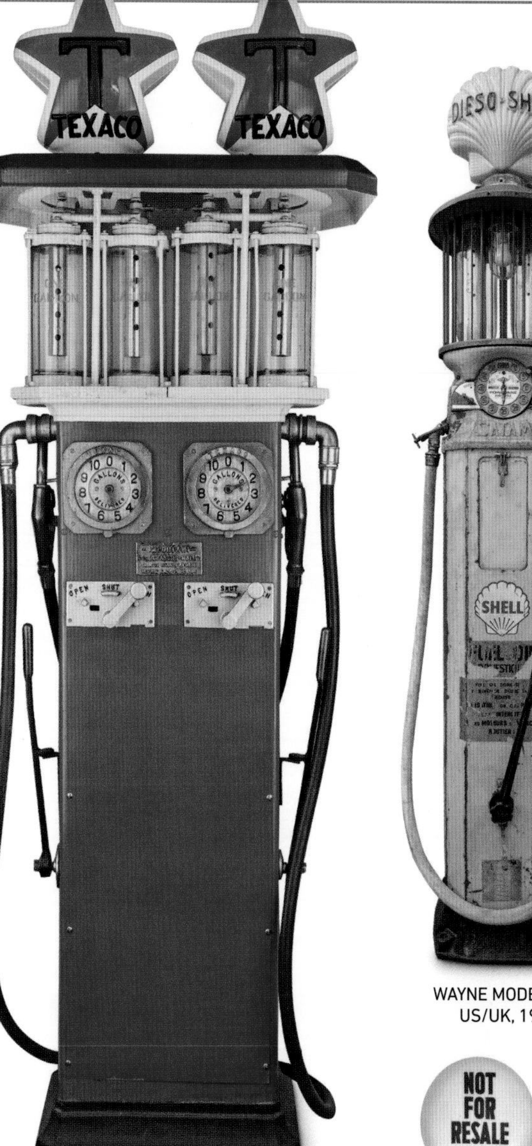

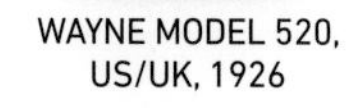

WAYNE MODEL 520,
US/UK, 1926

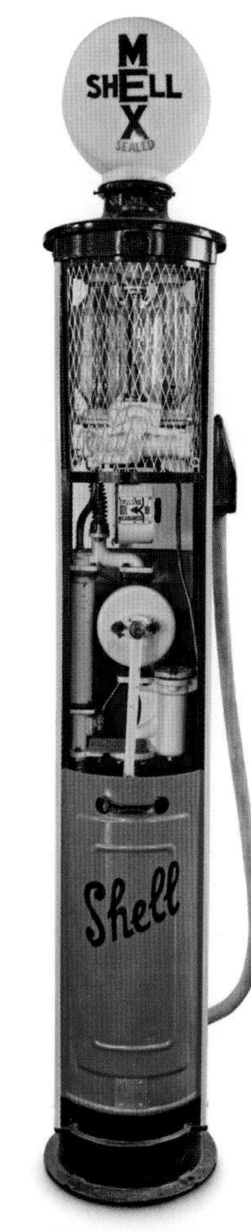

SHELL VICKERS,
UK, 1929

HAWKE DOUBLE PUMP, UK/US,
USED IN AUSTRALIA, 1930s

KEY DEVELOPMENT

The spread of filling stations

Fuel pumps were originally provided where petrol was sold as a cleaning fluid or as a fuel for lamps. The growth in the number of pumps mirrored the development of motoring worldwide, so the earliest purpose-built filling stations were in the major car markets, such as the US and UK. Russia also had hundreds of petrol stations by the outbreak of World War I. Smaller or less-developed countries had to wait: petrol stations did not become common in Greece until the 1950s, for instance, and the first self-service petrol pumps in India were opened as recently as 2011.

SPOILT FOR CHOICE, **THIS GERMAN MOTORIST COULD CHOOSE FROM 12 DIFFERENT PETROL BRANDS AT A SINGLE PETROL STATION IN THE 1930s.**

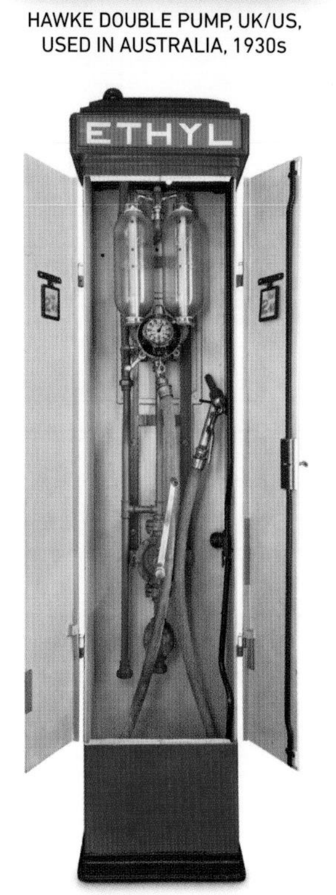

SASSO CABINET PUMP,
SWITZERLAND, 1932

SATAM TWIN-DOOR,
FRANCE, 1930s

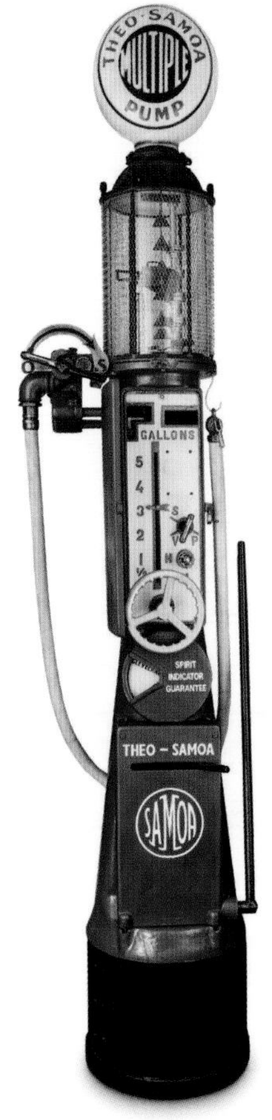

THEO-SAMOA,
UK, 1932

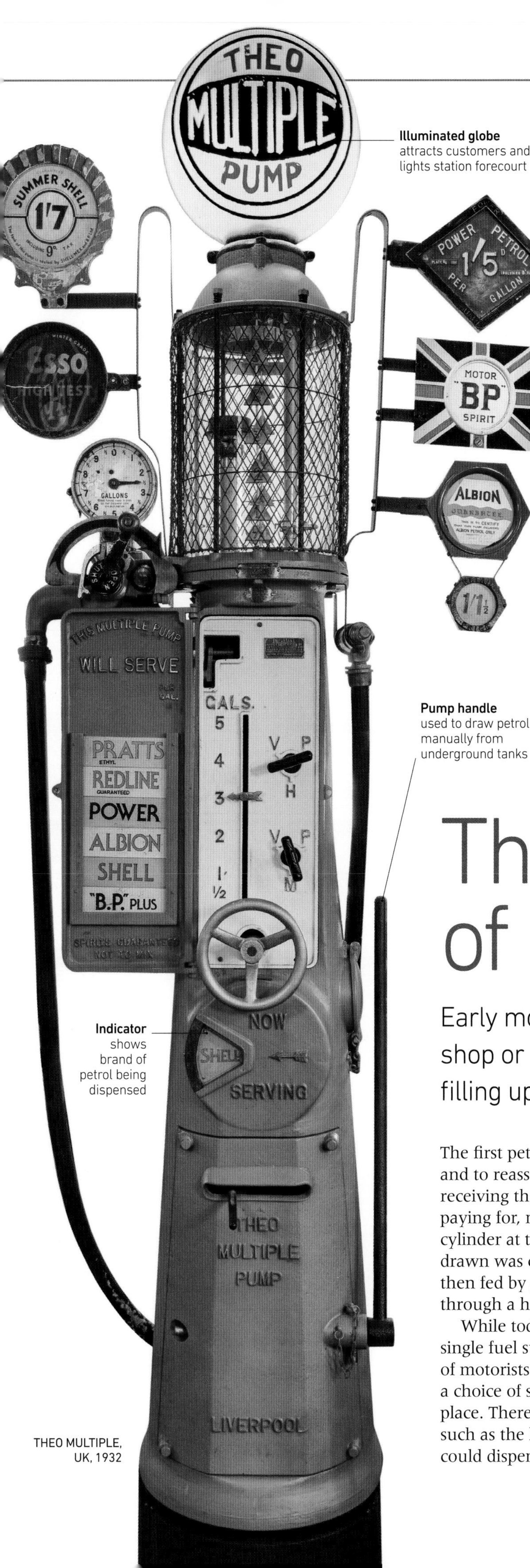

Illuminated globe attracts customers and lights station forecourt

Pump handle used to draw petrol manually from underground tanks

Indicator shows brand of petrol being dispensed

THEO MULTIPLE, UK, 1932

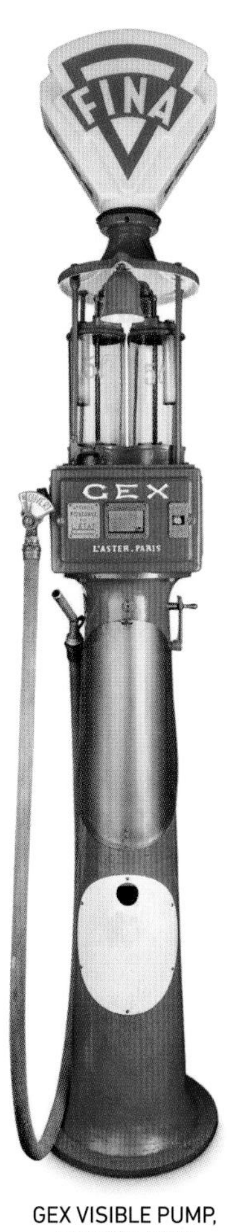

GEX VISIBLE PUMP, FRANCE, 1932

MULLAR "BIG BEN", GERMANY, 1930s

THEO ELECTRIC, US/UK, 1936

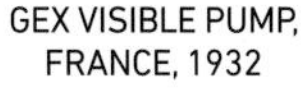

The golden age of petrol pumps

Early motorists often bought petrol in a can from a chemist's shop or hardware store, but the arrival of petrol pumps made filling up easier, quicker, and safer.

The first petrol pumps were hand-operated, and to reassure motorists that they were receiving the quantity of petrol they were paying for, many featured a glass measuring cylinder at the top, where the amount of fuel drawn was clearly shown. The petrol was then fed by gravity into the car's fuel tank through a hose.

While today's filling stations are tied to a single fuel supplier, for previous generations of motorists it was common to be offered a choice of several different brands in one place. There were even "multiple pumps", such as the large Theo Multiple (left), which could dispense several brands of fuel. As motoring became more popular, double pumps were also created to enable filling stations to serve more motorists at once.

While the purpose of the pump for the motorist was to dispense fuel, for the competing petrol producers it was an opportunity to promote their products. Hence, pumps were commonly brightly painted, and featured enamel signs and illuminated glass globes to advertise the various brands. At a time when road networks and infrastructure were still being developed for cars, these colourful pumps must have seemed like beacons of progress and modernity to 1920s and '30s drivers.

At the carwash

Once the motorcar began replacing horses, it did not take long for entrepreneurs to develop a support network catering for early motorists' maintenance needs. One such service was the carwash, which first appeared in Detroit in 1914.

In the early 20th century, paved roads were rare outside cities, and in most cases a drive in the country resulted in the car becoming covered in dirt, mud, and more.

In that filth, two Detroit businessmen, Frank McCormick and J.W. Hinkle, saw money. Inspired by Henry Ford's moving assembly line process, they created what is widely believed to be the world's first automated carwash – the Automobile Laundry, located at 1221 Woodward Avenue in Detroit. Their slogan was: "Everything back but the dirt".

The Automobile Laundry's "automation" did not consist of any kind of exotic machinery – like Ford's assembly line, it required a lot of human effort. According to the Detroit Historical Society, the customer exited his car, taking with them anything breakable. Several men then pushed the car down a line while brigades of bucket- and brush-wielding workers scrubbed the wheels and bodywork with soap and water. Then the car was hand-dried and the brass components polished. The process took 30 minutes and cost $1.50 – a princely sum in those days.

True automation came in increments and took decades. First came conveyor belts that automatically moved the car. Next, in the 1930s, came overhead sprayers, dispensing soap and water – workers still manually washed and dried the car. In the 1940s, automatic brushes were introduced to soap the vehicles. Finally, in 1946, less than a mile from where the Automobile Laundry once stood, Paul's Auto-Matic Wash added the last ingredient: a 50–horsepower blow drier. It is estimated that the Automobile Laundry could handle 100 cars per day – Paul's Auto-Matic Wash could clean 180 cars per hour.

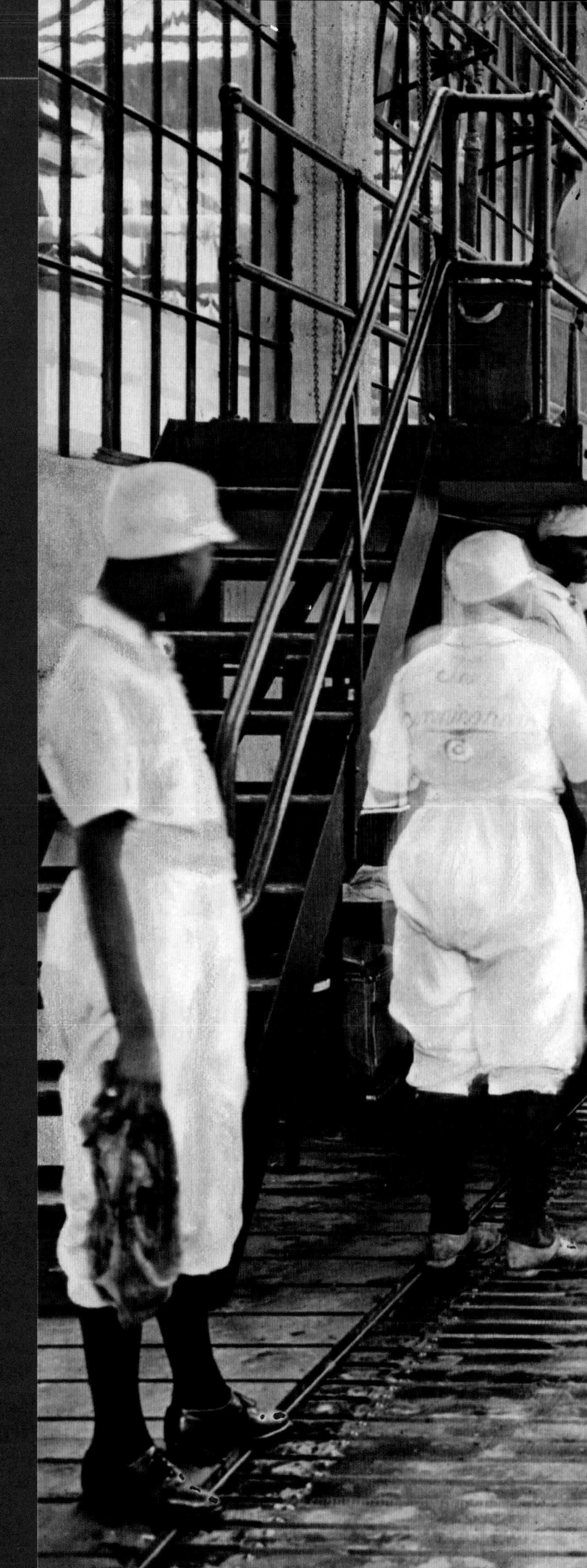

▷ **Manual carwash in Detroit, 1920s**
Before the development of carwashes with conveyor belts, the vehicle was physically pushed past the team of workers by hand. Having taken 30 minutes to complete each car in 1914, by the 1920s the process had been streamlined to just five – from rinsing through to drying the bodywork.

34-882

Streamlined style

Creativity and engineering converged in the 1930s as heavyweight marques on both sides of the Atlantic competed in designing cars that combined thrilling looks with technical mastery.

△ **Alfa Romeo 8C 2300 Berlina Sport, 1933**
Inspired by its successful line-up of race cars, Alfa Romeo produced a high-specification, coach-built touring model powered by a straight-eight, twin-cam engine. Only 249 of the elegant fastback Berlinas were produced.

In less than a decade, automotive design underwent a revolution, spurred on by advances in technology and a new approach to coachwork styling. Streamlining, as it was called, was a phenomenon with wide-ranging influences. The European art movements of Futurism and Constructivism, which glorified speed and engineering, filtered into the consciousness of car designers, along with the Art Deco design trend.

The science of design

Although influenced by Art Deco, streamlining stripped away ornamentation in favour of a scientific approach, with an aesthetic that expressed speed and motion. The sharp geometry of Art Deco was replaced with curving forms and elongated lines. Aeroplane design also had an impact, with newly launched airlines advertising their dynamic silver machines on billboards, and in newspapers and magazines.

Designers were excited by the idea that science and industry were reshaping the world, and they looked for a visual language to express this. One visionary who proved highly influential was Austrian engineer Paul Jaray. He pioneered the streamlining concept, starting with aircraft and moving on to cars in 1927. His wind tunnel experiments helped to hone the engineering aspects of aerodynamic design. Jaray created cars for many big makers, including Mercedes-Benz, Audi, Chrysler, Maybach, and Ford. Another prominent streamliner was Michigan-born designer Norman Bel Geddes, whose 1932 book *Horizons* included a futuristic vision of automobile design. He was later contracted as a consultant by Graham-Paige, Chrysler, and General Motors.

The technical side

Engineering innovations helped to radicalize car design. Pillarless windows allowed smooth lines, with no frame to interrupt the eye as it travelled down the car. Ingenious hidden door hinges also helped create a seamless look, and an underslung chassis meant that seats could be lower, allowing a better flow of the roofline. Adding to the stylish appearance, two-tone colour schemes were now achievable, thanks to the advent of quick-drying cellulose paints.

△ **Tatra T87, 1936**
This brochure for the T87 with its distinctive fin emphasized the eight-cylinder car's speed.

British coachbuilder Albany Carriage was an early adopter of the streamlined look, showcasing its aircraft-inspired Airway saloon in 1927. Czechoslovakian maker Tatra was also quick off the mark. Applying the principles of Paul Jaray, Tatra started work on a prototype car in the early 1930s. It had an air-cooled engine at the rear, allowing aerodynamic coachwork due to a small frontal area. The highlight was the Tatra T87 of 1936 – with a top speed of 100 mph (160 km/h), it was one of the fastest production cars of the era. However, the Tatra paled compared to the Pierce-Arrow Silver Arrow shown at the

RAYMOND LOEWY **PICTURED WITH EXAMPLES OF HIS STREAMLINED AUTOMOTIVE DESIGNS.**

BIOGRAPHY

Raymond Loewy

A passion for functional styling underpinned Raymond Loewy's career, and his work had a huge influence on streamlining in the car industry. He started as a fashion illustrator in 1919 for *Vogue* and *Harper's Bazaar* in Manhattan, and freelanced as a window designer for Saks Fifth Avenue and Macy's. A decade later, he switched to industrial design, and was hired by the Hupp Motor Car Company in 1930. He went on to design locomotives, refrigerators, the Coca Cola bottle, and the Shell logo. The highlight of his contribution to the automotive world was the 1953 Studebaker Starliner Coupé, which the Museum of Modern Art in New York called "a work of art".

> "We **enter a new era**. Are we ready for the **changes that are coming**?"
>
> NORMAN BEL GEDDES, *HORIZONS*

Chicago World's Fair in 1933. Its radical design featured a wide-angle V12 in a low engine bay, and hydraulic tappets. At a staggering price tag of $10,000, it was lauded as the car of the future. The same year, Cadillac unveiled its V16 Aerodynamic Coupé. Meanwhile, in Italy, Alfa Romeo released the 8C Berlina Sport, and in Germany Mercedes-Benz was at work on its own stunning aerodynamic saloon.

Rolling off the production line in 1934, the Mercedes-Benz 500K Autobahn Kurier Sport was heavily influenced by the theories of Professor Wunibald Kamm, who had worked with Mercedes-Benz on race cars for the land speed record cars. It set the standard for body shapes for years to come. Rivalry came in the form of Chrysler's Airflow of 1934. Although it was a sales flop, its sleek external features, space-frame design, and superior handling influenced a new wave of European cars from lowly Fiats to exclusive Talbots.

▷ **Chrysler Airflow on display in the Chrysler Building, New York, 1937**
The futuristic Chrysler Airflow proved too extreme in a time of economic uncertainty. Not only was its streamlined look somewhat alien, its technical innovations made the build and retail price more expensive.

△ **Singer Nine Le Mans, 1933**
In 1928, Singer was the UK's third-largest carmaker, but despite the Nine's lively handling and good reputation, the firm eschewed innovation and eventually fell behind the times.

△ **Chrysler Airflow, 1934**
This American saloon projected modernity through a streamlined look (it was not actually very aerodynamic), but an increasingly poor reputation for quality hindered its popularity.

Unitary design

Popular cars underwent a fundamental change in the 1930s. It was not always obvious to the public, or to anyone admiring the new models at motor shows, but drivers and passengers could feel it when they were out on the open road.

Engineers devised clever ways to eliminate the need for a separate chassis frame, welding the body and chassis into a single "unitary" piece. The trend had begun in 1917, when Dodge introduced the first saloon car body as a single structure made from welded steel panels – this did away with the wooden inner frame that had previously supported the body panels. However, in 1934, Citroën modernized the concept with a welded body/chassis unit for its Traction Avant.

Why did it matter? Unitary, or "monocoque", design enabled factories to increase automation, and it offered drivers and passengers vastly better rigidity and strength, increased safety through improved roadholding, and a much more comfortable ride quality, with vibration and rattling reduced significantly.

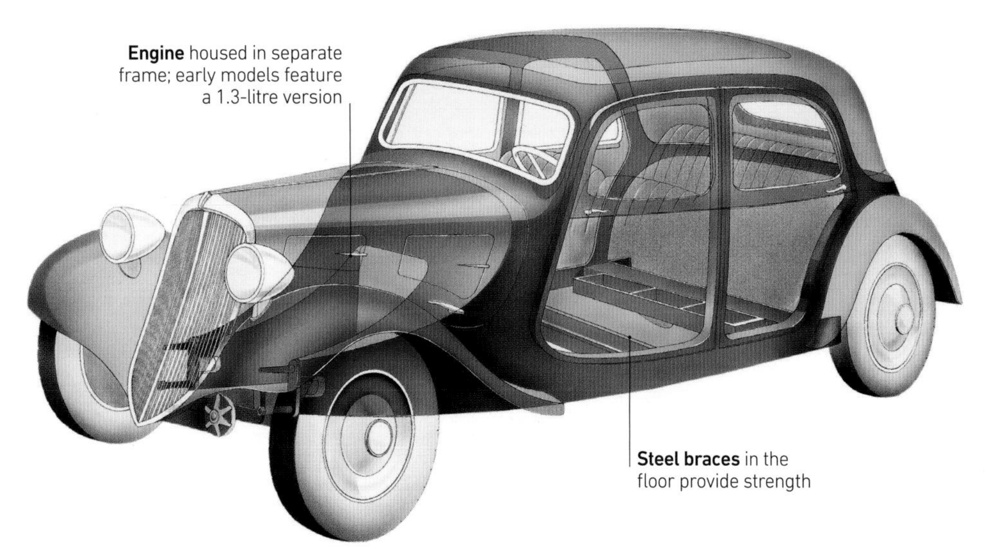

△ **Structure of the Citroën Traction Avant, 1934**
This diagram shows how the Traction Avant's body and chassis were combined in one torsionally strong unit. The engine was mounted in a separate subframe at the front of the car. It was also one of the first mass-produced models to be front-wheel drive.

▽ **Monocoque design in use**
An Opel Olympia and its three passengers pause during their journey through the Austrian mountains in the 1930s. The compact Olympia was Germany's first mass-produced car with a monocoque design, with 168,000 vehicles produced between 1935–40.

△ **Citroën 11 Large, 1935**
Better known as the Traction Avant (Front-Drive), Citroën's advanced family car design was a true standard-setter – W.O. Bentley owned one and praised its superior roadholding.

△ **Lancia Aprilia, 1937**
This ultra-modern small car embraced the latest unitary-construction technology; its body was so strong that no central pillar between the doors was needed. The sleek body was designed using a wind tunnel.

"**Nothing surpasses** the exclusive **Airflow** principle of **design!**"

CHRYSLER AIRFLOW ADVERTISING SLOGAN

Motoring in the fast lane

Multi-lane highways built for car use opened up entire countries and continents. The US term for them – "freeways" – embodies the freedom they have brought to generations of drivers.

High-speed roads reserved for motor vehicles are an essential part of modern life. Germany is often cited as the country that invented these roads. After all, the AVUS two-lane toll road near Berlin opened in October 1921. However, although it was open to public traffic, its main role was as a test circuit and race track.

In reality, Italy invented the motorway in 1924, with a car-only *autostrada* that connected Milan to the northern Italian lakes. The brainchild of engineer and entrepreneur Count Piero Puricelli, it was opened in September 1924 by the King of Italy. It had just one lane for each direction, but, crucially, the lanes were separated. Suddenly Italian drivers could access the northern mountains quickly and easily. It felt grand, too: special uniformed officers would greet each individual vehicle in full military style.

The idea was bold, for in 1924 there were just 57,000 cars in the whole of Italy. But, by 1938, over 2,000 cars were using the road each day and all construction costs had been paid off. Other nations quickly flocked to study and copy the idea.

Multi-lane motorways took longer to arrive. The first two-lane highway was completed in 1932 between Cologne and Bonn in Germany. The term *Autobahn* had not yet been coined – that arrived after the Nazi takeover, when Adolf Hitler ordered the first Autobahn to be built from Frankfurt to Darmstadt; it opened in 1935. Famously, sections of German Autobahns to this day have no upper speed limit; the record for the highest-ever speed on an Autobahn is a remarkable 268 mph (432 km/h), set by racing driver Rudolf Caracciola in 1938. The passion for multi-lane highways soon spread to the US (whose first freeway opened in 1940), Sweden (1953), France (1954), and the UK (1958).

▷ **A revolution in travel**
A German Autobahn in around 1935. First planned in the 1920s, Germany's new multi-lane highways allowed drivers to travel at high speed. The central reservation between on-coming lanes also improved safety.

Art Deco elegance

Art Deco revolutionized design in the 1930s – a re-working of art styles from the past, updated with materials such as stainless steel and chromium, and soft, curved edges. It influenced many areas of life, including the car industry.

△ **Advert for Morris Big Six, 1936**
Art Deco influences could be used to sell more conservatively styled cars. The colours in this Morris advertisement lend the car a typically American glamour.

Before Art Deco really took hold, motorcar style was fairly well established as a box for the people, a box for the engine, and mudguards over the wheels. Although the details might have been different, the majority of cars in any given class looked broadly similar. The advent of Art Deco influences allowed manufacturers to be rather more free with their concepts throughout the 1930s, '40s, and '50s.

Form following function

Manufacturers such as Cord used the new style to herald technological advancement. The Cord 810 was one of the first front-wheel drive cars to appear in the US, and the first to feature independent front suspension. It was only natural that an up-to-the-minute body design was conceived to match – with rounded edges, hidden lights, and elaborate strakes in place of a radiator grille, it embodied the historically driven future that Art Deco encapsulated so well.

Streamlined style

William Stout created one of the most overtly Art Deco vehicles of all time with his Scarab (see p.261). Intended as a cross between a car and the fuselage of an aeroplane, it did away with the traditional separate chassis, introduced third-row seating, and featured a swivelling second row and removable table. The Stout Scarab could be used as an office or family space, as well as mere transport, fifty years before the MPV craze truly took hold.

△ **Wakefield Trophy, 1929**
This trophy was awarded by oil tycoon Sir Charles "Cheers" Wakefield to Major Sir Henry O'Neil de Hane Segrave for setting a new land speed record of 231.362 mph (372.341 km/h).

This vision of the future deserved a forward-thinking body design – and it had one. Its slickly curved, futuristic ponton styling (featuring a full-width, slab-sided body with integrated

KEY DEVELOPMENT
The Art Deco movement

Art Deco effectively pastiched a number of eras of artistic thought, merging the best of them with new materials and technologies to create an unusually advanced take on what had gone before. A popular example of Art Deco design is the Chrysler Building in New York. Having originated in France just before World War I, Art Deco had been refined into the slightly less extroverted Art Moderne style by the time of the Great Depression in the US. This utilized curved surfaces, plastics, and chrome plating to mirror the streamlining that could be seen in aspects of technological culture. Elements of Art Deco design could be seen well into the 1950s, and it remains popular with antique enthusiasts today.

ROLLS-ROYCE PHANTOM 1 AERODYNAMIC COUPÉ **BUILT BY JONCKHEERE IN 1925.**

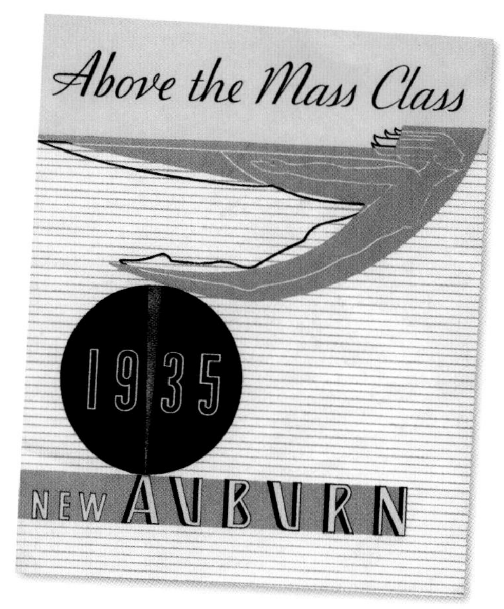

◁ **Auburn sales brochure, 1935**
Auburn, alongside its sister companies Cord and Duesenberg, was at the forefront of Art Deco car styling in the US. The Auburn Speedster was one of the most stylish cars of the 1930s.

wings, and dispensing with running boards) was revolutionary, and owed more to aircraft than car design. Just nine Scarabs were produced, of which five survive today.

In Europe, French manufacturers such as Delage and Talbot-Lago pioneered Art Deco automotive design. Several of their late-1930s designs, including the Talbot-Lago T150 coupé, showed to stunning effect the softened edges and chrome of the Art Deco era, and made the world of motoring an infinitely more varied place. Such style was not cheap, however – Talbot-Lagos were very much the preserve of the wealthy, while most other drivers had to stare on, enviously, from within their humdrum saloon cars.

But it was not just car manufacturers that offered Art Deco style to those who could afford it. Coach-builders, such as Jonckheere of Belgium, would create lavish Art Deco bodies for any chassis taken to them. Nothing was too extravagant for them to provide – including, in the case of a Rolls-Royce Phantom 1 (see left), an aerodynamic coupé body shape equipped with a tail fin and a pair of round doors.

Lingering influence

Art Deco style in cars endured through to the 1950s in the US, with models such as the first Chevrolet Corvette sitting well alongside the jet-inspired creations of Harley Earl on US roads. But soft curves were becoming a thing of the past, and while the use of chrome spiralled, the true age of Art Deco motoring was over.

▽ **Sunrise in an SS1, 1935**
SS Cars (later to become Jaguar) brought a touch of Art Deco style to the British middle classes. Cars like this SS1 Tourer were among the most elaborate of the day.

The Bentley Boys

The wealthy drivers of British sports-car marque Bentley were united by a love of speed, and their devil-may-care attitude brought them success on the racetrack. They also enjoyed excess, and fully embraced the playboy lifestyle of the 1920s.

It is not known who coined the most stirring alliteration in prewar British motoring, but the glamorous "Bentley Boys", as they became known, summed up all that was rousing, exciting, and daring about motor sport in the 1920s. To join this informal club, you needed an unshakeable belief that Bentley cars could be turned into race winners. Well-heeled aristocrats like diamond heir Woolf Barnato and former fighter pilot Sir Henry "Tim" Birkin epitomised this spirit. Exploiting the power and tenacity of a Bentley to the full fuelled their passion, and brought results: Barnato won the 24 Hours of Le Mans race three times, while Birkin created the supercharged Bentley "Blowers" and once held the record, at 137.96 mph (222 km/h), for the fastest lap of Brooklands race circuit.

Company founder W.O. Bentley served as a tolerant godfather to the group and savoured the glamour they brought to his marque. Dashing characters such as Captain Glen Kidston, jeweller Bernard Rubin, Canadian racing driver John Duff, and reporter Sammy Davis came together to produce an unbeatable team, winning four consecutive victories at Le Mans (1927–30).

After races, the Bentley Boys partied hard and became famous for their day-long, champagne-fuelled celebrations. Indeed, Barnato had a realistic mock-Tudor pub created in the basement of his Surrey mansion. W.O. Bentley himself once said of the media furore created by the Bentley Boys: "The public liked to imagine them living in expensive Mayfair flats with several mistresses and, of course, several very fast Bentleys, drinking champagne in nightclubs, playing the horses and the Stock Exchange, and beating furiously around racing tracks at the weekend. Of at least several of them, this was not such an inaccurate picture." They lived fast, and at the end of the 1930 season the team disbanded. With it a short but dazzling chapter in Bentley's motor-racing history came to an end.

◁ **The victorious Bentley Boys**
Glen Kidston and Woolf Barnato (second and third from left) celebrate in France after winning the 1930 24 Hours of Le Mans race in a Bentley Speed Six. With them are Dick Watney (left) and Frank Clement (right), who finished in second place.

Far-flung corners

In the early decades of the 20th century, most of the world was still making do with bicycles and placid livestock. And yet, within a few decades the motorcar had all but conquered the world.

△ **The GAZ-M1, 1936**
By the mid-1930s, Soviet car manufacturer GAZ had produced the GAZ-M1, which became an icon of its time. It had a maximum speed of 65 mph (105 km/h) and was used by the Red Army as a staff car.

The globalization of the car industry took decades to gain momentum, but it was catalyzed in 1907 by a competition set in motion by French newspaper *Le Matin* – the Peking to Paris race. This set out to prove that a car could take you anywhere.

Peking to Paris

Remarkably, given the novelty of cars at the time, 40 entrants signed up for the race, although in the end only five teams shipped their vehicles to Peking (now Beijing), ready to start their engines on 10 June outside the French embassy. From Peking, the route followed telegraph lines, enabling journalists to file reports along the way. The teams crossed deserts, mountains, and steppes to Outer Mongolia and Ulan Bator, before skirting Lake Baikal en route to Moscow, and then crossing Russia, Poland, Germany, and Belgium to reach Paris.

Overall, the competitors travelled 9,317 miles (14,994 km), mostly over land that had never been driven on before. In these areas, locals looked on in disbelief as the teams passed through

▽ **Stuck in the mud**
Victor Collignon of France at the wheel of his De Dion-Bouton, being pulled out of the mud in the Gobi Desert during the Peking to Paris race, 1907.

▷ **Osa Johnson in the Congo**
After the Peking to Paris race, adventurers were keen to make the most of the automobile. In 1930, Osa Johnson explored the Belgian Congo in a Willys-Overland, befriending the Mbuti people of the Ituri Forest en route.

in their jalopies. Nomads in the Gobi Desert had their first encounter with a car when they rescued the crew of the three-wheeled French Contal. The car itself was bogged down in the sands and was left to rust.

Italian journalist Luigi Barzini documented the reactions of the local people he met. He described the Chinese as being indifferent to what they called "chicho", or fuel chariots, while one group of Mongolians was convinced that the car's power came from an invisible winged horse.

In the end, it was Italian Prince Scipione Borghese who crossed the finish line first – in an Itala – despite falling through a bridge in Siberia.

Far-flung industries

Cars remained a rarity in Asia until at least the 1930s. The first car in India arrived in 1897, imported by a resident of Calcutta. By the following year there were four cars in Bombay, one of them owned by Jamshetji Tata, founder of Tata Motors, India's largest carmaker, and now owner of Jaguar Land Rover. In Madras, Samuel John Green caused a sensation with his steam car in 1903. Otherwise, cars remained scarce until 1928, when General Motors' Indian subsidiary began assembling cars in its Bombay plant.

In other remote places, especially those with little infrastructure, cars were increasingly in demand. Harry Tarrant made Australia's first petrol-driven car in 1897, followed by various improved models, and in 1909 he took on the country's first Ford franchise for assembly and sales. In South America, Peruvian engineer Juan Alberto Grieve made the continent's first car in 1908, but when he sought government funding he was told that the country needed quality foreign imports, not homegrown Peruvian "experiments".

At the other extreme, China restricted imports to encourage patriotism, and the nation's first vehicle – the Jiefang liberation truck – hit the road in 1956. Even into the 1980s, cars were rare outside the Chinese capital, and those that existed were typically Chinese-made limousines for bureaucrats or imported Soviet Ladas. In the Soviet Union itself, a car industry was well established by World War II, producing the iconic GAZ-M1 in 1936.

▽ **Sharing the road**
In 1930, cars were still scarce on the roads of Kolkata. Here, a zebra-drawn carriage shares the road with a pair of Morris saloons.

"...as **long** as a **man** has a **car**, he can do **anything**..."

FRENCH NEWSPAPER *LE MATIN*, 16 JANUARY 1907

Making roads safer

Car sales had rocketed by the mid-1920s, but with no effective speed limits, driving tests, or minimum driving ages, accidents were all too frequent. Legislation was needed to keep both drivers and pedestrians safe.

△ **Keeping order**
Demonstrators demand road safety improvments, such as safer crossing points, in the late 1920s.

In the UK, a national speed limit of 20 mph (32 km/h) had been in place since 1903, but it was widely ignored. To remedy this, all speed limits for car drivers were abolished by the 1930 Road Traffic Act. Car owners could drive as fast as they liked, anywhere, without fear of prosecution, but buses ,and lorries were restricted to a national maximum of 30 mph (48 km/h).

As a safeguard, the Act introduced the offences of careless, dangerous, and reckless driving, as well as fines for driving while drunk or drugged. Drivers were also required to have third-party insurance, while disabled drivers had to take a driving test. The first edition of *The Highway Code* was published in 1930, to standardize driving etiquette.

Dispensing with an upper speed limit led to an increase in fatalities on UK roads. There were 4,886 in 1926, but they soared to 7,305 in 1930, and five years later stood at 7,343, half of which involved pedestrians. The new Road Traffic Act of 1935 made the top speed in built-up areas 30 mph (48 km/h).

This has remained largely in place ever since, as has the 1934 Act's stipulation that all drivers need to pass a driving test. Another measure was the introduction in 1935 of designated pedestrian road crossings, with a bright orange globe on a pole on either side of the road. Nicknamed "Belisha beacons" after the British transport minister, Leslie Hore-Belisha, they remain a distinctive sight on the UK's roads.

The US and Europe

Around this time, road safety was also improved and standardized across the US and Europe, where free-for-all driving was also the norm. The American Automobile Association (AAA) developed curricula in driver education. The United Nations ratified driving standards and road signs, while the recently invented three-coloured, four-way traffic lights (see pp.66–67) were installed in cities around the world. Manufacturers also began to take safety more seriously, introducing features such as brake lights and indicators.

◁ ***The Highway Code***
This pocket-sized guide to the rules of British roads for all users, including cyclists and horse riders, was first published in 1930 and is still in print today. Knowledge of it is essential to pass a driving test.

KEY EVENTS

- **1920s** Canada, Italy, and Spain follow the US in driving on the right-hand side of the road. The UK, its colonies, and Japan opt for the left.
- **1926** An international convention on motor traffic is held in Paris, and a convention on the unification of road signs is held in Geneva.
- **1926** The international driving permit is launched.
- **1930s** Most of Eastern Europe adopts driving on the right-hand side of the road.
- **1930** The Road Traffic Act is introduced in the UK, abolishing speed limits.
- **1934** The first high-school class in drivers' education is taught in Pennsylvania, US.
- **1935** A new UK Road Traffic Act limits top speeds in built-up areas to 30 mph (48 km/h).
- **1935** Belisha beacons are introduced in the UK to indicate pedestrian crossings.
- **1935** The US government produces its *Manual on Uniform Traffic Control Devices.*
- **1936** A chemistry professor from Indiana, US, develops the Drunkometer. It is first used by police in Indianapolis in 1938.

BELISHA BEACON GLOBES **BEING PACKED FOR DISTRIBUTION IN THE UK IN THE MID-1930s.**

△ **Manually-operated traffic lights** on 5th Avenue in New York, US, pictured in 1929.

▽ **Aspirational advertising**
1930s car-industry marketing emphasized the latest styling in beautiful settings to fuel the aspirations of potential buyers. Here, glamorous ladies admire the latest Morris car in a 1934 advertisement.

Early Japanese cars

Nowadays, Japan is a powerhouse of car manufacturing, but in the 1920s and '30s, when cars were being produced commercially in the US and Europe, it essentially had no such industry of its own.

△ **Roads destroyed**
Japan's fledgling car industry was set back by the Great Kanto Earthquake of 1923. It killed hundreds of thousands of people, and wrecked countless rail and tram links, as well as roads.

Most cars in Japan in the 1920s and '30s were foreign, imported as status symbols and playthings by the wealthy. There were some notable exceptions, however, such as the four-seater Takuri (1907), the DAT 41 (1916), and the Mitsubishi Model A (1917).

The Great Kanto Earthquake of September 1923, which devastated Tokyo and Yokohama, caused critical transport and communication problems. As an emergency measure, Japan imported 1,000 Ford Model T truck chassis from the US and converted them into taxis. Being robust, they could handle Japan's poor roads. The Model T was such a hit that Henry Ford set up a local assembly plant, building cars from kits, in Yokohama in February 1925. Other US companies entered the market, and the number of American cars made in Japan rocketed, while the production of Japanese cars remained piecemeal. To rein in the Americans, the Automobile Manufacturing Industries Act was introduced in 1936, restricting vehicle production to companies with more than 50 per cent Japanese ownership, and higher import duties followed. By the onset of World War II, the US plants in Japan were shut down for good.

Domestic manufacturing

In this prewar period, Datsun was arguably the most proactive Japanese carmaker, producing the four-cylinder Datsun Type 10, followed by a string of small saloons and the open "torpedo" models (the 14, 15, and 16). The more upscale, US-style, six-cylinder Nissan Type 70 was unveiled in 1937, a year before Datsun production was stopped in favour of military/industrial vehicles. But perhaps Japan's true landmark prewar entry was the Toyota AA, a large, streamlined, six-cylinder saloon, modelled after the Chrysler Airflow.

The companies that went on to become the Suzuki, Daihatsu, Subaru, Mazda, and Isuzu of today existed at this time, but it was not until after World War II that they produced cars.

◁ **Royal approval**
Prince Chichibu, the eldest brother of the Japanese emperor, appears delighted seated in a Datsun in 1934. The tiny car was intended as Japan's challenge to Western manufacturers, selling for less than American cars.

KEY EVENTS

- **1902** Henry Ford exports his first car to Japan.
- **1907** The Takuri, regarded as Japan's first true home-made car, is produced.
- **1923** The Great Kanto Earthquake devastates Tokyo and Yokohama, prompting the import of 1,000 Model T truck chassis from the US.
- **1925** Henry Ford sets up a Model T assembly plant in Yokohama.
- **1927** General Motors opens an assembly operation in Osaka.
- **1929** Ford's brand-new factory for building the Model A opens in Yokohama.
- **1930** 98 out of every 100 cars on the road in Japan are American.
- **1934** The Nissan Motor Co is officially formed in Yokohama.
- **1935** Toyota launches its first-ever production car, the AA.
- **1936** Japan introduces the Automobile Manufacturing Industries Act to curb American car producers.
- **1941** Japan enters World War II. US manufacturing plants in Japan shut down.

AFTER ITS LUXURY MODEL A **(1917), MITSUBISHI STOPPED MAKING CARS UNTIL 1960.**

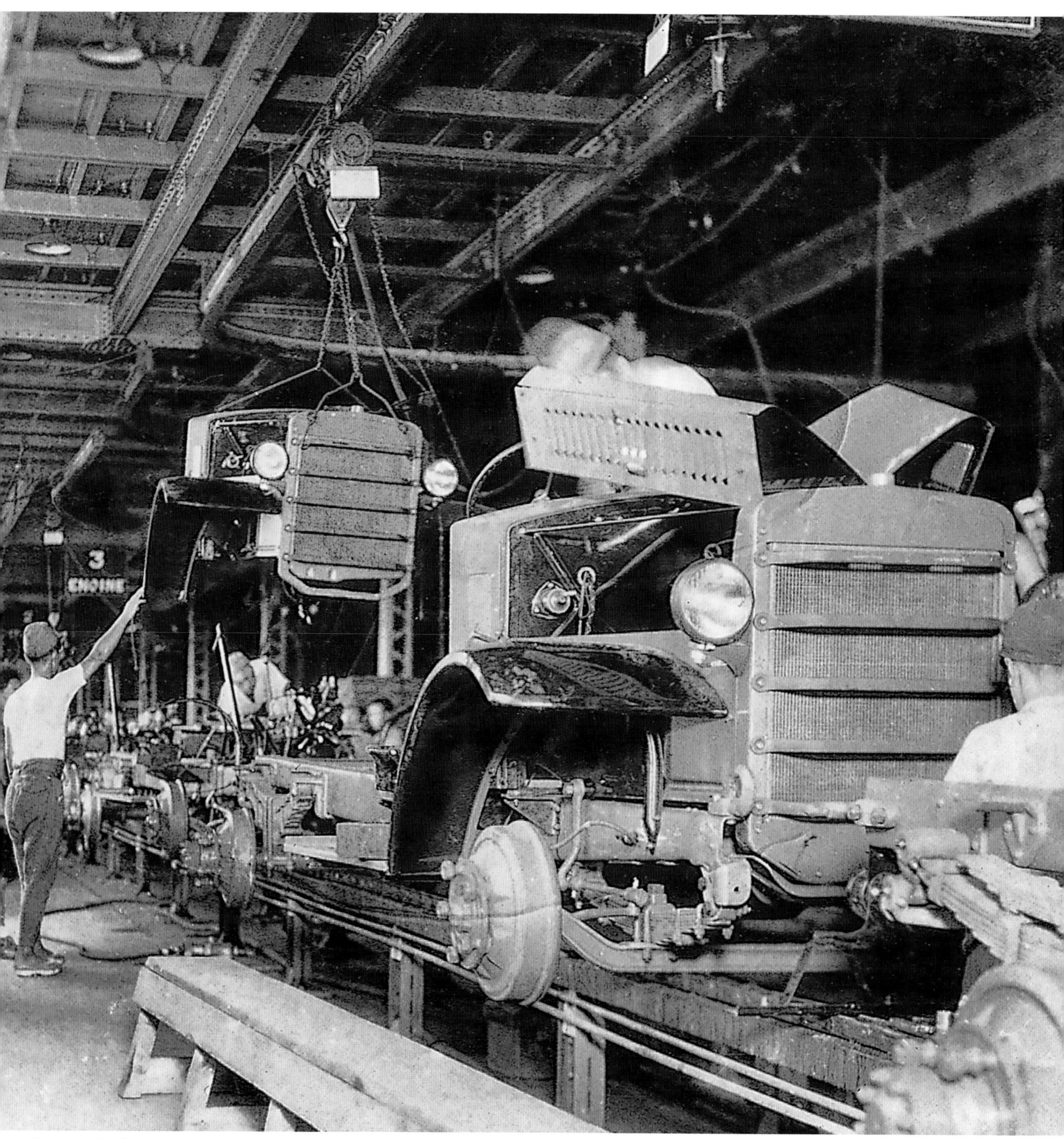

△ **The production line** at the Toyota Motor Co factory in Koromo, Aichi Prefecture, in the 1930s.

Fast and affordable

MG became synonymous with the affordable sports car between the wars, when it produced the little M-type, J-type, and P-type Midgets, and the larger six-cylinder Magna and Magnette models. These cars were fun to drive, and for weekend thrills enthusiastic owners often entered them in circuit races or off-road sporting trials. Many budget roadsters followed, such as Morgan's quirky V-twin three-wheelers, and, from Italy, the Fiat Ballila 508S, but MG had the niche for mass-market fun sewn up. Stiff springs and a "whippy" chassis were typical, but BMW changed all that with its sophisticated 328, featuring a stiff structure and well-controlled suspension. It was the future blueprint for sports cars, but unlike MGs, BMWs were not cheap.

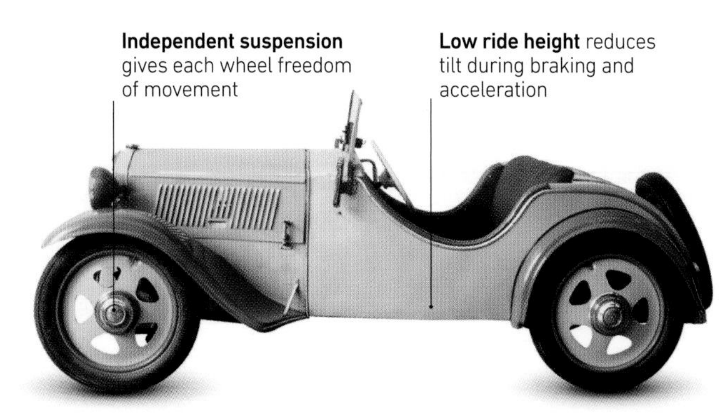

△ **DKW FA, 1931**
Powered by a motorbike engine, the cheap and fast DKW FA was the first of a line of front-wheel-drive DKWs assembled at Zwickau, Germany. The two-stroke engine was mounted sideways behind the gearbox.

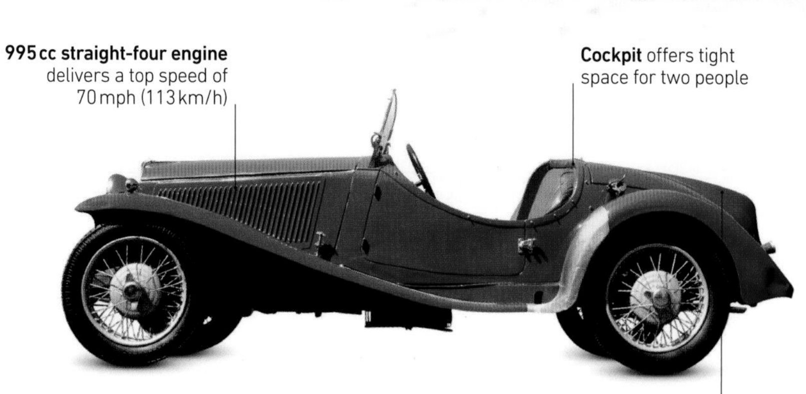

△ **Fiat Ballila 508S, 1933**
A year after Fiat launched its Ballila family car, this sports version was made available. It was based on a design by coach-builders Carrozzeria Ghia, and soon became the quintessential small sports car.

◁ **Taking the corner**
A pair of Austin Ulsters are pursued by an MG at Brooklands race course, UK, in 1931. The leading driver is Victoria Worsley, who came in seventh in the event and later switched to driving an MG.

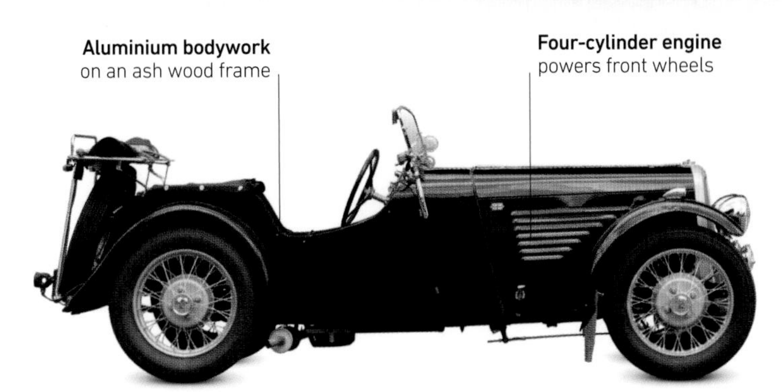

△ **BSA Scout, 1935**
BSA branched out from rifles to motorcycles and then cars. This was a four-cylinder sports car or tourer with front-wheel drive, independent front suspension, and a transmission brake on the front axle.

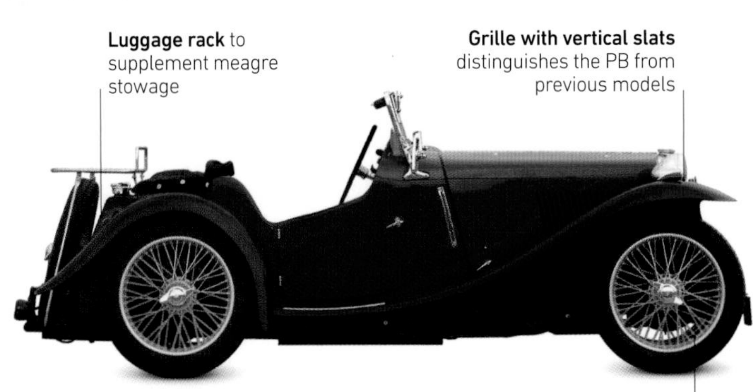

△ **MG PB, 1935**
P-type MG Midget superseded the J-type in 1934, with a longer and stronger chassis. The updated PB of 1935 had a big-bore 939-cc version of the Wolseley-sourced engine generating 43 bhp.

1936–1945

THE CAR COMES OF AGE

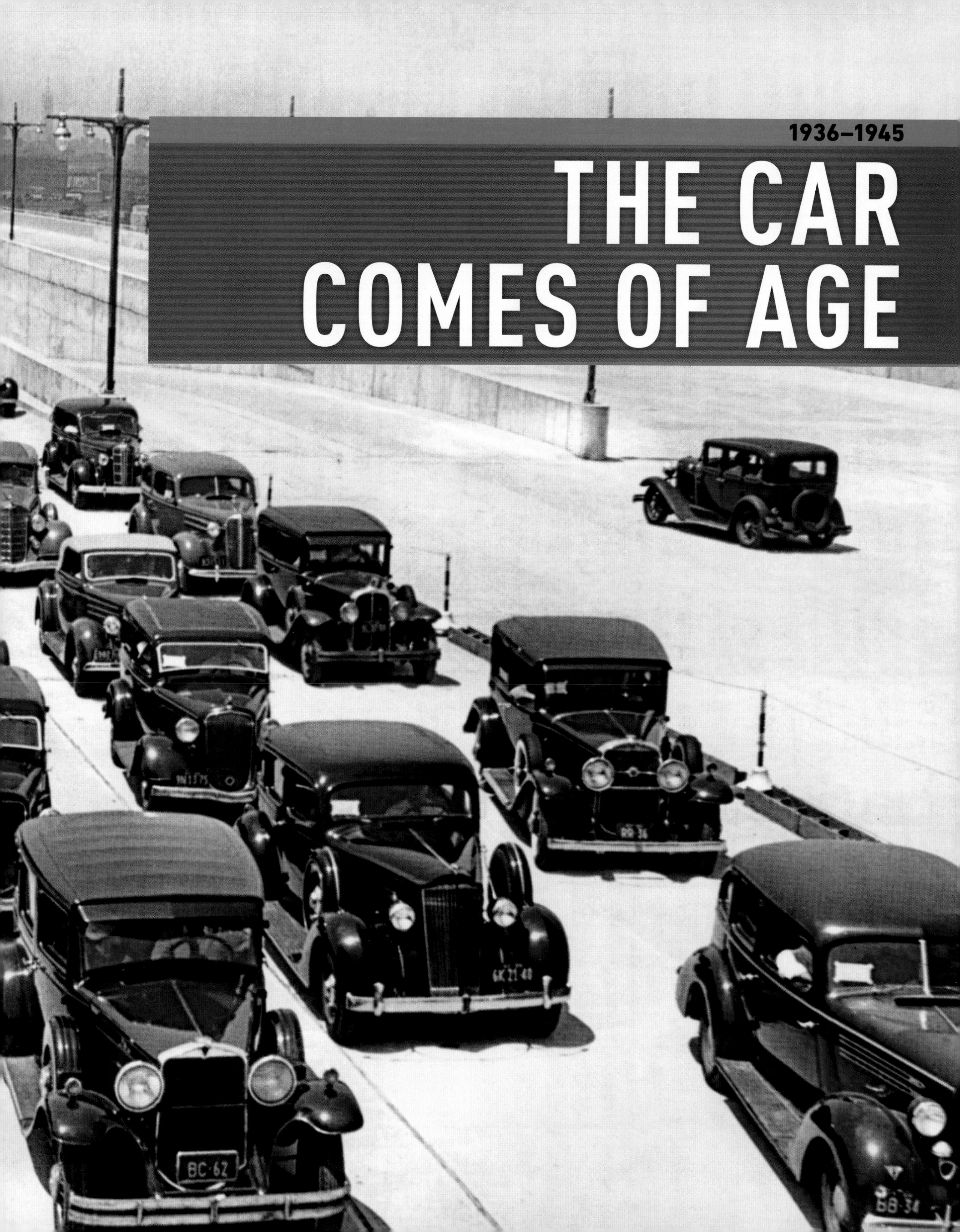

1936–1945

The car comes of age

World War II effectively halted the evolution of the car for five long years. In Europe, shortly after the conflict began, virtually all car manufacturing stopped, and factories that had previously met demand from motorists now did the bidding of ministries, dictators, or – in the case of France – occupying forces. Overnight, these factories became part of their countries' war machines, their goal no longer to make cars, but to build tanks, planes, and munitions. After the bombing of Pearl Harbor in 1941, the US car industry followed suit.

Cars for the people

The freedom of the road – a flourishing thrill of the 1930s – was another immediate casualty. Fuel shortages rendered private motoring almost impossible, and many cars were laid up and stored in the hope of better times. Nevertheless, the war did bring some automotive benefits. The most important of these was the development of a lightweight four-wheel-drive vehicle in the shape of the versatile Jeep – the grandfather of all of today's sport utility vehicles. Industrial and military vehicles with all wheels driven were one thing, but the Jeep – planned as a battlefield taxi and reconnaissance vehicle – was in a league of its own. Another vehicle whose fate was intertwined with the politics of the period was Germany's *Volks wagen*, or "people's car" a product that, before the war, Adolf Hitler promised would be available to almost every German citizen. It was a promise he could not keep.

The impressive factory in which the VW was built was all but destroyed by Allied bombing, and the car purchase scheme endorsed by the Nazis

THE GREAT DEPRESSION IN THE US ROCKS THE GLOBAL CAR INDUSTRY

WORLD WAR II BRINGS DESTRUCTION TO CITIES AROUND THE WORLD

"...most **middle-class people** expected to **own a car** when **finances allowed**."

was soon in tatters. However, the VW, along with Fiat's 500, and temptingly priced models from Opel and Renault, was in the vanguard of a "cars-for-all" movement that pointed to an era that was soon to dawn.

Making a myth

By the 1940s, most middle-class people expected to own a car when finances allowed. Showrooms were palaces of desire in which aspirations were stoked. You could sit in the driving seat and familiarize yourself with the dashboard – even though the ergonomics were terrible, the controls were often heavy and irksome, and the maintenance duties were potentially unending. The widespread marketing of everything to do with car ownership – from petrol and oil to tyres and picnic sets – sparked an explosion of passion for the car. In the cosy darkness of the cinema, glamorous metal models filled the silver screen alongside Hollywood's top talent, all pointing to a life that was not complete without a car. To be a man, in particular, was unthinkable without one – or so the myth suggested.

However, in spite of so much advertising, there was still a surprising number of car-free countries, such as Spain. Factories were humming in Detroit, in North America, but just over the border, Canada had only the most basic assembly plants. Innovations varied enormously in the impact they made. As soon as car radios were made to work well on the move, everyone wanted one, but the world's first diesel-engined production car – from Mercedes-Benz – remained one of the very few to embrace this fuel that unprecedentedly low running costs.

MANUFACTURERS CONTINUE TO ADVERTISE THEIR CARS DESPITE THE WAR

THE JOY OF MOTORING SLOWLY RETURNS IN THE 1940s

Trading up

Car ownership boomed in the 1920s. However, as more people bought new cars, second-hand dealers began to sell the used models, often with new bodywork, which generated problems for the motor trade.

The mass production of cars, led by Henry Ford's refinement of the assembly line system for his Model T, lowered prices and massively expanded car ownership. By 1929 there were already more than 26 million cars in the US, and California had an automobile for every 2.1 people. In 1913 three-quarters of all cars built were for first-time buyers, but by 1924 two-thirds of new cars were being bought to replace an older machine. As a result, the motor trade had to deal with increasing numbers of used cars.

Then, as now, buyers of shiny new cars paid little or no attention to what happened to the old machines they traded in. Some of these vehicles were lined up at second-hand car dealers and sold on to buyers who could not afford a new car. But mass production was ramping up all the time, driving down the costs of new cars, and so new cars were becoming more and more attractive to buyers.

The "used car problem"

It became increasingly difficult for dealers to dispose of the used cars that were traded in. Worse still, in a cut-throat market many dealers were offering generous trade-in prices to encourage customers through their doors, which meant that the second-hand cars were often sold at a loss. It was common for a new body to be built for a second-hand chassis in an effort to maximize a car's re-sale value, so an old tourer might be given a sports body to sell it on, or a saloon re-worked as a pickup truck for farm use. To solve the "used car problem", US carmakers

△ **Sent to the breakers' yard**
The glut of older vehicles led to an increase in the scrap metal business, collecting cars left behind by manufacturers' trade-in schemes.

▽ **Affordable payments**
The market for used cars was driven strongly by price concerns, as seen in this 1930s lot in Greenwich Village, New York, advertising payment instalments.

Chevrolet and Ford introduced schemes to buy back and scrap trade-in cars. Ford's scheme involved a full-blown recycling operation at its Detroit factory, which disassembled up to 600 cars a day, of all makes, which had been traded in to its dealers in Michigan. Every usable component or material was removed, then recycled to make everything from aprons to window panes.

In 1930, an industry-wide programme was introduced by the manufacturers' trade association, the National Automobile Chamber of Commerce. Hundreds of thousands of serviceable used cars were scrapped to make way for more new ones. However, the scheme ultimately proved to be unsustainable. In part this was due to low steel prices, which made recycling a used car uneconomical – although that would change when steel was in short supply in World War II and during the Korean War.

There were efforts to regulate trade-in allowances in the US by common agreement and by federal law, but these also proved unworkable. Instead, car dealers started to share information on trade-in prices, and price guide books were published to help traders avoid over-offering: the *Kelley Blue Book* was first published in 1926, and the famous *Glass's Guide* followed in the UK in 1933.

> "Always consider **what you get** for the **price you pay!**"
>
> CHEVROLET ADVERTISING SLOGAN

KEY DEVELOPMENT

Suburbs demand car ownership

Despite the economic depression in the West during the early 1930s, home ownership increased in many countries, driven by inner-city slum-clearance programmes and low interest rates. In the UK and the US, new suburban housing estates gave more people than ever a chance to own their own home, with a down payment of as little as £1. Many now had the space to accommodate a car, and for people who lived outside the city, cars became an essential part of everyday life.

HOUSES WITH BUILT-IN GARAGES **WERE HIGHLY DESIRABLE AS THE CAR BECAME IMPORTANT IN PEOPLE'S WORKING LIVES.**

Driving on a budget

In the wake of the Great Depression of the 1930s, significantly fewer people were able to afford a new car, and among those who could, many chose the cheapest model available on the market.

△ **Chevrolet Mercury, 1933**
Launched as a budget model in 1933, the six-cylinder Mercury was renamed the following year as the Chevrolet Standard Six.

Small, simple cars met the demand for motoring on a shoestring all over the world at this time, but the US had already started to develop a preference for larger vehicles. As a result, producers of small family cars, such as American Austin (later reformed as American Bantam) struggled to gain a foothold there. Most American car buyers instead looked to entry-level models from the big manufacturers. Chevrolet was the market leader with its Mercury, Standard, and Master models, powered by the stovebolt six-cylinder engine. Ford's rival saloons featured the flathead V8, with a low-cost 2.2-litre version available from 1937.

Economy of style

Ford's cheapest European cars were very different. There were no V8 engines – a 933-cc, side-valve four-cylinder powered the tiny Model Y, produced from 1932 to 1937. The beam axles were suspended from transverse leaf springs, Ford's usual practice, and the gearbox was three-speed with synchromesh on the top two ratios. In its cheapest two-door form, the Model Y was the first saloon car available in the UK for less than £100. It was also built in Germany as the Ford Köln.

European rivals

In the UK, the biggest rival to Ford's Model Y was the home-grown Morris Eight. The two were well matched: the Morris had more power but more weight; it also had better brakes but was slightly more expensive. Gradual developments in the design of the Morris resulted in steel wheels as well as more modern styling on the 1938 Series E, while the Ford was restyled as the 7Y, which became the Anglia in 1939. Both were competing with the long-running Austin Seven, which had sold 290,000 cars from the start of production in 1922 to its end in 1939.

Fiat engineer Dante Giacosa designed a car for Italy that was just as small as the Austin Seven but much more technically advanced. First produced in 1936, the Fiat 500 (nicknamed *Topolino*, meaning "little mouse") had a novel layout, with the radiator positioned behind the engine, which improved space inside the car and allowed a low, raked front end in place of a conventional vertical grille. Sliding sidescreens instead of roll-down windows meant that the doors could be concave, providing more elbow room. The Topolino also had independent front suspension, hydraulic brakes, and a four-speed gearbox at a time when many cars made do with three. Half a million were sold by 1955.

In Germany, the Dixi company built Austin Sevens under licence, and the DKW motorcycle company offered the two-stroke DKW F1 saloon, the

▽ **Austin Seven**
Advertising for the Austin Seven highlighted its credentials as an economical and reliable saloon. It was one of the most popular cars in the UK at the time.

▷ **1939 Plymouth convertible**
Plymouths were one of the cheapest – and most popular – cars in the US. The advertising for its 1939 convertible boasted a "thrilling performance and exceptional economy".

first successful front-wheel drive car, from 1931. Meanwhile, Opel's cheapest offering was the Olympia (named in honour of the 1936 Olympic Games held in Berlin), which was available from 1935 until World War II ended production in 1940. It had novel monocoque construction, and was soon joined by the even cheaper Kadett. The war also delayed production of the Volkswagen Beetle, which was planned to sell for under 1,000 Reichsmarks – less than half the Olympia's price.

Inspired by the Opel Olympia, Louis Renault launched the Juvaquatre in France from 1938. Although replaced in 1948 by the 4CV, estate and van versions were available until 1960.

KEY DEVELOPMENT

Imported motor industries

Most countries in the West had indigenous mass-market carmakers early in the 20th century, but Spain's most famous car brand was the high-end Hispano-Suiza, which disappeared soon after World War II. It was not until the 1950s that the SEAT brand emerged, its cars heavily based on Fiats (see pp.194–95).

Although Canada had no major car brands of its own, it did possess a significant motor industry. The three biggest US carmakers – Ford, General Motors, and Chrysler – all had car assembly plants in Canada, which for a short time was the second-biggest car producing country in the world.

SEAT WAS THE FIRST **SPANISH MANUFACTURER TO PRODUCE AFFORDABLE CARS.**

"**Motoring** at its **lowest cost!**"

AUSTIN SEVEN ADVERTISING SLOGAN

Hollywood glamour

With their gleaming grilles, elongated bonnets, and exaggerated curves, the voluptuous American cars of the late 1930s and early '40s oozed sex appeal and symbolized success.

As the Great Depression ended, the age of Hollywood glamour propelled cars into the spotlight. They were cast as the seductive partners to immaculately dressed stars, working their magic on an audience weary of austerity. Fuelled by the Art Deco movement, car design was reaching a creative peak, and the timing could not have been better for positioning cars as the ultimate style icons.

Few brands epitomized the Hollywood dream like the luxury marque Duesenberg. Although the company folded in 1937, that same year the producers of the hit comedy *Topper*, starring Cary Grant, paid homage to its kind by making over an ordinary Buick to resemble a Duesenberg/Cord hybrid. Duesenberg's Model J, or Doozy, was driven by industrialist Howard Hughes, gangster Al Capone, and a raft of celebrities, including Greta Garbo, Mae West, Gary Cooper, and Clark Gable. It also appeared on the silver screen with Fred Astaire and Ginger Rogers in Oscar-winner *The Gay Divorcee* (1934), and again in *The Great McGinty* (1940).

It was not long before Hollywood and the car industry were promoting each other. A Buick advert from 1935 proclaimed that "Hollywood – creator of style – chooses Buick for its own". Carole Lombard featured in a 1938 advert for DeSoto, which was a tie-in for her upcoming appearance in David O. Selznick's *Made For Each Other* (1939). A Buick Phaeton was the centrepiece of the farewell between Humphrey Bogart and Ingrid Bergman in *Casablanca* (1942).

Opulent American cars often featured in the pages of fan magazines with their famous owners, including Bing Crosby's 1939 Oldsmobile Coupé Convertible, Cary Grant's 1941 Buick Century, Rita Hayworth's 1941 Lincoln Continental, and film mogul Cecil B. DeMille's 1937 Cord. Although Hollywood's glamour cars were out of reach for the hard-pressed cinema-goer, they did help to feed their fantasies that they, too, might one day take the wheel of an impressive car of their own.

▷ **Rita Hayworth with a Lincoln Continental**
The Hollywood actress poses beside her famed Continental in 1941. It was an iconic car, but such links to screen idols also made it seem attainable.

GA 41
184 49F
PEACH STATE

ATLAS OIL CAN

SHELL OIL CAN

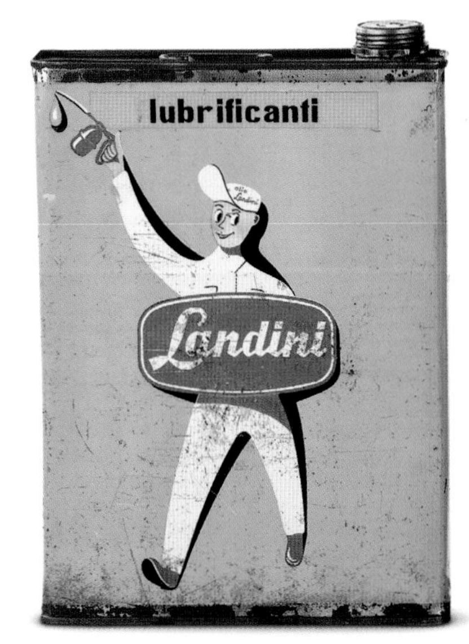

LANDINI OIL CAN

MOBILOIL OIL CAN

NEAL'S OIL CAN

CASTROL OIL CAN

REDLINE GLICO OIL CAN

PENN HILLS OIL CAN

KEY DEVELOPMENT
Synthetic oils

As the stresses within engines increased, more effective lubricants were needed. Synthetic engine oils, which are manufactured chemically rather than refined from crude oil, can offer reduced friction and greater durability. They were first produced in the 1930s for use in aero engines, but it was not until the 1970s that synthetic motor oils for cars were widely available from brands including Mobil, Amsoil, and Motul.

Semi-synthetic oils, which blended conventional mineral oil and synthetics, were also created to provide many of the performance advantages of synthetic oils but at a lower cost.

MOBIL 1 SYNTHETIC OIL **WAS LAUNCHED IN 1974. MOBIL SPONSORED MOTOR SPORT.**

LISTROIL OIL CAN

DUCKHAM OIL CAN

BRITISH AERO LUBRICANTS OIL CAN

Engine oil for all

Oil companies vied with each other for motorists' loyalty. Bright, distinctive branding featured on oil cans and signage.

Dr John Ellis was reputedly the first to create a successful lubricant from mineral oil, initially for steam engines. One of its important properties was that it avoided deposits on valves, leading to its brand name, Valvoline. Another motor oil pioneer, Sir Charles "Cheers" Wakefield, founded his own company in London in 1899 to produce a lubricant with added castor oil – from which its brand name, Castrol, was derived. The company was one of the first to sponsor motor racing and record-breaking events to help publicize its products. As motoring spread, more companies adopted eye-catching branding to compete for drivers' business.

Many early oil brands were swallowed up by large corporations, and while Shell, BP, Texaco, Gulf, and other long-serving brands still exist, others such as Duckhams, Vacuum, and Powerlube are no longer marketed.

SPEEDOIL 960 OIL CAN

DRAGOIL OIL CAN

LUBROL OIL CAN

SHELL OIL DISPENSER CART

▽ **Oldsmobile 76 Club Saloon, 1946**
This Oldsmobile's fastback "Futuramic" lines are displayed in a vast showroom in 1946. Spacious retail space in cities became essential to showcase new cars to potential buyers.

After the Crash

The Great Depression began in 1929 and reshaped the automobile industry. In the early 1930s, as unemployment rose and car sales shrank, manufacturers cut jobs, closed plants, and introduced lower-priced cars.

The global economic depression triggered by the October 1929 US stock market crash caused the biggest upheaval the world's auto industry had ever experienced. By 1932, motorcar sales had fallen 75 per cent in the US, and the effects were also felt worldwide. The US industry went from making a profit of $413 million in 1929 to losing $191 million in 1932, roughly equal to $2.9 billion in today's money. Small firms selling mid-priced or luxury cars were forced to merge or close. Larger, better capitalized companies, such as General Motors (GM) and Ford Motor Company, made radical changes in the way they operated their businesses to adapt to the severe economic conditions – moves that would set their course for the next 50 years. Firms were obliged to adapt or die. When the stock market crashed in 1929, GM and Ford controlled two-thirds of the US new-car market, while smaller firms, such as Packard, Studebaker, Nash, Hudson, and Chrysler – as well as some lower-volume luxury brands – accounted for the rest. However, by the early 1940s, when the industry switched to war production, GM was larger than ever. Chrysler, founded in 1925, had moved into second place in sales, and Ford was third, hanging on weakly.

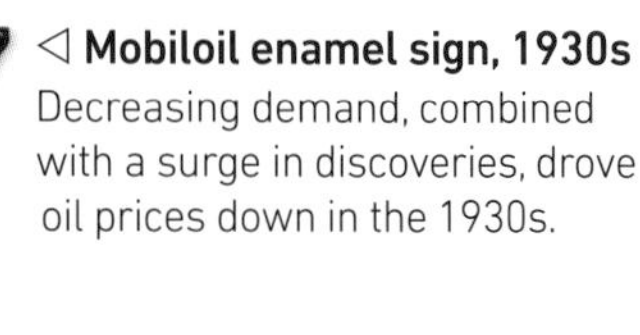

◁ **Mobiloil enamel sign, 1930s** Decreasing demand, combined with a surge in discoveries, drove oil prices down in the 1930s.

◁ **Citroën factory, France, 1930s**
Citroën survived the 1930s by offering popular models, such as the Traction Avant (pictured here on a production line) with its ground-breaking unitary body and chassis (see p.100).

Elsewhere, revered brands such as Duesenberg – considered one of the world's greatest high-performance luxury cars – along with Cord, Franklin, Pierce-Arrow, Graham, and Marmon were forced out of business.

Ford and GM survive

GM succeeded by cutting costs, using common parts across its brands, and boosting production of its less expensive Chevrolet brand. Additionally, while banks may not have been lending money, GM was, offering low-cost car financing loans to customers. Throughout the Depression, GM never lost money, and it gained 15 per cent of the market share, mostly at the expense of Ford. Meanwhile, Chrysler's success during the Depression came mostly from Plymouth, its low-cost brand. Chrysler increased Plymouth production by 50 per cent and opened new dealerships in the 1930s. Ford, dealing with high costs, loose accounting practices, lack of new products, and its stubborn boss (see pp.52–53), struggled during the Depression, and sales of the Model A, which had replaced the Model T, fell by 50 per cent by 1931. Ford's expenses were higher because it made parts for its overseas operations in Dearborn and shipped them to its plants around the world. As a result, increased tariffs made Ford's operations in the UK and Germany unprofitable.

◁ **"Crash"-damaged car being sold by owner, 1929**
The 1929 Wall Street Crash made the world re-evaluate its assets, from single cars to entire car companies.

The Depression in Europe

The European car industry also suffered in the 1930s, but the effects on British, French, and German carmakers were not as severe. British car production grew in the 1930s, rising from 239,000 in 1929 to more than 500,000 by 1937. At the start of the decade, Austin, Morris, and Singer controlled 75 per cent of the British market. By 1940, the UK had six major carmakers: Morris, Austin, Standard, Rootes, Ford, and Vauxhall. During the Depression years, the British Midlands and industrialized south prospered, but areas that depended on mining and shipping suffered.

The German economy, still burdened from the effects of World War I, was in turmoil even before the Crash. In the 1930s, Ford and GM expanded into Germany and the 1926 Daimler/Benz merger began to prosper. In France, Peugeot, Citroën, and Renault gained a new competitor, Simca, which had been set up by Italy's Fiat: the first French-built Simcas were based on designs by Fiat.

LIFE BEHIND THE WHEEL
Duesenberg, a short-lived luxury

Rolls-Royce has been known as the leading luxury car for nearly a century, but there was an American brand that many view as equal or better. Brothers August and Frederick Duesenberg, self-trained engineers, started the company in 1920. The first car, the Model A, was one of the most expensive on the market at $6,500, but it featured advanced engineering such as four-wheel hydraulic brakes and a straight eight-cylinder engine. The cars were hand-built and production was only about 150 per year. The Model J had an eight-cylinder, 265-hp engine; a supercharged version had a top speed of more than 100 mph (161 km/h). Duesenberg folded in 1937, due to the Depression, and only 481 Model Js were built.

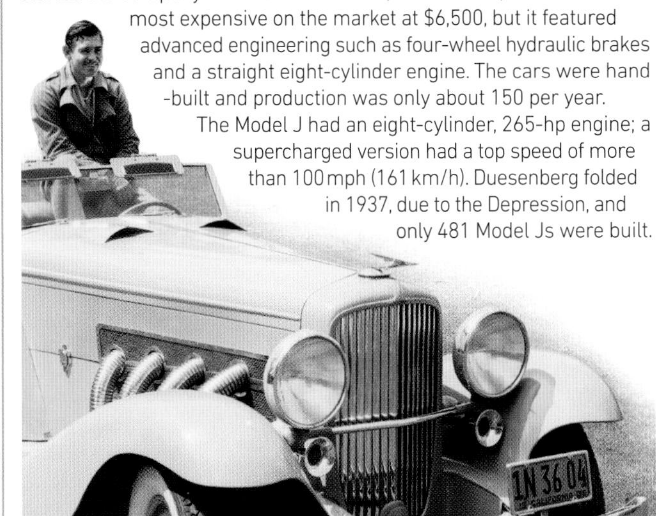

CLARK GABLE **POSES IN HIS CUSTOM DUESENBERG MODEL J DURING THE 1930s. THE MARQUE WAS A BYWORD FOR HOLLYWOOD EXCESS.**

Thick armour protects the two-man crew

Tyres virtually solid to avoid punctures

Six-cylinder Daimler petrol engine mounted at the rear

△ **Daimler Dingo Scout Car Mark III, 1940s**
This British armoured car was designed in 1939. It had a preselector gearbox with five speeds forwards and backwards, and four-wheel steering for exceptional manoeuvrability. They were still in use in the 1970s.

Fold-down hood provides cover

Engine-driven propeller can be lowered for water propulsion

△ **VW Schwimmwagen, 1941**
The engine and suspension of the German KdF-wagen, which became the Volkswagen "Beetle", were used to create the Kübelwagen. The Schwimmwagen was an amphibious version and 15,000 were built.

"It's a **Power-House** on wheels."

WILLYS JEEP ADVERT, 1948

▽ **Willys Jeep at Omaha Beach, France, 1940s**
The Jeep became the ubiquitous light transport machine of World War II, seeing action all over the world.

△ **Willys MB Jeep, 1941**
The definitive World War II Jeep, combining the best features from designs by Bantam, Willys, and Ford. Willys built more than 350,000, and Ford built nearly 280,000 of its licenced version, the GPW.

△ **GAZ-61, 1938**
The Red Army was slow to equip itself with a go-anywhere vehicle to handle harsh Russian winter conditions. The GAZ-61 was a stopgap four-wheel-drive saloon favoured by generals as a command car.

The first four-by-fours

War prompted the development of versatile light utility vehicles that could be used for a wide variety of military tasks. Some were equipped with four-wheel-drive transmissions to provide extra traction in slippery conditions. The best known was the American Jeep, developed by Bantam, Willys, and Ford – but there were many more. In Germany, the Volkswagen was adapted into the rear-wheel-drive military Kübelwagen – and a remarkable amphibious version called the Schwimmwagen. After the war these vehicles went on to be adopted for non-military use and created a whole new market sector. Purpose-built civilian 4x4s like the Jeep CJ series and the British Land Rover were soon developed to meet rising demand.

△ **Willys Jeep Jeepster, 1948**
After World War II, Jeep sought to capture the civilian market with this two-wheel-drive fun car. Although it failed to catch on, the Willys Jeep inspired many manufacturers to develop road-going four-wheel drives, leading to the modern SUV.

Cars in World War II

As the 1930s drew to a close and the shadow of World War II loomed, the scene was set for a radical change to the automobile industry – and a severe curtailing of motorists' freedom.

△ **Carpooling in the US**
A propaganda poster from 1942 encouraged civilians to share cars as a way to save fuel.

One of the first strategies employed by the UK and its allies after war broke out was to destroy enemy industrial sites – especially car factories appropriated for ammunition, tank, and aircraft production. The Renault factory in Paris became a key target after it began to serve Vichy France in 1940, under the direction of personnel from Daimler-Benz in Germany. In March 1942, it was devastated by Allied bombs in what was then the biggest drop in the war. Remarkably, the Volkswagen factory in Germany continued to make utility vehicles based on a modified Beetle chassis until August 1944, when the threat of bombing called a halt to production. Another German carmaker, Borgward in Bremen, produced tanks and trucks throughout the war, and was not destroyed until 1945.

The war effort

In the UK, as on the continent, carmakers were called on to contribute to the war effort. The Austin Motor plant in Longbridge, Birmingham, began producing parts for tanks, mines, depth charges, and ammunition, while Ford's Manchester plant churned out Rolls-Royce Merlin engines for military aircraft. Almost every automotive facility was working overtime for the war effort.

◁ **Scrap metal drive**
Rita Hayworth starred in a poster campaign that called on civilians to donate any metal items, especially tin, copper, and steel, to further the war effort.

Petrol was the first commodity to be rationed, within a few months of war being declared, followed by tyres. Most of the world's natural rubber supply came from Southeast Asia, which had been occupied by Japan early in 1942. From then on, tyres were in short supply. Industrial plants needed all the rubber they could get for military production, and civilians were called on to donate rubber items, from Wellington boots to garden hoses. Although civilian motoring was restricted, high demand for people to drive ambulances and other essential vehicles saw legions of women recruited and trained to take up positions behind the wheel.

Mirroring its British counterpart, the auto industry in North America retooled, installing new manufacturing equipment virtually overnight and taking on huge contracts from the government to produce military vehicles, ammunition, bombs, and helmets. Car sales were suspended on 1 January 1942, and production ground to a halt: no cars, commercial vehicles, or parts were made between February 1942 and October 1945.

The remaining stockpile of half a million cars was rationed for "essential drivers", such as war workers and medical staff. Fuel and tyre supplies were also strictly controlled after a failed attempt at voluntary rationing, and the national speed limit was reduced to 35 mph (56 km/h). Civilians were encouraged to share cars to save fuel, and to contribute: paper for packing weapons; silk and nylon for parachutes; and metal for making bombs.

Civilian models revived

Despite their absence from domestic car production, the big brands continued to advertise, promoting their contribution to the war effort. Finally, in autumn 1944, the War Production Board granted permission for Ford, Chrysler, Nash, and General Motors to begin preliminary work on new civilian models. Meanwhile, German and Japanese automotive companies emerged crippled from the war, albeit to become powerhouses of global car design and production within a few decades.

◁ **London streets during the Blitz**
Bombing raids during World War II turned the world of motoring upside down. Factories on both sides were destroyed, as were countless privately owned vehicles, such as this Humber, wrecked on Pall Mall, London, in 1940.

LIFE BEHIND THE WHEEL
Petrol rationing

When rationing was introduced to the UK in September 1939, petrol was the first commodity on the list. Buses were converted to gas, and some private drivers installed coal-fuelled gas producers on the backs of their cars. In July 1942, petrol was further limited to essential users only. In the same year, rationing was introduced in the US, but more to save on precious tyre wear. The UK pressured Australia to stop buying petrol in order to support the sterling exchange rate. Coping with 50 per cent less petrol than British or New Zealand motorists, Australians eventually rebelled and began hoarding, pushing the country into chaos. In France, petrol rationing hampered farming, and food production fell by half.

CARS SURROUND THE PUMPS **OF A FILLING STATION IN NEW YORK IN 1943, A YEAR AFTER RATIONING WAS INTRODUCED TO THE US.**

Building the People's Car

A Nazi initiative created the single best-selling model of car anywhere in the world. However, it was thanks to the British Army that the design and the original concept endured.

Nazi leader Adolf Hitler had a dream: "strength through joy", a vision for a happy, strong, powerful Germany. Alongside holiday camps, trips, concerts, and plays, Hitler conceived of a car that every German could afford. Dr Ferdinand Porsche created designs for this state-sanctioned car, inspired by the work of Tatra designer Hans Ledwinka. The motor industry could not meet Hitler's price demands, so the project was taken over by the German Labour Front and a factory was built using misappropriated funds.

The government offered a savings scheme: 5 Reichmarks per week bought stamps that were stuck in a book, and a complete book could be used to buy a car. However, nobody received their cars in the end, as the funds had been diverted to the war effort. A legal battle that dragged on until 1965 led to a small number of savers receiving discounts on a new Volkswagen, but the debt was never fully paid. In the aftermath of World War II, the factory was offered to British motor manufacturers, but none was interested in the car. A British Army officer, Major Ivan Hirst, was given the factory instead – the British Army needed cars, and Germans needed jobs. He recognised the virtues of the car and, following a demonstration, the British Army placed an order for 20,000 vehicles.

Former Opel production manager Heinz Nordhoff took over in 1949, and set the company on its present, well-respected path. The Volkswagen brand expanded with several variations on the theme, before launching the Golf as a true Beetle replacement in 1974. German production continued until 1979, while the Beetle was built in Mexico until 2003.

◁ **A model of efficiency**
Dr Ferdinand Porsche (far left) presents Adolf Hitler with a model of the Volkswagen on Hitler's 49th birthday in 1938. The Beetle (as it became known) went on to become popular across the world.

Home from home

Want to see the world, or more of your own country at least? Caravans and travel trailers satisfied the wanderlust of intrepid drivers and holidaymakers seeking home comforts on the road.

Once the preserve of Gypsies and travelling tradesmen, the first caravan built for recreational purposes was commissioned by proud Scotsman William Gordon Stables in 1885. Stables used his caravan, which was pulled by a horse and named The Wanderer, to travel the length of the UK.

The idea of caravanning for pleasure slowly spread among those with the means, and The Caravan Club was established in the UK in 1907. It organized meets, rallies, and events across the country as its membership steadily grew. By 1912, the UK had 450 dedicated caravan parks.

△ **Eccles caravan, 1926**
A proud family of holidaymakers poses alongside their new four-berth Eccles De Luxe caravan and Morris Oxford.

As prosperity and peace returned after World War I, caravanning resumed under a new form of propulsion – the motorcar. Just one year after the end of the conflict, Eccles Motor Transport Ltd developed the first prototype caravan designed to be pulled by car.

▷ **Airstream trailers**
US travellers take a break and park up their Airstreams at the side of the highway in 1956.

> **"How lovely** it must be to live in a **house** that has **wheels...**"
>
> ENID BLYTON, *MR GALLIANO'S CIRCUS*

Priced at the princely sum of £300, it could accommodate two occupants, and featured a stove and washbasin, with felt insulation between its exterior and mahogany-panelled interior. It was promoted as "the holiday problem solved" and sales grew strongly during the 1920s. To promote the durability of its products, and the flexibility that caravans offered, the company entered the 1932 Monte Carlo Rally – its caravan towed by a Hillman. In a further promotional stunt, Eccles travelled through Africa and across the Sahara Desert with one of its caravans, again pulled by a Hillman.

Home on the road

In the US, the first commercially-produced "travel trailers" also arrived in the 1920s. Always on a much bigger, bolder scale than their European cousins, few captured the imagination more than the polished magnificence of the Airstream Clipper, introduced in 1936. With room for four, an onboard water supply, and electric lighting, it was a totem of American freedom.

Caravans also became popular in mainland Europe, but there, as in the UK and US, their manufacture was suspended with the outbreak of World War II. With the combined effect of fuel rationing, material shortages, and economic hardship, it would take until the 1950s for caravans to become a regular sight on roads again. When they finally did, they flourished. Modern, lightweight materials allowed caravans to be made more cheaply with many ingenious design features. No longer the preserve of the wealthy, caravanning entered a golden age that lasted until the 1980s.

◁ **Family outing, 1934**
Dedicated caravan sites sprang up quickly during the 1930s, and allowed families to enjoy different parts of the country.

LIFE BEHIND THE WHEEL

Rise of the camper van

A converted delivery van appeared a logical solution for providing mobile bed-space that could be driven anywhere. Most standard vans, however, were simply too cramped for comfort. This led British conversion specialist Martin Walter to design an extendable canvas roof for its 1954 Bedford Dormobile that could be erected when the vehicle was parked. With the additional head-height, occupants could actually stand upright in the van – to cook a meal on a small stove, for instance. The idea was adopted by Westfalia in Germany, which included a similar pop-up roof on its Camper conversion kits for Volkswagen Kombis from 1956.

BEDFORD DORMOBILE ROMANY **ADVERTISEMENT FROM 1962 DEMONSTRATING THE VERSATILE POP-UP ROOF MECHANISM.**

BGH 627

△ Style over speed
Built by SS Cars Limited (renamed Jaguar in 1945), the SS1 was known for its stylish appearance rather than its performance, although it proved a capable rallying car. Here it competes in Scotland in 1936.

The rise of diesel

Rudolf Diesel's engineering breakthrough – an engine that burned fuel more economically – made him a household name. Diesel-powered cars caught on slowly, but eventually found favour with motorists.

△ **Diesel pump**
Although it would be several decades before diesel cars were widely available, diesel trucks were common on the roads during the 1930s. They could be refuelled at dedicated pumps.

The year 1893 holds particular significance for motorists. On 24 December of that year, Henry Ford completed his first functioning petrol-fuelled engine on his kitchen sink, just a few weeks after the electric car was invented in Canada. Four months earlier, in August, France had introduced the world's first driving test. But, outshining these events, on 23 February the same year, Rudolf Diesel patented the engine that was named after him.

Efficiency and economy

Seven times more efficient than the steam engine, the diesel engine was also significantly more efficient than the petrol engine (invented 20 years earlier by Nikolaus Otto) because the fuel contained more energy. Additonally, unlike existing petrol engines, Diesel's device required no external spark to ignite the fuel. Instead, the diesel was compressed to a high temperature so that it ignited within the cylinder. Combustion moved the piston, which activated the motor. The engine also used a different type of fuel, called distillate, which was less refined than petrol and therefore cheaper. By the time Rudolf Diesel invented his engine, oil refining had been underway for more than 40 years, mainly to extract paraffin for lamps, and kerosene. Distillate was a byproduct of this process – essentially waste matter that was thrown away. Distillate was renamed "diesel" in 1894, after the new type of engine that finally provided a use for it. Rudolf Diesel had experimented with various substances, including coal powder, ammonia, and peanut oil before deciding on distillate as the best option.

◁ **Diesel powers utility vehicles**
By the 1930s, diesel was the fuel of choice for most utility vehicle users. This German advertisement shows a Shell diesel delivery truck making its rounds.

In the early 1900s, the diesel engine was super-sized to power locomotives, tractors, trucks, and eventually ships. It began to replace cumbersome coal engines. Continued refinements widened the diesel engine's application, especially for marine transport. On their own, the existing diesel engines were too fast to power a ship's propeller; if the prop rotated too quickly, it lost thrust. However, by installing an electric motor between the diesel engine and the propeller, the rotation speed could be controlled and kept at the desired level. Eventually, diesel engines with higher horsepower and slower speed evolved to drive cargo ships and military vessels. Diesel was a mainstay of World War II transportation on land and sea.

Smaller diesel engines were also developed for use in yachts and cars, but diesel-fuelled motoring did not really take off until the 1950s and 1960s, when diesel's fuel efficiency started to make sense in the devastated post-war economy of Europe. Mercedes-Benz had wowed car enthusiasts with its 260 D production car at the Berlin Motor Show in 1936, and Hanomag previewed its diesel car

◁ **The Mercedes-Benz 260 D, 1936**
The Mercedes-Benz 260 D was one of the first mass-produced diesel-engined passenger cars, the other being the diesel version of the Hanomag Rekord. Some 2,000 were built until 1940, after which Daimler-Benz devoted itself to manufacturing military vehicles.

△ **Engineer Karl Haeberle with the Hanomag diesel car, 1939**
Grabbing headlines in February 1939, German engineer Karl Haeberle set four world records in the 1.9-litre diesel Hanomag. The car had caused a sensation three years earlier at the Paris Motor Show because of its diesel engine.

in the same year in Paris. Diesel engines grabbed the headlines several times in that decade, when British army captain George Eyston set land speed records in diesel-powered cars (see box), reaching speeds close to 160 mph (260 km/h) – but diesel engines were slow to catch on for general motoring. A genuine advancement came in the 1990s, when the turbocharger was improved and popularized, delivering better performance and economy. Then, in the late 1990s, common rail fuel injection technology brought diesel engines a giant leap forward.

Developed by Bosch in Germany, the common rail system fed fuel to all the cylinders in the engine at a constant pressure and enabled multiple injections in a single injection cycle. The end result was an engine that was quieter and generated fewer raw emissions. Thus, at the end of the 20th century, the combined effect of the turbocharger and common rail technology elevated the diesel car to a new level of efficiency and desirability.

"The **automobile engine will come**, and then I will consider my life's work **complete**."

RUDOLF DIESEL

BIOGRAPHY

George Eyston

After serving in the Royal Field Artillery during World War I, George Eyston studied engineering at Cambridge, which helped to fuel his passion for racing and record-setting. In 1932 he became the first driver to exceed 120 mph (193 km/h) in a 750-cc car dubbed the Magic Midget.

While working for the Associated Equipment Company, Eyston hatched the idea of a race car with a diesel engine. In 1934 he claimed a diesel speed record of 115 mph (185 km/h) in a Chrysler fitted with an AEC London bus engine. Eyston set three land speed records between 1937 and 1939, achieving a top speed of 357.50 mph (575 km/h). He later worked with Stirling Moss and other drivers to break further records.

GEORGE EYSTON **PICTURED IN HIS MG MIDGET AT BROOKLANDS RACETRACK, 1931. HE WENT ON TO PIONEER DIESEL RACING ENGINES.**

"There was... the **newly-found excitement** of **driving a car** where it was said **cars could not go**."

RALPH A. BAGNOLD, *LIBYAN SANDS: TRAVEL IN A DEAD WORLD*

Automobile adventures

The increasing reliability and sophistication of cars in the 1930s, coupled with the daredevil spirit of the decade, kick-started a new European craze for far-flung motoring adventures.

Many of these exploits were based in exotic locations, especially in Africa and the Middle East. Desert driving was a challenge, but thanks to the experiments of Brigadier Ralph A. Bagnold in his Ford Model T, it became viable in the 1930s. Bagnold developed ladders of wire and bamboo for crossing dunes, a dash-mounted sun compass, and a system for conserving radiator water. He also introduced the simple ruse of lowering tyre pressures for driving on sand. Swedish rally driver Eva Dickson used these ideas to become the first woman to cross the Sahara in 1932 – her prize a crate of champagne from Baron von Blixen of Sweden, with whom she'd made a bet.

One of the greatest motoring achievements of that era was the round-the-world trip undertaken by Czech drivers Bretislav Jan Procházka and Jindrich Kubias in 1936. Beginning in Prague, the pair drove 27,700 miles (44,600 km) across 15 countries and three continents in 97 days – 53˚ of which were spent at sea. The car was a slightly modified Škoda Rapid, which had a four-cylinder 1.4-litre engine, independent suspension, and hydraulic brakes. The car was state-of-the-art, but still the drive was punishing, the pair averaging 390 miles (630 km) per day on mostly bone-jangling gravel tracks.

The route took them through Germany, Poland, Latvia, and Azerbaijan to the Russian city of Kaluga, where today's Rapid is assembled. After sailing the Caspian, they crossed Iran, weathering a sandstorm en route. To make up time they drove from Quetta, Pakistan, to Mumbai, India, in three days, then sailed for Japan via Hong Kong and Shanghai. After passing through Honolulu they reached San Francisco and crossed the US in a record-breaking time (100 hours and 55 minutes). The last stage of their journey took them from the French port of Cherbourg to Paris, and then to Nuremberg, Germany, and back to Prague.

◁ **The route and the car**
Bretislav Jan Procházka and co-pilot Jindrich Kubias's modified Škoda Rapid is exhibited beside a map showing the route of their 1936 pan-continental journey.

The other Mr Ford

Unlike his controlling father, Henry Ford's only son Edsel was a quiet, shy man. However, as president of the Ford Motor Company from 1919, he won key battles with his father to help cement Ford's global success.

△ **Greeting the new Ford**
"Room for everybody, fore and aft" was among the descriptions of the new Ford V8 Saloon De Luxe in a magazine advertisement from c.1937.

Edsel Ford, who died of stomach cancer at the young age of 49 in 1943, made major contributions to the company that his father founded in 1903, and became its president when he was only 25 years old.

Edsel's influence

Overshadowed by his father, Edsel is perhaps best remembered for the car that bears his name: the Edsel, which was released by Ford in 1957, and was a spectacular flop. However, Edsel can be credited with far-sightedness and a good head for business. In 1922 he recognized the opportunity to expand on Ford's brands by purchasing the bankrupt Lincoln Motor Company, the luxury carmaker, for $8 million, and later, in 1938, by founding the style-orientated Mercury brand of mid-priced cars.

One of Edsel's main concerns as Ford's president was the look of its cars. Unlike his father, he recognized that styling and design were going to be just as important to Ford customers as performance, reliability, and technology. With that in mind, he appointed E.T. Gregorie as head of design in 1931. The appointment was timely: Ford needed to catch up with General Motors, which had created its own design studio, Art and Colour, four years previously and was already rolling

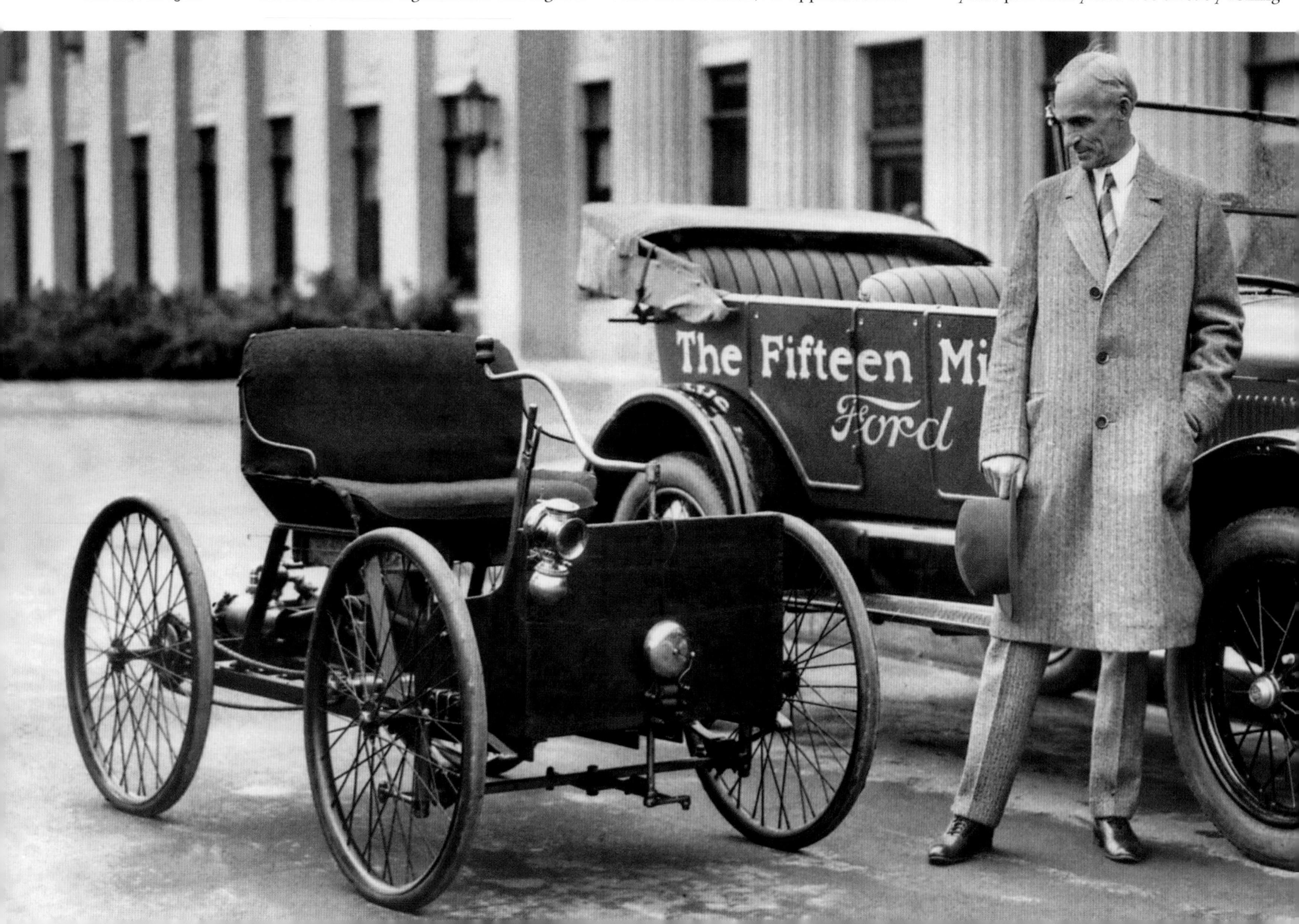

▽ **Car evolution**
Henry Ford (left) and his son Edsel, in 1927, compare the original 1896 Quadricycle, the first vehicle developed by the elder Ford, and the 15-millionth Model T to come out of the Ford plant in Detroit, Michigan.

▷ **The 20-millionth Ford, 1931**
After being in production for 18 years, the Ford Model T was finally replaced in 1927 by the Ford Model A. Here, a fresh Model A – the 20-millionth Ford – leaves the assembly line.

out new eye-catching models. However, Edsel did not overlook the safety angle, and pushed for such innovative features as shatterproof glass and hydraulic brakes. He also created the company's credit arm, which allowed customers to purchase Ford cars by paying in manageable instalments.

Competitive edge

Edsel's vision for the company's future was not restricted to the US. He saw the potential in Ford's flagging operations in Europe and succeeded in turning them around. In the UK, success came with vehicles designed specifically for British roads and drivers. However, Edsel's greatest achievements were the two cars that saw Ford prosper after production of the Model T ended in 1927: the Model A in the US, and the Model Y in the UK. With the Model Y, Ford finally had a car that could compete with Austin and Morris.

The Model A won approval from the elder Ford after Edsel convinced him that the Model T was not only technically obsolete but a company liability. He argued that Ford needed a more powerful car if it was to compete with GM's Chevrolet and others.

Edsel determined the Model A's basic proportions, while Henry directed his engineers. The car broke new ground in several areas, and was the only non-luxury car in 1927 to have hydraulic shock absorbers and shatterproof safety glass. While the 20-hp Model T had a top speed of about 40 mph (64 km/h), the Model A could cruise comfortably and economically at 45 mph (72 km/h). By January 1928, orders for the car, which was available in eight body styles, had reached 600,000.

> "He was responsible for **many good things** in the **Company's history**."
>
> HENRY EDMUNDS, DIRECTOR OF FORD'S ARCHIVES

Paying homage

In 1970, Henry Edmunds, director of Ford's archives, said of Edsel: "He was responsible for many good things in the Company's history – insistence on verve and dash in product styling, on a reliable and safer product, on fair and courteous relationships with dealers and the public... He did all the essential things that Henry refused to do, and consequently held the Company together... Without him, the Company might never have attained the solid image it has today".

KEY DEVELOPMENT
Ford Model Y

The 1932 Model Y is the car that made Ford's overseas operations in the UK and Germany profitable. The Model A was a blockbuster in the US, but not so in Europe, especially in the UK, where it was classified as a luxury car and was taxed accordingly. The Model Y was the first Ford car aimed specifically at the company's overseas operations. It was designed and engineered in Dearborn, Michigan, in just four months, and first shown in Europe in February 1932. It went on sale in October, but was now being built at Ford's Dagenham plant in the UK and in its Cologne factory in Germany. Descendants of the Model Y would remain in the Ford line-up for a remarkable 27 years.

THE MODEL Y **WAS THE FIRST FORD VEHICLE SPECIFICALLY AIMED AT MARKETS OUTSIDE THE US.**

▽ **Peugeot 402, 1935**
This large Peugeot, from France, was one of the first mainstream cars to adopt streamlined styling. Forward-thinking customers were just as impressed by its standard equipment, which included semaphore indicators, twin sun visors, and a dashboard clock. There was also a huge choice of 16 body styles.

Taking control

Although car controls were being standardized, many cars retained individual quirks. These required considerable skill to use – particularly sports models, which had a wide array of attention-demanding gauges.

Starting a modern car usually demands no more than pressing one button, but a driver of the 1920s or '30s would have been familiar with a more complex process. First the cold start control, or choke, would be set to enrich the fuel/air mixture entering the cylinders to make starting easier. On some cars an ignition control lever, usually mounted on the steering wheel, would be set to retard, or "lower", the spark to avoid backfiring, but the difficult job of setting the spark timing was increasingly being handled automatically. Often a hand throttle was provided so that the engine could be set to a fast idle as it warmed up.

Starting and changing gear

Electric starter motors, first seen in the US in 1911, were now a common feature on cars. The motor was usually operated by a button mounted either on the dashboard or on the floor next to the pedals. However, manufacturers still made provision for manual starting using a crank handle in case the car battery was flat or the motor failed to start the engine.

◁ **A Bentley receives its dashboard, 1930s**
High-performance cars, such as Bentleys, had extensive instruments to inform the driver.

Moving up through the gears was easier now that synchromesh was common. However, synchromesh was not always fitted to the bottom gear, so drivers had to learn to "double declutch" when changing down. The trick was to pause with the gear lever in neutral, to engage the clutch for a moment, and then declutch again and select the lower gear. Doing it correctly prevented noisy clashes of gears, but it took some skill.

Automatic gearboxes had not yet been invented, but one alternative to a conventional manual gearbox was the Wilson "preselector", invented by British engineer Major W.G. Wilson. This system enabled the driver to move the gear lever to any position at will, but the gear change was only made when a pedal was pressed.

Speed and turning

The accelerator and brake pedals worked the same way as they do in a modern car – but the response to either pedal was not as fast as a modern driver would expect. Steering was unassisted, and often heavy when the wheels were turned. As a result, steering wheels were large and drivers had to sit close to them.

Flashing indicator lamps were as yet a novelty, but many cars had semaphore "trafficators": arms that sprang out from the sides of the body to indicate the direction of travel at junctions. The trafficator switch was mounted on the dashboard or the steering wheel – the steering column stalk had yet to arrive.

▽ **Bugatti Type 51, 1930s**
To keep their weight down, cars built for racing in this period featured simple controls, including an externally mounted handbrake lever.

△ **The interior of a 1930s SS1**, with windscreen wiper motor at the top of the screen. Note the large steering wheel.

1946–1960

REBUILDING THE WORLD

1946–1960

Rebuilding the world

The pent-up demand for new cars after the deprivation of World War II was enormous. However, potential customers had to be patient, as the industry needed time to recover. As a result, the first new peacetime cars were prewar models revived with a few improvements. This, combined with restricted supplies of fuel and raw materials, added to the prevailing atmosphere of austerity.

A burgeoning market

By 1950, however, a fascinating variety of new small family cars had become available to the public. Renault and Volkswagen favoured rear-engined designs, while the British Morris Minor was slow but sophisticated. Fiat's revamped 500 was familiar and cheap, Citroën's 2CV embodied minimalism, and Swedish newcomer Saab aimed for aerodynamic efficiency. Further up the scale, solid workhorses that also doubled as excellent taxis were soon to come from Peugeot and Mercedes-Benz. Australia got its own, tough new saloon, the Holden; estate cars became an emerging market sector that combined business with pleasure, and the Land Rover delivered four-wheel drive to the civilian world.

In the US, meanwhile, aircraft-inspired style and V8 power had become the watchwords as the Detroit giants roared into the 1950s – a decade that would see a peak in confident automotive style and engineering as fins and engine capacities soared to dramatic new heights. However, at the same time, European vehicle manufacturers were making quiet inroads into the US, with success notably greeting the French Renault Dauphine, German VW Beetle, and nimble sports cars from the UK and Italy.

CAR PRODUCTION FINALLY REGAINS PACE AFTER WORLD WAR II

MANUFACTURERS LAUNCH A WEALTH OF NEW DESIGNS

"... **aircraft-inspired style** and **V8 power** had become the **watchwords**."

Long distances, high speeds, and mechanical strength were also evident in the motor sport boom, which created demand for sports cars and made international heroes of race and rally winners. Even at the lowly end of competition, worn-out jalopies fit only for the scrapyard could enjoy one last chaotic hurrah in stock car races.

New global challenges

Life behind the wheel became the new normal. In the US, motorists enjoyed drive-ins, from cinemas and restaurants to banks. In Europe, the emphasis was on expanded foreign travel, with tunnels, bridges, and ferries opening up new routes and promoting new horizons.

While the expanded freeway network in the US promoted driving over huge distances at sustained speeds, cities in the old world struggled to cope with the demands that traffic imposed. Motorways in the UK and autoroutes in France progressed slowly. Only a small number of cities were reshaped around the needs of cars – and most of those suffered problems as a result. Fortunately for Japan, its infrastructure could expand at the same rate as its burgeoning car industry. Meanwhile, for South America and Russia, from the pampas to the steppes, the relationship between driving and open space created its own challenges.

A further problem for motorists arose from the volatile geopolitics of a changing world order. Fuel crises arising from post-colonial ructions in the Middle East made themselves felt at the petrol pump from 1956 onwards. But, even then, the car industry responded imaginatively, supplying a new breed of tiny cars with motorbike engines and bubble profiles.

EUROPEAN SPORTS CARS FIND A READY MARKET IN THE US

IMPRESSIVE AMERICAN V8s PROVE COSTLY TO RUN WHEN FUEL BECOMES SCARCE

The world gets back on the road

As the world's leading carmakers emerged from the daze of World War II, they had to adjust to new realities, including petrol rationing, steel shortages, and governmental dictates.

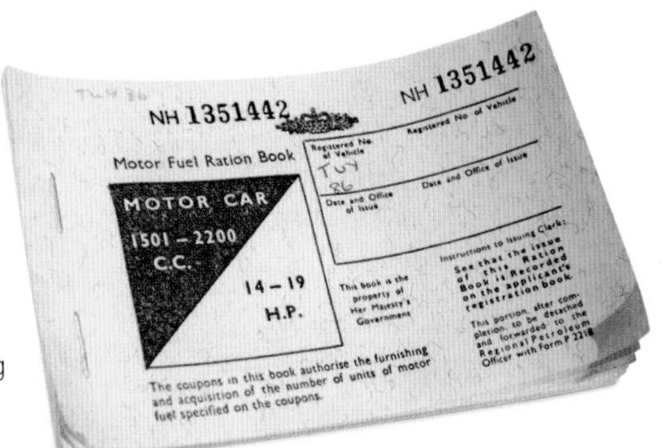

▷ **Ration book for fuel, 1950**
Petrol rationing continued in the UK after the end of World War II until 1950, although it was adopted briefly again in 1957 during the Suez Crisis.

In the face of the austerity measures that continued in the UK and Europe after World War II, automobile manufacturers sprang into action to meet the build-up of demand for new cars. Ford was well positioned to take advantage of the post-war years, as it was equipped with sophisticated machinery used to produce military vehicles for government contracts, thanks to taxpayer funding. Other manufacturers, such as Morris, were at a disadvantage, having converted their facilities for aircraft production, and had to retool.

Post-war US

Ford was also one of the first car manufacturers to bounce back in the US, launching its 1946 model ahead of its main competitor, Chrysler. Externally the car was similar to the 1942 prewar model, but it was powered by a larger 3.9-litre, 100-bhp V8 engine. Trailing behind the two market leaders, Plymouth, Buick, Dodge, and Pontiac also launched post-war models, contributing to a car-buying boom. Meanwhile, the US government initiated a road-building programme to service burgeoning cities and suburbs.

Roads were also a priority for the British government, which had drawn up plans for a new highway network during World War II. Desperate for foreign currency, the British government also ordered carmakers to prioritize export sales, setting aside 50 per cent of all units for this purpose, at the expense of domestic customers. Of the half a million cars produced in 1950, for example, more than two-thirds were shipped to the US. The government of France went one step further in its control of the automotive industry, nationalizing the country's main car producer Renault in 1945.

Raw industrial materials in all countries were scarce, with strict quotas in place for steel. In the UK, steel was freely available for export production, but the domestic quota allocated to each car manufacturer depended on how successful their exports were. Domestic demand was boosted when petrol rationing was lifted in 1950 although at that time there was only one car for every 16 people. It was not until 1955 that supply of materials caught up with demand.

Flooding the market

In the rush to populate the post-war market with new cars, many makers attempted to release too many models, rather than simply focusing on a few key ones. In the UK, Morris pinned its hopes on the Morris Minor, unveiled at the 1948 Motor Show. Although it quickly became the UK's favourite small car, Morris failed to capitalize on its potential in the export market, instead investing resources into other domestic models, including the Morris Oxford, Cowley, and Morris Six. Rover made a similar strategic move, spreading resources across several models instead of focusing on the new Land Rover, which was launched in 1948.

◁ **The need to rebuild roads, 1940s**
As well as the destruction of buildings and property, World War II left huge numbers of roads in ruins across Europe, as seen in this photograph of urban devastation in Germany.

△ **Paris Motor Show, 1948**
Manufacturers display their newest models in an effort to recover from wartime disruption. Here visitors inspect a glamorous Delahaye 135, although virtually no-one could afford to buy it.

Austin also diversified with a number of large saloons when it could have pushed the new A40 as a rival to Volkswagen in overseas markets – after the war, the latter had started from scratch in a bombed-out plant in Germany, but it made up for this handicap with the manufacture of its Beetle (see pp.138–39). French mass-market manufacturers mostly prospered, but the country's luxury marques – Bugatti, Delage, Delahaye, Hotchkiss and Talbot – all struggled because of punishing domestic purchase taxes levied on any car with an engine with a 2-litre capacity or over. While much of the developed world was still recovering from World War II, there were few available export markets to offset this restriction.

> **"Unchallenged** for sheer **sterling worth."**
>
> MORRIS MINOR ADVERTISING SLOGAN, 1950s

KEY DEVELOPMENT
Manufacturers merge

For some smaller US car companies, survival after World War II was difficult. The 1950s saw a series of mergers take place, notably the American Motors Corporation (AMC), combining Hudson and the Nash-Kelvinator Corporation, with the Studebaker-Packard Corporation. Although Studebaker pioneered the overhead valve V8 engine, its cars sold poorly, and the company merged with Packard in 1954. The company continued to sell cars under the Studebaker name until 1966.

A STUDEBAKER-PACKARD MAGAZINE ADVERT **FROM 1959 DEMONSTRATES THE VARIETY OF MODELS AVAILABLE.**

Easy drivers

The pioneers of motoring were hardy folk. Exposed to the elements, their comfort came a distant second to maintaining control of the machine. However, as time passed, the car became essential, and life behind the wheel became more comfortable.

△ **Chrysler Imperial** The 1957 Chrysler Imperial, with its air-conditioning, was a vehicle in which the driver could truly stay cool behind the wheel. This model was owned by film director Howard Hughes.

Many of the basic systems in cars had been tried and tested by the 1940s, and manufacturers now felt confident enough to start offering enhancements aimed at making the driving experience easier.

Automatic gear changing had been in development for decades before General Motors adopted a hydraulic system in its Hydramatic transmission, which made its production debut in 1940-model Oldsmobiles and Cadillacs. There were four forward speeds responding to accelerator position and the car's speed, allowing the driver to relax, using a single foot to operate the two pedals – accelerator and brake. At first it cost $25 extra on the Cadillac and $100 more on an Oldsmobile. It spread to the Buick range, too, in 1948, when it was first allied to a torque converter. Rival firm Chrysler was late catching up on this tempting new option, not releasing its Torqueflite automatic transmission until 1953. It was keen not to be caught out again when it came to offering customer-pleasing features.

In 1939, Packard had become the first marque to introduce air-conditioning into its cars. However, the Bishop & Babcock Weather Conditioner unit that it used was discontinued in 1941, the same year that Cadillac abandoned its own attempt to introduce air-conditioning. It took until 1953 for another carmaker to investigate the concept. Walter Chrysler had pioneered Airtemp air-conditioning for the Chrysler building, after which the company made the option available on its 1953 Imperial. In the same year, Cadillac, Buick, and Oldsmobile licensed air-conditioning systems from Frigidaire.

Assisted driving

Inventions such as power-assisted steering, the brake servo, automatic choke, and cruise control all helped make driving easier. Cruise control emerged in 1948, when Ralph Teetor devised a system for electrically adjusting the position of the throttle cable in order to maintain road speed based on the rotation of the speedometer cable. This system, called Autopilot, was first fitted to the 1958 Imperial. By 1965, AMC had developed a simpler vacuum-based system, which has since been replaced by electronic and adaptive control.

◁ **The Oldsmobile 88, 1955** Featuring an automatic transmission, power-assisted steering, a radio, defroster, and heater, the 88 was one of the most well-appointed standard cars of its day.

KEY EVENTS

- **1939** Packard introduces the Bishop & Babcock Weather Conditioner into its cars. The system features a heater.
- **1948** Cadillac, Lincoln, and Daimler all offer electrically-operated windows for the first time, only 29 years after the first wind-up/down windows had been introduced.
- **1951** Chrysler becomes the world's first carmaker to offer power-assisted steering, which takes the effort out of parking manoeuvres.
- **1951** Werner Armstrong patents the first automatic choke, which enables an engine to be started at any temperature.
- **1953** The Chrysler Imperial offers the Airtemp – the most advanced air-conditioning system of its day.
- **1955** Electrically-controlled, self-levelling suspension is introduced on Packards, although it is too complex to become an industry norm.
- **1955** The first boot lock that could be remotely operated from the driver's seat is launched on the latest Cadillacs.
- **1959** Several companies start selling in-car record players, such as the RCA Victrola.

***CHRYSLER'S HIGHWAY HI-FI*, LAUNCHED IN 1956, WAS THE FIRST RECORD PLAYER MADE FOR CARS.**

△ **Drivers in the 1950s** could increasingly enjoy the convenience of sunroofs, cruise control, a radio, and air-conditioning.

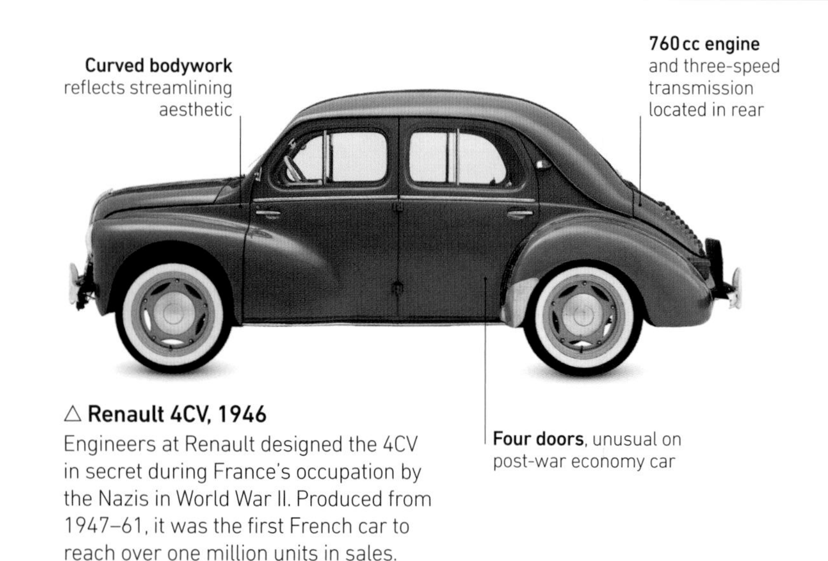

△ **Renault 4CV, 1946**
Engineers at Renault designed the 4CV in secret during France's occupation by the Nazis in World War II. Produced from 1947–61, it was the first French car to reach over one million units in sales.

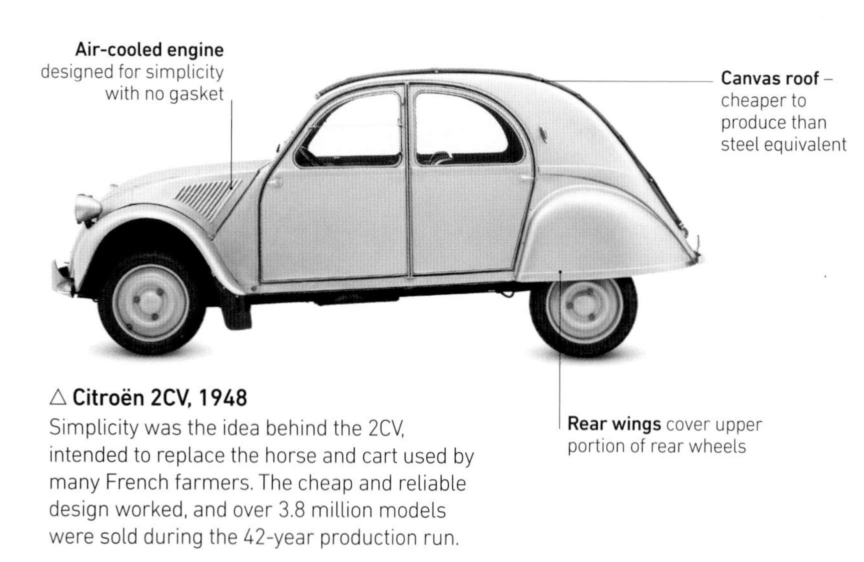

△ **Citroën 2CV, 1948**
Simplicity was the idea behind the 2CV, intended to replace the horse and cart used by many French farmers. The cheap and reliable design worked, and over 3.8 million models were sold during the 42-year production run.

"The **first words** that a baby should **learn to pronounce** are Mummy, Daddy, and **Citroën**."

ATTRIBUTED TO ANDRÉ CITROËN

▽ **Renault factory, 1957**
Brand new cars line up outside the Renault factory in France. The manufacturer recognized the need for affordable, reliable vehicles for ordinary people in the years following World War II.

Aerodynamic body meant small engine could be used, saving fuel

Green paintwork applied to all 1,246 Saab 92s built in first year of production

△ **Saab 92, 1949**
The Saab 92, made from 1949–52, was the company's first production car and aimed at ordinary Swedes. Economical methods of manufacture made the car light without compromising on strength.

Canvas fold-back top appealed to younger consumers

Four-cylinder engine capable of 60 mph (96 km/h)

Chassis just under 9 ft 8 in (3 m) long

△ **Fiat 500C, 1949**
This little car, offered as a two-seater convertible and a tiny estate, was a thorough update of Fiat's original 500 "Topolino" of 1936. It was enormously, and deservedly, popular.

Mobilizing the masses

Motoring started out as a pastime for the wealthy. Cars were very expensive, and it would have taken most people several years to save enough for a second-hand car, let alone a new one. But in the aftermath of World War II, several European countries launched inexpensive basic vehicles for the masses. Some – including the Volkswagen Beetle and Citroën 2CV – had been developed prior to hostilities, but it took until the end of the war for civilian output to begin. Others, such as the Austin A30 and the Saab 92, were developed in the post-war climate. Although none of these cars possessed many creature comforts, they nevertheless perfectly met the needs of a society recovering from six years of war and hardship.

◁ **The Volkswagen plant, Germany, 1953**
A line of Volkswagen bodies (see pp.138–39) await their chassis on the factory production line. Then known as the 1200 (the Beetle title followed much later), the Wolfsburg plant was then, and remains, the world's biggest car factory.

Everyone wants a car

The peace of the post-war period allowed a huge expansion in private car use. As more and more people found they could afford to run a car, whole new road networks were needed to carry them.

As the chaos of World War II receded, an economic revolution shook the world – one that resulted in a vast increase in private car ownership. With a surge in the number of popular new cars such as the Volkswagen Beetle, Morris Minor, and Renault 4CV, roads had to be transformed to cope.

A huge programme of new road building and upgrades was undertaken across Europe and the US. Most notable was the idea of motorways (or freeways) – multi-lane roads designed for long-distance travel at high speeds, usually with few access points and no road crossings.

△ **Masses of cars for the masses**
This brochure for the Morris Minor in the 1950s entices ordinary people to welcome the car into their family, satisfying a craving for the freedom of independent travel.

Miles of motorways

Early experiments – including the world's first dual highway, the "Autostrada A8" in Italy in 1924 (see pp.102–03) and the "Autobahns" in Germany in 1935 – spread to many more countries in the following years. In the US, the first freeway, the Pennsylvania Turnpike, opened in 1940, although a full nationwide programme did not start until 1956. Sweden's first motorway was built in 1953, followed by France in 1954, and the UK in 1958 with the Preston Bypass in Lancashire (see p.213), which later became part of the M6 motorway.

Motorway interchanges were often vast and complex. In the UK, the term "spaghetti junction" was coined for the Gravelly Hill interchange on the M6 motorway in Birmingham, a name that has been used for dozens of intersections worldwide ever since.

△ **Motorways connect**
Motorways enabled more and more drivers to travel rapidly over long distances. Although there were collisions, such as this one on an Autobahn in 1957, these were few and far between and never deterred drivers.

KEY DEVELOPMENT
Complex interchanges

The problem with big, fast-moving roads was that traditional intersections, with their stop signs, traffic lights, and roundabouts, severely affected traffic flow. Expansive junctions of criss-crossing access roads were the answer, as they allowed vehicles to drive on and off these fast roads with little reduction in traffic speeds.

No fewer than a dozen major freeways intersect the city of Los Angeles in California. Major interchanges between freeways were built that had no local traffic access points, easing congestion.

TRAFFIC ON A HIGHWAY **WITH INTERCHANGES IN LOS ANGELES, CALIFORNIA, US.**

Suburban dreams

City centres were often unsuitable environments for car owners. However, as the 20th century progressed, an increasing number of urban dwellers embraced a fresh start in suburbia – a slew of purpose-built residential zones on the edges of cities. With the need to travel greater distances, a new class of car-loving commuter was born.

Car ownership among suburbanites exploded and new roadscapes were increasingly designed specifically around cars. Such networks often encouraged car use at the expense of walking or cycling, and even short trips started to be carried out in cars.

Huge new car parks sprang up to manage this new influx, including multi-storey ones. Vast shopping malls increasingly encouraged out-of-town

▷ **Heaven or hell?**
Huge multi-lane roads offered the promise of freedom, but became a victim of their own success. This view of the four-lane Pasadena Freeway in California, taken in 1958, shows lines of traffic barely moving.

shopping by car, while drive-in food outlets meant people could even eat while remaining in their cars.

Cities such as Los Angeles and Detroit in the US and Melbourne in Australia were modelled and remodelled to cater for cars above all else, which led to the new phenomenon of "car dependency".

Spiralling demand

This rise in cars on the road also increased levels of traffic congestion, which led to demands for more and bigger roads. Equally, there was pressure for "impediments" to traffic flow – such as pedestrians, cyclists, and trams – to be edged out of the system. A whole new road infrastructure incorporating bridges and tunnels started to become the norm.

Changes such as these only served to encourage even higher traffic volumes in a seemingly never-ending cycle of intensifying car use. Not for the first time, motoring had become a victim of its own success; the effects were to be felt for decades ahead in ever-expanding territories around the globe.

△ **Suburban bliss**
This advert for Chevrolet's Corvette and Bel Air models from 1957 shows the cars in their natural setting: outside an ultra-modern suburban home with its own garage to house the cars.

CRÈME
LION NOIR

Demolition derby

By the 1950s, increasing sales of new cars meant there were more and more worn-out vehicles reaching the ends of their lives. Smashing up older cars on the racetrack became a thrilling new form of motor sport.

Stock car racing – competing in production saloon cars – migrated from the US to the UK and Europe in the mid-1950s. The events mainly took place on shale-surfaced oval tracks that had been built for greyhound racing, and which were also used for motorcycle speedway races. While mainstream racing circuits were generally in the countryside, these shale ovals were often in cities and pulled in huge crowds to watch the action. Early venues sprang up across the UK, and in Ireland, Belgium, and the Netherlands.

Although stock car racing was theoretically a non-contact sport, clashes were inevitable when up to 40 cars were crammed on to a ¼-mile (½-km) track, and it was quickly obvious to promoters that the crashes were just as popular as the racing – perhaps even more so. In the US, the easy availability of worn-out Ford Model T vehicles led to "demolition derby" races in which smashing up the opposition was the objective, with the last car running declared the winner. In the UK a whole new class of "banger" racing was invented, still based on a sprint to the chequered flag but with contact between cars permitted to add extra excitement.

The cars used in these races were production models that had been stripped of their interior trim and glass for safety, and were often painted in outlandish colour schemes, although underneath this they were usually irretrievably rusty. Most of the cars involved were the larger mass-produced family saloons of the 1950s and '60s that were now worthless. However, some drivers took delight in racing rare machines, putting them at odds with the trend for saving and restoring classic cars that began to emerge in the early 1970s.

◁ **Drawing a crowd**
Spectators watch a stock car race at Buffalo Stadium, near Paris, France. Like many stock car venues, the stadium was previously used for other sports – in this case, for cycle races, and boxing and football matches.

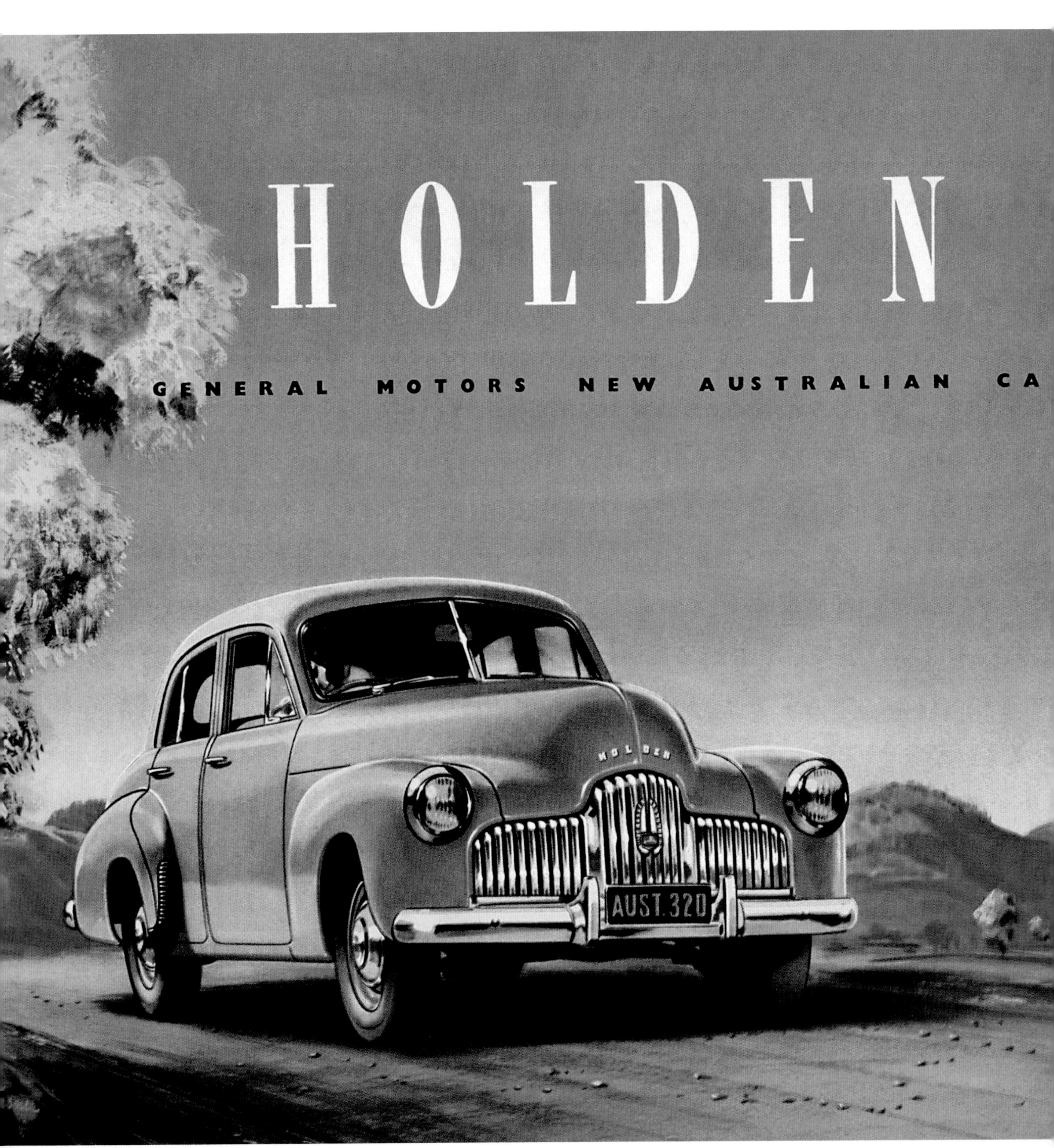

△ **The 1948 sales brochure** for the Holden 48-215, the first car developed and produced for the Australian market, shows the rough roads that the Holden was able to handle.

Australia's Holden

Before 1948, Australian drivers had to make do with cars that were designed for other territories. But then General Motors-Holden (GM-H) brought out the Holden 48-215, the first all-Australian automobile.

KEY EVENTS

- **1931** General Motors in Australia merges with Holden Motor Body Builders, creating General Motors-Holden's Ltd (GM-H).
- **1948** The Holden 48-215, known as the FX, is successfully launched, based on a previously rejected Chevrolet design for the American market. It becomes Australia's first indigenous mass-produced car.
- **1951** The 50-2106 Holden pickup utility vehicle, or "ute", is launched.
- **1953** The FJ replaces the FX, which results in Holden gaining a 50 per cent share of the Australian market.
- **1960** Ford unveils its Falcon XK, an American import, but it proves unsuccessful.
- **1972** Ford Australia launches its first home-grown car, the Falcon XA, which captures the public's imagination and increases the pressure on Holden.
- **2016** Ford Australia ceases manufacturing.
- **2017** The last Holden is built.

HOLDEN'S MAIN ASSEMBLY LINE **AT FISHERMANS BEND IN MELBOURNE, AUSTRALIA.**

The Holden 48-215 struck a chord with Australians and effectively sparked a whole new market. The car had been developed originally as a Chevrolet by General Motors but was deemed too small for the American market in the wake of World War II. The decision was taken to develop it exclusively for Australia, which resulted in it becoming the country's first mass-produced car. Alongside the 48-215, GM-H (usually referred to as Holden) produced the 50-2106, a utility vehicle designed to satisfy local interest in "utes".

Australian cars needed to be hardier and sturdier than their European or American counterparts to cope with the poor roads and challenging terrain. The vast size of the continent also meant that vehicles had to be capable of covering long distances without breaking down. The Holden's simple mechanics made it the perfect design.

The 48-215, popularly known as the FX, was an immediate success. It was followed by the FJ in 1953, which ensured that Holden attracted a 50 per cent share of the home market. Its rivals were forced to take note.

Competing for the market

Holden's main competitor was Ford, which first assembled and then manufactured American models; by the 1970s its Falcon XA was available exclusively in Australia. These cars competed head-on with the likes of the Holden Kingswood and Torana, and led to fierce debate as to which was the superior marque. Rivalry in Australia between Ford and Holden fanatics was intense – not just at sporting events, such as the Bathurst 1,000 endurance race, historically dominated by both marques, but day to day on ordinary roads.

For Australian motorists not wishing to compete in this one-upmanship, Japanese cars offered a reliable and inexpensive alternative to Fords and Holdens. Many Ford products were rebadged Japanese cars – such as the Telstar, based on a Mazda model – and Japanese used cars were readily available in a right-hand drive configuration.

△ **Holden FJ**
First produced in 1953, the FJ saloon was a huge hit with the public. Although little had changed from the FX of 1948, its styling was more upbeat and the ride more comfortable. Production continued until 1956.

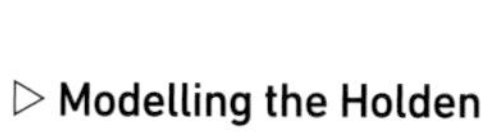

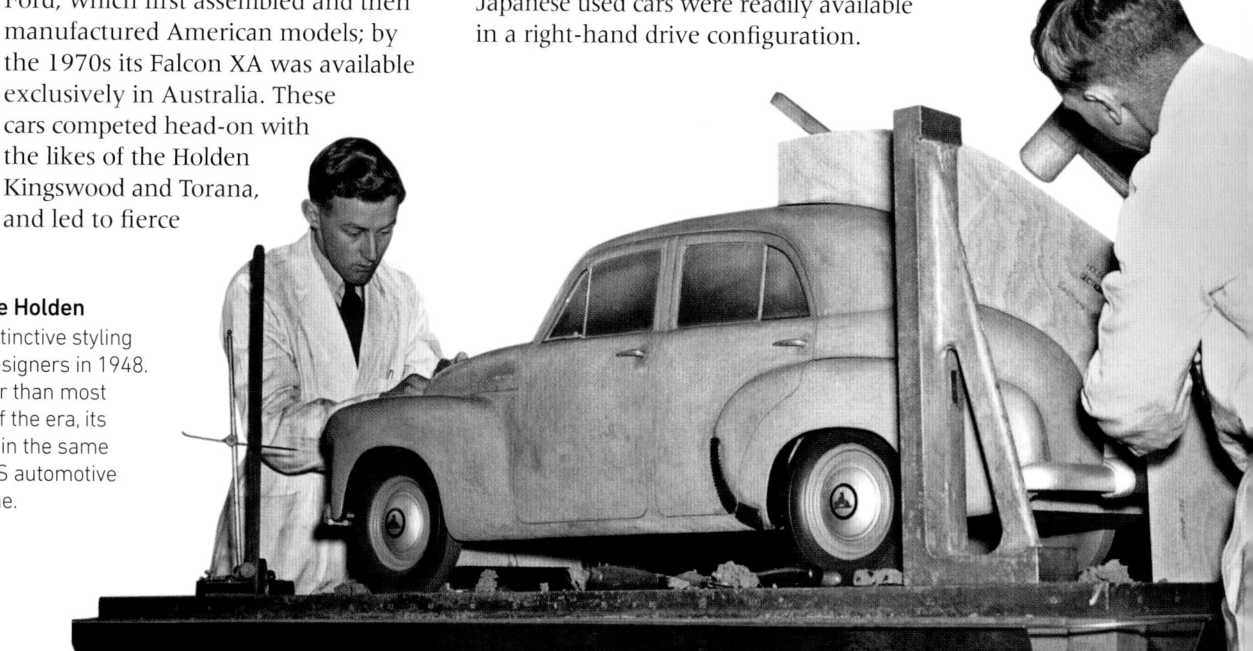

▷ **Modelling the Holden**
The Holden's distinctive styling is finalized by designers in 1948. Although smaller than most American cars of the era, its curved lines are in the same style as much US automotive design of the time.

NO ENTRY, FRANCE

NO STUDDED TYRES, SWEDEN

ALTITUDE MARKER, US

NO-VEHICLE PASSAGE, EUROPE

HIGHWAY ROUTE 66 MARKER, US

STOP SIGN, MOROCCO

Motorway signs in many countries have white or yellow type and symbols on a green background

DIRECTIONAL SIGN, SYRIA

LOCATIONS AHEAD, SAHARA DESERT

BEWARE OF ANIMALS CROSSING, AUSTRALIA

Sign language

Standardized signs for directions, information, and hazard warnings were adopted across many countries to make driving easier and safer.

The growth of cycling in the 19th century led to the first large-scale adoption of road signs in the US and Europe. Early signs were posted by cyclists' and motorists' groups until governments started to take over in the first years of the 20th century.

In Europe the standardization of road signs between countries began as early as 1908. European signs were developed between the world wars, using symbols rather than words where possible to avoid problems with different languages. The UK devised its own signage system, formalized by the Road Traffic Act of 1930. Following the introduction of motorways in the UK in the late 1950s, signage was overhauled along European lines. The new signs, designed by Jock Kinneir and Margaret Calvert, were introduced in 1965 (see pp.212–13). The US had a system defined by the *Manual on Uniform Traffic Control Devices*, first published in 1935, although states were not required to fully implement these policies until the 1960s. Even then, some local variations were allowed. Now, countries around the world follow many of the same conventions, such as red borders for warnings or green backgrounds for directions.

Triangular sign with red border has been a standard warning sign in Europe since 1968

BEWARE OF ELK, SWEDEN

BEWARE OF TANKS CROSSING, US

Yellow or orange diamond signifies person, animal, or vehicle crossing road

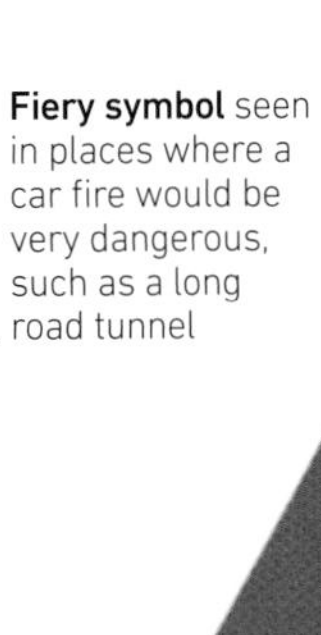

Fiery symbol seen in places where a car fire would be very dangerous, such as a long road tunnel

DO NOT CARRY FLAMMABLE LIQUIDS, FRANCE

Simple symbol gives essential information without needing words

CONSTRUCTION IN PROGRESS, US

BEWARE OF THE QUAYSIDE, UK

Wildlife warnings around the world feature animals from deer to camels, bears, snakes, and penguins

BEWARE OF PENGUINS CROSSING, NEW ZEALAND

BEWARE OF SNAKES CROSSING, CANADA

BE ALERT FOR AMISH NON-MOTORIZED VEHICLES, US

DRIVING TECHNOLOGY

Roadside speed cameras

The first speed limit enforcement cameras were built in the US in the 1960s. Since then, developments in sensors and image processing, and the introduction of digital cameras, have made modern speed cameras more accurate and cheaper to operate.

The cameras operate in two ways. Some use RADAR or LIDAR (laser) sensors aimed at a vehicle, while others calculate speed from the time taken for the vehicle to travel a known distance.

SPEED CAMERAS **PHOTOGRAPH VEHICLES EXCEEDING THE GIVEN LIMIT EITHER DIGITALLY OR ON TRADITIONAL FILM.**

△ **Morris Oxford MO, 1948**
The Morris Oxford looked much like a scaled-up Minor – and to a degree it was. Torsion bar front suspension, unibody construction, and a sidevalve engine were effectively taken from the Morris Minor concept and simply enlarged to create this model.

△ **Mercedes-Benz Ponton, 1954**
Mercedes' first major post-war models, the Ponton range included the four-cylinder-engined 180 and 190 models, and the longer, more upmarket 220 six-cylinder. These formed over 80 per cent of Mercedes' sales from launch until their replacement by the flamboyant Fintail.

European workhorses

The 1950s were an austere time in Europe – fresh out of the war, money was tight and people were far more concerned about essentials than they were about fripperies, such as an abundance of chrome. While American car manufacturers pushed forwards with glitz and glamour, in Europe the motor industry favoured worthy cars that got the job done simply, reliably, and with a degree of toughness vital in something that represented a significant investment to the people of nations getting back on their feet.

Marques such as Mercedes-Benz, Peugeot, and Morris forged reputations for sturdy, reliable cars during this era, while American giant Ford tried to add style into the mix with its so-called "Three Graces" – the Consul, Zephyr, and Zodiac. On the other side of the Iron Curtain, the GAZ-21 combined scaled-down American lines with simple mechanicals and reliability that could withstand a Siberian winter.

△ **Ford Fairlane brochure, 1958**
While the US market was driven by glitz, glamour, and a constant clamouring for next year's model, things in Europe were very different. Reliability and thrift were the bywords driving the European market.

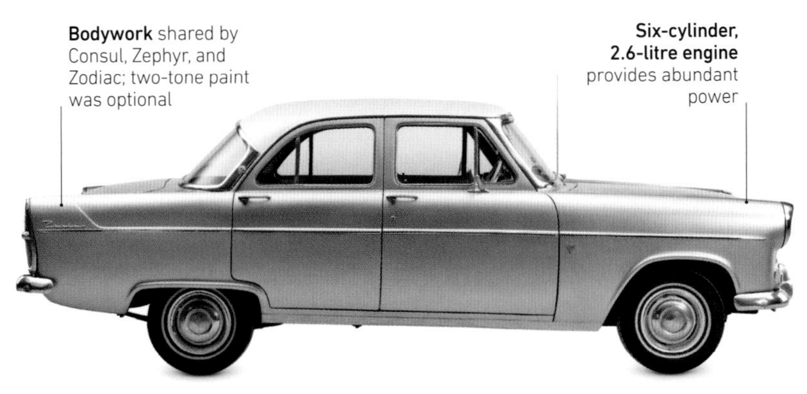

△ **Ford Zephyr MkII, 1956**
Part of Ford's "Three Graces" range, the Zephyr was an attempt to marry British scale and solidity with American style. Along with General Motors rival Vauxhall, the Zephyr ensured that US-style glamour was a familiar sight on UK roads.

△ **GAZ-21 Volga, 1956**
A replacement for the Pobeda, the Volga was intended to provide fast and spacious transport for Soviet officials. Named after a Russian river, it was well beyond the means of ordinary Russians, although it was widely used as a taxi.

"These were **tough cars, fast for their day**..."

LEISURE IN POST-WAR BRITAIN, STUART HYLTON, ON THE FORD ZEPHYR AND ZODIAC

▽ **Peugeot 403, 1955**
The Peugeot 403 was ideal for austerity Europe – a large, solid car with simple mechanicals and a no-nonsense approach. It was the preferred car for much of middle-class France.

Traffic grows in cities

In the post-war period, the boom in popularity of cars – often encouraged by local authorities – led to cities becoming newly overwhelmed with traffic. Whole new road systems needed to be built to cope with the demand.

△ **Modern life**
As this *Life* magazine cover from 1960 shows, heavy traffic (seen here in Los Angeles, California) was increasingly a part of everyday life for citizens. Whole cities were remodelled to cope with the extra pressures.

Following World War II, dramatic increases in the volume of traffic on roads began to have an impact on cities around the world.

In the UK, for instance, the annual distance travelled by cars, measured in passenger miles, increased almost five-fold between 1952–69, from 36 billion miles (58bn km) to 178 billion miles (286bn km). In Europe, cities that had been reduced to rubble during the war were now presented with what many saw as a new menace: the car. In rebuilding bombed-out cities, authorities took the chance to modernize their transport systems via a huge explosion in road building, with cities across Europe remodelled around car use. Road transport was increasingly viewed as "the future".

The new gridlock

Cities found themselves grinding to a halt under the burden of heavier traffic. Problems started when traffic entering junctions blocked traffic coming from other directions. In New York, the ensuing chaos gave rise to a new term: "gridlock". Ways to prevent motorists from blocking intersections – such as yellow box junctions in the UK – had some success, but other solutions were needed to ease congestion.

Changing roadscapes

Boston in the US changed its roadscape completely. By 1948, more than 100 highways fed traffic into the metropolitan area, and over two million people every day were taking to roads built for just a handful of wagons and horseback riders. Boston adopted a system of carefully channelled "arterial" traffic, bypasses, elevated sections, and limited-access expressways. Whole neighbourhoods were flattened in compulsory land purchases to make way for the 330-ft (100-m) wide new roads.

Road planners also increasingly adopted the idea of "ring roads" (known as "beltways" in the US). These had started to appear in the 19th century – for instance, around the Austrian capital, Vienna – but the concept of bypasses around large towns and cities took off in the 1960s.

One of the most famous ring roads, the Paris Périphérique that encircled the old city, was begun in 1958 and would take 15 years to complete. When it was finished, one-quarter of all Parisian traffic flowed along it, making it easily the busiest road in France. A victim of its own success, the Périphérique was – and remains – terribly congested. London's North and South Circular roads suffered a similar fate, as did the huge "beltway" around Washington, D.C. in the US.

▽ **New York, 1953**
Traffic jams were a menace to cities, especially at intersections. This is a view of 42nd Street and Fifth Avenue in New York, where the term "gridlock" was first coined.

KEY DEVELOPMENT
Parking on the next level

High-rise buildings were a 20th-century phenomenon that transferred to car parking. To cope with the huge rush of cars into cities in the post-war period, vast multi-storey car parks were built, swallowing hundreds of vehicles into a relatively small "footprint" of land.

Plans to link roof-top car parking with high-level "overpass" roads proved fanciful, however. It was much more cost-effective to build multi-level car parks accessed by ramps.

***AUTORIMESSA PIAZZALE ROMA*, MULTI-STOREY CAR PARK IN VENICE, ITALY (1953).**

In Coventry, a British city all but levelled by the Luftwaffe during World War II, the whole centre had to be rebuilt, including its roads. A young town planner called Donald Gibson created one of Europe's first traffic-free shopping precincts and surrounded it by an inner ring road – a very early example of such a road in the UK. It was carefully designed after extensive research into traffic flow and the new science of junction planning. One ambitious scheme involved linking high-level junctions to new roof-top car parks, but these turned out to be too costly to build.

◁ **Raising the road**
One solution to urban traffic jams was to build roads in the air, known as overpasses. Here the final span of the Hammersmith Flyover in London is put into place (1961–62).

Parking up

To accommodate the huge numbers of cars pouring into cities, multi-storey car parks were built (see box, opposite). In the UK, for example, National Car Parks (NCP) was founded on the basis of buying up bombed-out city sites and transforming them into car parks.

> "After publishing my **'Gridlock Prevention Plan'** I became known as 'Gridlock Sam'."
>
> SAM SCHWARTZ, CHIEF TRAFFIC ENGINEER, NEW YORK CITY

▷ **Route 66, US, 1950s**
Established in 1926, and gradually made redundant by modern interstate highways, the 2,500-mile (4,023-km) Route 66 from Chicago to California remains the most evocative and inspiring long-distance road journey across the US. Made famous by "beat" novelist Jack Kerouac and others, the route has enduring appeal.

US
66

Cars for the jet age

In 1938 Buick presented the Y-Job, the automotive industry's first concept car. Well ahead of its time, its styling paved the way for a new look in American automotive design, which, by the 1950s, featured aircraft- and rocket-inspired fins.

△ **Buick Y-Job publicity image**
Earl created the Y-Job to test public opinion, when competition for customers was intense at the end of the Great Depression.

The Buick Y-Job was revolutionary in 1938 – as big a leap forward in car design as the Citroën DS would be in 1955. The electric windows, power-operated hidden headlamps, and wrap-around bumpers were innovations that the car industry would later adopt around the world, and the Y-Job's styling would still look modern 15 years later. Following World War II, the Y-Job's designer, Harley J. Earl, and his contemporaries set about transforming the look of American cars. Earl was head of design for General Motors, Buick's parent company, and his vision was clear: to modernize car design, with a nod to the aerospace industry. Earl's era would be defined by cars that people desired, and a gradual shift towards a form of consumerism in which the public was enthralled by the idea of next year's model. Although the Y-Job remained a concept car and was never mass-produced, Buick's Roadmaster, brought out in 1942, did resemble it, and it was not long before other manufacturers joined in trying to design America's boldest car.

The trappings of success

By the 1950s, tailfins had become the way forward, originally inspired by World War II fighter planes, and increasingly by the concept of rocket travel. The 1948 Cadillac, whose design Earl had approved, is widely credited with introducing tailfins, and the next decade saw a battle between Earl and his counterpart at Chrysler, Virgil Exner, to see whose cars could become the most outrageous. The Chrysler 300 and Chrysler Imperial started the trend for bigger and bolder fins, with each year's design more fantastic than the last. The 1959 Cadillac had the tallest fins of them all, at 45 in (114 cm), but it also had overriders and tail-lights resembling afterburners, giving the illusion that they could propel the car faster than physics would allow.

Nevertheless, some manufacturers chose to ignore the trend, seeking a less brash aesthetic. While Ford's Thunderbird had Stateside scale on its side, its tailfins were modest and its overall appearance was too refined for Detroit. Similarly, the early Corvettes had a somewhat European flavour. Conceived as "America's sports car",

△ **Advert for the Edsel, 1958**
The Edsel was heavily marketed as a car of the future prior to its launch, yet its design was immediately considered old-fashioned once revealed. Early models were also plagued with performance and build quality issues.

they were bigger and more powerful than their European rivals, including Austin-Healeys, MGs, and Alfa Romeos, and featured toned-down lines in a bid to appeal to those who felt the fascination with fins too frivolous.

Reactionary design

Ford never took the idea of fins seriously, and so when, in 1957, they tried to design a car that would compete with GM in the marketplace, the look was "so old it was brand new". The Edsel, named after Henry Ford's only son, exhibited few features of contemporary jet-age styling (see pp.148–49). Rather, it seemed to be styled so as to suggest that the 1950s had never happened – an evolution of prewar thinking that accepted double headlamps and a pontoon body as its reluctant future. Unsurprisingly, the Edsel failed miserably and Ford dropped the brand in shame within three years. Its next attempt at toning down Detroit's styling, with the tastefully understated 1963 Lincoln Continental, was more successful.

LIFE BEHIND THE WHEEL
The Motorama show

In the 1950s General Motors held its own US motor show, the Motorama, which visited such locations as New York, Miami, Los Angeles, San Francisco, and Boston. A total of 10.5 million visitors saw Motorama over its 12-year life span. Most years, GM used it to display not only current models but also concept cars, such as the 1956 Pontiac Club de Mer. More than 100 trucks were needed to move Motorama around the country, and they arrived in a set order to minimize assembly time.

***THE WALDORF-ASTORIA* IN NEW YORK HOSTED THE FIRST OF GM'S ANNUAL TOURING MOTORAMA SHOWS. ATTENDANCE WAS BY INVITATION ONLY.**

> "**Our job** is to **hasten obsolescence**."
>
> HARLEY J. EARL, HEAD OF DESIGN, GENERAL MOTORS, 1955

◁ **1959 Cadillac Eldorado Seville**
Cadillac's luxury Eldorado model line was made between 1953 and 2002. This version was notable for its fins, bullet-shaped rear lights, and sweeping jet-age styling.

“The **rolling egg**.”

POPULAR NICKNAME FOR THE BMW ISETTA MICROCAR

The rise of the bubble car

The reintroduction of fuel rationing during the Suez Crisis of 1956 led to a sharp rise in the popularity of microcars.

Microcars had been popular on the continent since the end of World War II – in particular in France, home of the *voiture sans permis*, (a car that did not require a license) and in Germany. German microcars were the result of former armaments manufacturers looking to diversify – if Messerschmitt could not build planes, it could build cars. Some larger motorbike or car companies sought to use the market to increase post-war productivity. While the UK had its own microcars, including the Bond Minicar and Meadows Frisky, the best-remembered microcar is BMW's Isetta.

The BMW Isetta was a licence-built copy of the ISO Isetta, a tiny, small-wheeled car manufactured in Italy. BMW re-engineered much of the car, to the point where there were no interchangeable parts. It used a BMW motorbike engine, and early models were below an engine size threshold that allowed them to be driven in Germany using a motorcycle licence.

BMW also built the Isetta in the UK. In order to take advantage of the UK laws surrounding three-wheeled vehicles, British Isettas only had three wheels, and were classified as a motorcycle and sidecar for licensing purposes. With three wheels and no reverse gear, British buyers could drive an Isetta without passing a driving test for a conventional car. British newspapers nicknamed them "bubble cars" because of their striking ovular shape.

Although fuel rationing ended in 1957, there remained a solid market for microcars into the 1960s. Buyers were attracted by legislation allowing them to be used on a motorcycle licence and by their economy. However, British motor mogul Leonard Lord hated microcars – and commissioned a team to create a vehicle that would usurp them in popularity. That car became the Mini.

◁ **Two women pose in a BMW Isetta microcar**
The BMW Isetta was the archetypal microcar for most in the late 1950s. Its entire frontage was hinged on one side and served as the vehicle's single door. This picture was taken in Munich, Germany.

Sports cars flourish

Success in motor sport and the growth of amateur race clubs and events increased demand for road-going sports cars, boosting the reputations of marques such as Ferrari, Porsche, Jaguar, and MG.

△ **"Safety fast!", 1953**
Consumers were increasingly able to purchase high performance cars – this advert for MG's TF series emphasizes safety as well as speed.

After gaining experience as a race driver and team manager with Alfa Romeo, Enzo Ferrari went into business for himself in Maranello, Italy. He built the first car under his own name in 1947, and the Ferrari marque quickly became a force in top-level motor sports, both in Formula 1 and in endurance racing. Ferrari won the first post-war 24 Hours of Le Mans race in 1949, and dominated the World Championship for Sports Cars from its inception in 1953. Ferrari also took over from Alfa Romeo as the team to beat in the new Formula 1 World Championship, powering Alberto Ascari to back-to-back championship titles in 1952 and 1953, and the Argentine ace Juan Manuel Fangio to his fourth world title in 1956. But the famous team did not have everything entirely its own way: it had some powerful rivals.

Maserati was based just up the road in Modena, and had made its name as a producer of high-class racing machinery in the 1920s and 1930s. Under Orsi family ownership Maserati again became a force in motor racing, and its 250F was the definitive front-engined Grand Prix car.

Another rival came from further afield: Germany's Mercedes-Benz, which had been a player in Grand Prix racing between the wars, built a series of sophisticated racing cars, under the leadership of engineer Rudolf Uhlenhaut, which beat all comers. The spaceframe 300SL, with its characteristic upward-opening "gullwing" doors, won at Le Mans in 1952. Rather than attempt to repeat this success the

▷ **The dangerous romance of racing, 1959**
Sports cars line up in the pits before the start of the 1959 24 Hours of Le Mans race. Despite the dangers of racing, such spectacles fuelled the public's appetite for road-going sports cars.

following year, Mercedes instead switched to Formula 1. The team was quickly winning Grands Prix, and helped Fangio to his fifth and final World Championship in 1954. But by 1958 Mercedes and Maserati had both withdrawn from racing in response to tragic accidents at Le Mans and the Mille Miglia.

Jaguar also made its mark in motor sport in this period, with five wins at Le Mans for the C-type and D-type models in the 1950s, first in the hands of works team drivers and later by Ecurie Ecosse. As with the on-track successes of Ferrari and Maserati, Jaguar's achievements increased the brand's prestige and made more racers want to buy their competition cars. But Jaguar's race victories also translated into more sales of its road-going XK sports cars.

Aston Martin followed a similar path, toiling through the 1950s to make its DB3, DB3S, and DBR1 competitive and reliable – eventually winning both the 24 Hours of Le Mans race and the World Sportscar Championship in 1959.

Further down the hierarchy, Porsche's 356 coupé and 550 racing car, Chevrolet's Corvettes, the MGA, and the Triumph TR sports cars might not have had the speed to beat their larger-engined competitors, but they competed with honour in their respective classes. The growth of amateur, or club, racing also fuelled the popularity of sports cars. This enabled enthusiasts to drive their cars to work during the week, and then race them at the weekends – living the lifestyle the entire time.

"**Racing** is the **only time** I feel **whole.**"

JAMES DEAN

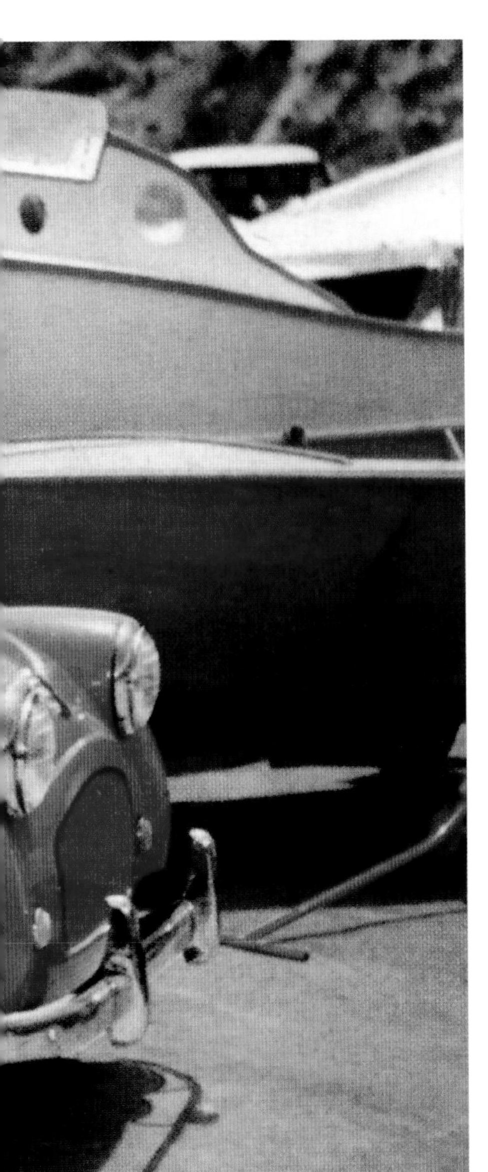

◁ **1955 Triumph TR2 car advertisement**
The TR2 was the result of British company Triumph's bid to make a modestly-priced sports car, and its combination of speed and affordability made it a hit with US drivers.

LIFE BEHIND THE WHEEL
James Dean's "Little Bastard"

Fifties actor and teen idol James Dean was one of many celebrities fond of fast cars and motorcycles. He traded in an MG for a Porsche 356 Speedster, which he raced in the spring of 1955; he then looked to buy a proper racer. After considering a Lotus Mark IX, he bought the car with which he will always be associated – a Porsche 550 Spyder. Painted on the tail was a nickname that he himself had acquired at Warner Brothers: "Little Bastard". It was while driving this car to a race meeting at Salinas, California, in October 1955, that Dean crashed and was killed at the age of just 24. Following Dean's death, some claimed that the remains of "Little Bastard" were cursed; more prosaically, the wreck was displayed at car shows as a safety warning.

JAMES DEAN **(RIGHT) DRIVES HIS PORSCHE 550 SPYDER WITH ENGINEER AND RACER ROLF WÜTHERICH, PICTURED IN 1955.**

▽ **At the drive-in, 1958**
Charlton Heston as Moses in *The Ten Commandments* holds drivers and passengers spellbound at this drive-in cinema in Utah, US. In 1958, there were almost 5,000 drive-ins across the US – marking the peak of their popularity.

Roll on, roll off

The advent of the roll-on/roll-off car ferry, connecting British motorists with Ireland and the rest of Europe, made foreign travel easier and more affordable. It also sparked an enduring passion for continental food, sun, and snow.

When the UK's first roll-on/roll-off car ferry, known as the ro-ro, was launched at the Port of Dover in 1953, it marked the beginning of a fundamental change in the holiday habits of British drivers. Until that point, all vehicular traffic on and off the island had to be loaded by crane. Lift-on/lift-off, or lo-lo, was not only time-consuming – it could take up to an hour to load 15 cars – but also expensive and sometimes risky, as cars could be damaged in the process.

A boom in car sales in the UK in the 1950s, coupled with a growing base of consumers with enough disposable income to afford foreign holidays, prompted transport companies to develop a lo-lo alternative: the ro-ro.

▽ **All aboard**
Before ro-ro ferries, vehicles were hoisted onto ships but only after their fuel tanks had been emptied and their batteries removed. This car is being loaded onto the *St David* at Fishguard Harbour in 1935.

The first ferries

The origins of the ro-ro ferry go back to the mid-19th century, when specially designed ships – such as *Leviathan*, the Firth of Forth ferry in Scotland – were used to take steam trains across waterways that had no bridges. These train ferries were fitted with railway tracks so that trains could simply roll onto them, then easily roll off them on arrival at their destination. Train ferries were also used during World War I to take munitions and tanks from the UK all the way to the front lines in France and Belgium, and then return after the signing of the armistice in 1918. During World War II, which relied far more heavily on the use of tanks, ships were adapted for the first time to take military vehicles across the English Channel.

Commercial usage

After World War II, shipping companies used these tank landing ships as a template for the building of ro-ro vessels to transport cargo, cars, and trucks. Vessels that only took cars were called pure car carriers, or PCCs. These were made up of several levels connected by ramps, which allowed motorists to drive their cars onboard and then drive off within minutes of landing. Since cars were relatively light compared to traditional cargo loads, the dividing floors and outer shell of the ship could be made thinner and lighter, thereby improving speeds and fuel economy. With ferry crossings cheaper and more convenient, ordinary Britons now had access to the balmy weather of Spain, Portugal, and southern France, and excellent alpine skiing in winter.

In the first year of operation, the Dover ro-ros transported 100,000 vehicles to France, a huge increase from the 10,000 previously handled by lo-los. However, before 1971, when the UK joined the EEC (European Economic Community), ferry crossings could still be a lengthy process, with customs procedures, as well as car searches on the return leg for alcohol and cigarettes, on which duty would be charged.

△ **Dream holidays**
The Austrian Alps became a real holiday destination for many ordinary British motorists in the 1950s. Thanks to the invention of the ro-ro ferry, European travel became cheaper and simpler.

Following the UK's membership of the EEC, motorists no longer had to queue for hours at border crossings on the continent, and the linking of the European motorway network to ferry ports began to offer motorists a more seamless experience on continental roads. All these factors helped to fuel a boom in ro-ro driving holidays. By 1985, the number of cars crossing the English Channel on PCCs had risen to more than 2.5 million; by 1994, it had reached over 4.5 million.

A new generation of high-speed cross-Channel ferries with reduced travel times was introduced in the 1990s, in part to compete with the growing popularity of low-cost airlines. But the ferry companies also faced tough competition following the opening of the Channel Tunnel between the UK and France in 1994.

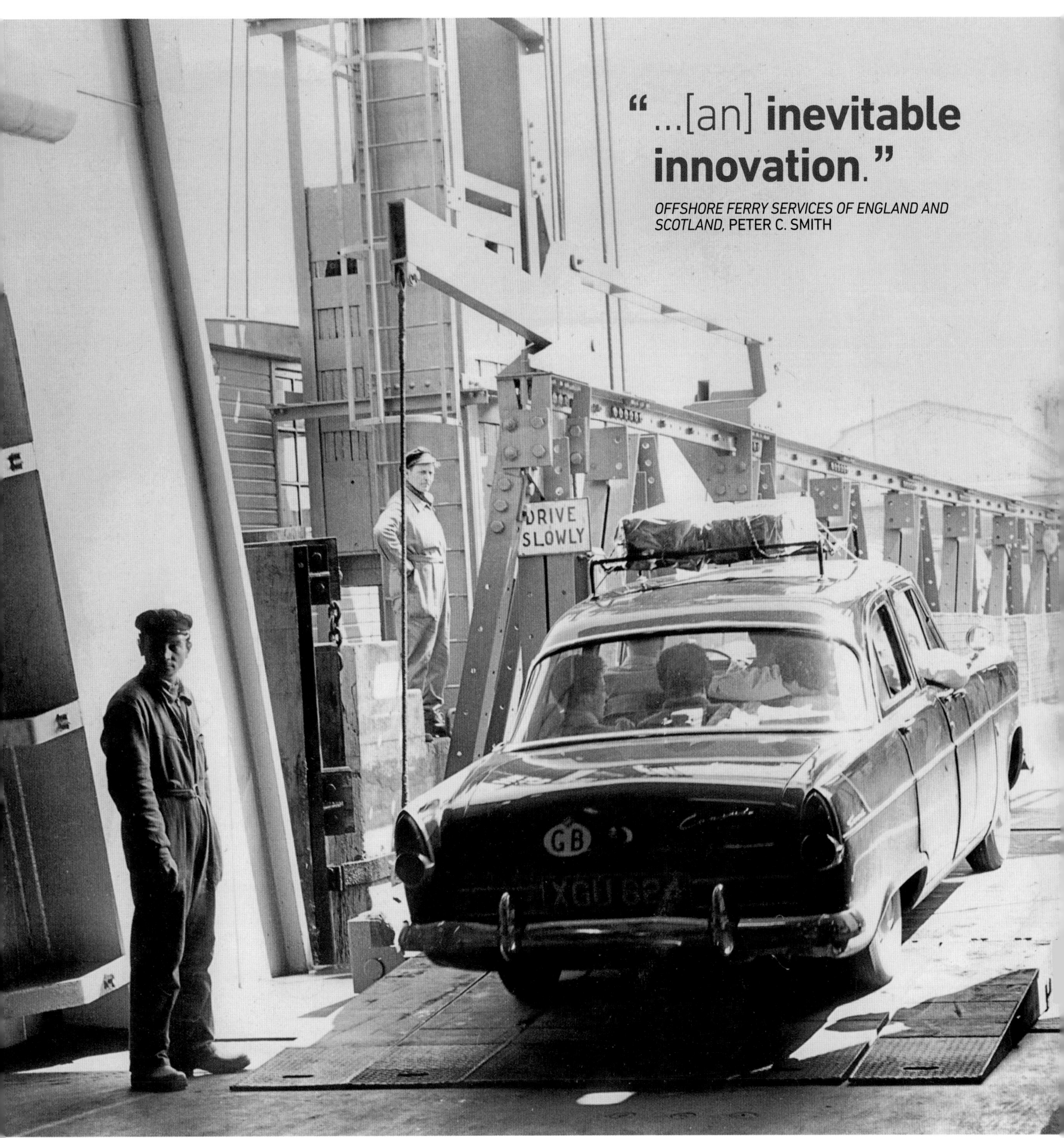

> "...[an] **inevitable innovation**."
>
> *OFFSHORE FERRY SERVICES OF ENGLAND AND SCOTLAND*, PETER C. SMITH

△ **A British motorist** – as seen by the "GB" sticker – drives off a cross-channel ferry in the 1960s.

だんぜん元気が出る
グロンサン
VICTOR
酒は
テイジンサービス
テイジンサービス
オリンピック

Motoring in post-war Japan

As Japan began post-war reconstruction, local car production tie-ins started with European and US companies. However, some enterprising Japanese manufacturers were also on hand to help get Japan back on the road.

As Japan recovered from World War II, cars and driving were very much secondary concerns for much of its population. On the streets of Tokyo, the vast majority of cars were still American, as they had been before the war, and when vehicle production picked up again under the 1945–52 US occupation, light trucks and three-wheelers were the main focus. Very few Japanese citizens owned a car, and road traffic mostly consisted of taxis and trucks. Around this time Mazda was producing simple three-wheelers – the motorcycle-based Mazda Go truck had first appeared before the war. The Honda Motor Company Ltd was officially formed in 1948, initially producing innovative motorized bicycles, with great success.

In the 1950s, to get the industry on its feet, some Japanese manufacturers signed production deals with foreign makers. Nissan began to make the Austin A40 under licence, while Isuzu built the Rootes Group's Hillman Minx, and Hino set up a production line for Renault 4CVs. Mitsubishi, meanwhile, began Japanese manufacturing of the iconic US Willys Jeep. Toyota stood out among the domestic car companies by determinedly following its own path, and in 1955 created the Crown, now seen as Japan's first truly significant post-war car. By 1958 Toyota was exporting the Crown to the US (as the "Camry"), and the line is still in production today.

Japan was also home to some unusual homegrown microcars, such as the wild, aerodynamic Fuji Cabin, launched in 1955, and the lightweight Flying Feather from the same era. Although few of these tiny vehicles were made – and they are now extremely collectable – they were noble attempts to get post-war Japan's motorists back in the driving seat.

◁ **The streets of Tokyo**
This photograph, taken in the 1950s, captures traffic on a busy Tokyo street. By this time Japanese car manufacturers had begun to form the nucleus of the domestic car industry.

The home mechanic

The car ownership experience evolved rapidly throughout the 1950s, and as motoring became commonplace, drivers often took matters into their own hands for better value and more longevity.

▷ **Dunlop tyre poster, 1960s**
Makers of car consumables, such as tyres, sought to promote their brands when drivers started servicing their own cars at home.

By the mid-1950s, there was no doubt that cars had crossed the line from being something special for a fortunate few to "consumer durables" owned by many. With the growth in ownership, motorists were increasingly courted by manufacturers in the hope of them becoming loyal repeat-customers.

The car ownership experience was cumbersome by today's standards. Although a 12-month/6–12,000-mile (9,600–19,000-km) warranty was becoming the industry norm, typical intervals between services were 2–3,000 miles (3,200–4,800 km). The main reason for this was that automatic lubrication was in its infancy, and mechanics would spend hours greasing and oiling a car's moving parts. Even a small car, such as the Ford Popular 100E, needed lubricating manually in 13 places each time it was serviced.

Home maintenance

To avoid the cost of such frequent servicing, drivers everywhere embraced do-it-yourself maintenance once their car's warranty had run its course. This in turn led to a surge in marketing from brands that offered the "consumables" of driving. Manufacturers and retailers were not just selling to mechanics any longer – they were also targeting the ordinary driver on a tight budget, offering tyres, batteries, oils, windscreen wipers, lighting, brake parts, filters, and cleaning fluids.

In turn, this powered the growth of several national chains of parts and accessories stores around the world, where the driver could shop for everything he or she needed. Mail order parts was big business too, and publishers showered owners with a wide variety of "how-to" books to take the mystery out of home servicing (see p.241).

Among the many new products was WD-40, the revolutionary light oil compound that freed up parts and protected against moisture (named "WD-40" as it was chemist Norm Larsen's 40th attempt at a "water displacement" product). Home mechanics were now also buying Swarfega, the gritty skin cleaner that helped get oil off dirty hands after working under the bonnet.

Over time, a throwaway culture developed, since it was very often cheaper to replace an item (such as a punctured tyre) than to replace it.

△ **Mobiloil sign**
Born of the Vacuum Oil Company, Mobil remains one of the world's top producers of motor oil.

Technical developments

Around this time, cars needed improved power supplies, as manufacturers included more electrical equipment as standard, and to start the high-compression V8 engines that appeared in the US from 1949 onwards. Systems were upgraded from 6- to 12-volt to cope. By 1960, Chrysler began replacing the generator with the alternator, which was smaller, lighter, and provided more current, especially at lower engine speeds.

◁ **Ford Popular 100E, 1959**
Affectionately known as the Ford "Pop", this car required servicing after every 1,000 miles (1,600 km). It was mechanically simple, however, allowing for easy home maintenance.

△ **Regular servicing** Mechanics work on a Ford Anglia 105E. Launched in 1959, this was a fourth generation of Anglia, yet it still required regular servicing.

Nonetheless, many popular models of the 1950s were still sparsely equipped, so car accessory manufacturers seized the opportunity to enable owners to install items for themselves, including wing mirrors, rear-screen demisters, sun visors, and wind deflectors.

Many cars in this period lacked rust protection. Well before carmakers themselves tackled the issue of corrosion, US entrepreneur Kurt Ziebart came up with his patented rustproofing process in 1959 that added a thick sealant to the car's vulnerable underside. It was yet another example of third-party involvement in keeping cars going that filled the chasm between the dream and the reality of car ownership.

DRIVING TECHNOLOGY

Radial tyres

Beneath a tyre's outer layer, or tread, lies a reinforcing structure known as the carcass. This structure is composed of a network of cords, which are made chiefly of polyester and steel and were traditionally laid in plies at 45 degrees from the direction of travel. This arrangement, known as "cross ply", was superseded by Michelin's "radial" design of 1946. In radial tyres, the cords are laid at 90 degrees to the direction of travel and so do not overlap. This avoids friction between the plies, which in turn reduces tyre wear and enhances fuel economy. To shore up the whole structure, the tyre is surrounded by additional steel or Kevlar belts, which further enhance performance.

THE FIRST CAR **TO FEATURE MICHELIN'S RADIAL TYRES WAS THE CITROËN 2CV OF 1948. AT THE TIME, CITROËN WAS OWNED BY MICHELIN.**

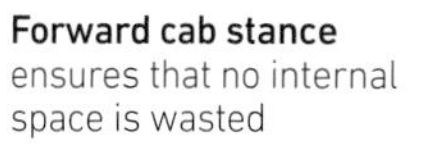

△ **Volkswagen Kombi, 1950**
While the VW Kombi is not strictly an estate car, large numbers of families bought and used them as daily transport. Ample space ensured that they could carry more than enough to meet an average family's needs.

△ **Morris Minor Traveller, 1953**
The Morris Minor Traveller was one of the last estate cars made in the UK to utilize a wooden frame. The front end and chassis were built at one factory; the estate rear section was added at another.

"... both **family standard** and **status symbol**."

NORM AND ANDREW MORT, *AMERICAN STATION WAGONS*

▽ **Chevrolet Bel Air Nomad, 1956**
In the US, cars like the Chevrolet Bel Air Nomad and Pontiac Safari proved popular. Two-door estates with plenty of space, but even more style, they were less of a family car than a personal statement.

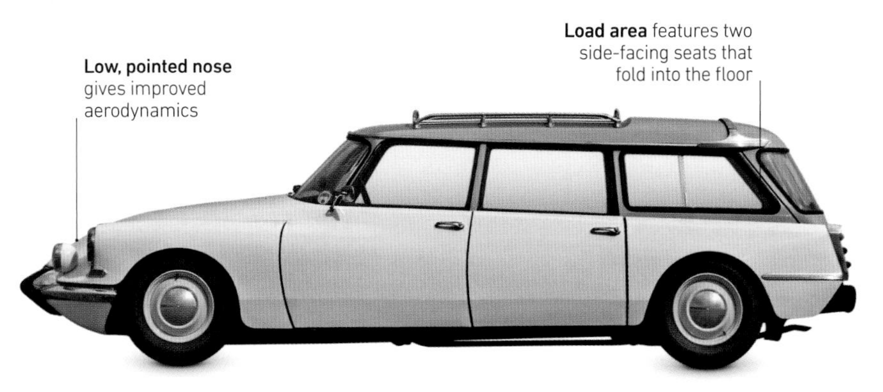

△ **Citroën DS Safari, 1958**
Citroën's DS Safari was based on its cheaper and simpler sister, the ID. It was a popular car among antiques dealers and motorcycle racers, who appreciated its combination of comfort and space.

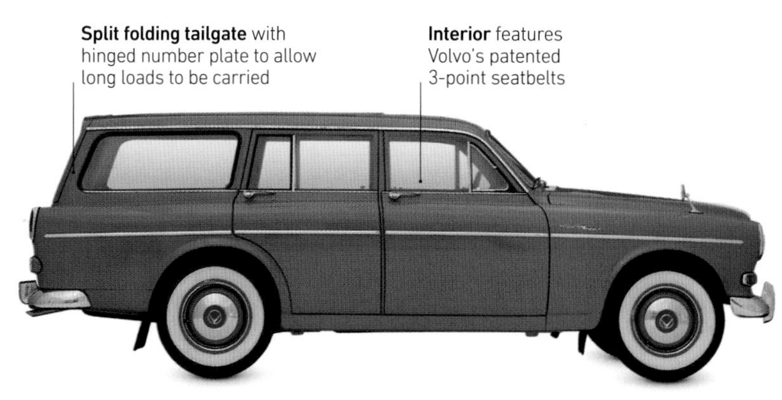

△ **Volvo 221, 1962**
Volvo is famed for its estate cars, a reputation which began with the 221. With its spacious load area, roomy cabin, and easy access – thanks to having five doors – it proved popular with families and tradesmen.

Family load-luggers

For as long as there have been cars, there have been people who want to carry large loads in them. However, the design of cars that could easily accommodate passengers and/or cargo really began to take off in the post-war period – in particular the 1950s and '60s. This was the point at which manufacturers realized that estate cars, or station wagons, could be desirable as well as practical, if equipped with an upper-hinged tailgate and rear seats that folded flat to increase load capacity. Key to this was Chevrolet in the US, which launched the Bel Air Nomad after the trial of its Nomad concept car in 1954. The Nomad was based on the Corvette, and its rakish yet capacious rear end was easily adapted to the new Bel Air body. European manufacturers were quick to follow suit.

△ **Land Rover Station Wagon, 1949**
While the Land Rover might have been developed for off-road use, the long-wheelbase Station Wagon variant made excellent transport for adventurous families.

Car production in the Latin world

In the mid-20th century, the Latin markets of Spain, Portugal, Argentina, and Brazil adopted strategies of channelling foreign investment into car production, which served as an effective catalyst for economic progress.

Industrialization was the goal of many countries in the 1950s and '60s, particularly the Latin-speaking nations in Europe and South America, where sizeable populations provided large potential markets for domestically produced cars. With low wage costs and government policies designed to attract investment, these countries also provided appealing locations for foreign producers. Establishing national automobile industries was viewed as the key to building economic prosperity and became a priority for Spain and Portugal, and their respective former colonies Argentina and Brazil, which were at the forefront of industrial progress in Latin America.

The Spanish revolution

Spain in particular underwent a dramatic turnaround, transforming itself from an emerging country with a modest national car industry in the early 1950s to one of the top ten automobile makers in the world. After years of neglect through the Civil War years, Spain began the process of economic development under Franco. Among the policies of the General and his technocrats was the decision to establish a national car brand to be made wholly in Spain. With investment and technical expertise from Fiat in Italy, the Sociedad Española de Automóviles de Turismo (SEAT) was established in 1950. Within four years, the company had gone from relying on imported parts to using 96 per cent locally-made components. Then, in 1957, the launch of the SEAT 600, with its basic but sturdy build, helped to make Spain a nation of drivers.

Unlike Spain, Portugal made no popular national car of its own – although there were a few domestic specialists such as military vehicle maker Bravia, and the Alba sports car (above) remains a rare classic. However, Portugal did become an important production centre for car manufacturers from other countries. In 1963, the government adopted a strategy of banning imports of fully built-up cars, and requiring a proportion of at least 25 per cent local components to be included on assembly lines of international companies. Several of the big makers, including Ford, General Motors (GM), Renault, and Citroën, established subsidiaries in Portugal, and by 1973 there were some 30 assembly lines producing both passenger and commercial vehicles.

△ Alba sports car, 1952
With a 90-hp engine and a top speed of around 125 mph (200 km/h), the sporty Alba was the only car to have been entirely manufactured in Portugal at the time.

Latin America

In South America, Argentina and Brazil led the way in car production, with both countries focusing on manufacture for foreign auto companies. After the overthrow of General Perón, who had made foreign investors wary, Argentina's government signed a deal with US manufacturer Kaiser to begin production in the city of Santa Isabel in 1956. Ford also set up a plant in 1959 on the

◁ Volkswagen factory, São Paulo, Brazil
Volkswagen started building Kombis in Brazil in 1957, followed by the Beetle – renamed Fusca - in 1959. With its reliable air-cooled engine, the Fusca proved highly popular and remained the country's top selling car for 24 years.

outskirts of Buenos Aires, initially to manufacture trucks; by 1962 the US carmaker was also producing the Falcon car. By then, there were around 20 different companies in Argentina making commercial and passenger vehicles.

As early as 1925, GM had built a factory in Brazil for the manufacture of trucks and utility vehicles, but it was not until 1956 that the nation's fledgling assembly industry began under President Kubitschek, who launched a five-year plan under which imported cars were banned, forcing transnationals to choose between abandoning the market or starting local manufacture with 90–95 per cent locally produced parts. Eleven companies committed to the scheme, which offered heavy subsidies. With over a million miles of roads to tackle, of which less than ten per cent were tarred, Brazilian drivers demanded robustness. Volkswagen offered solid German engineering with its Beetle and Kombi, and began production near São Paolo in 1959. Chevrolet was the first to launch a home-grown model, the Opala, in 1969 – its reputation for reliability made it the choice of the Brazilian police force and legions of taxi drivers. By 1970, Brazil had grown to become the world's tenth largest car producer, and was mid-way through an economic boom.

▷ **Kaiser Bergantin, 1960**
The Kaiser Bergantin was manufactured in Argentina by IKA, a joint venture with Kaiser Motors of the US. IKA was the country's first car company.

"Fifty years of progress in **five."**

CAMPAIGN SLOGAN OF BRAZILIAN PRESIDENT JUSCELINO KUBITSCHEK

▽ **SEAT factory, 1961**
In this promotional photograph, a fleet of SEAT cars, including 600s, line up in front of the company's former factory in Barcelona.

Glittering prizes

For every race there is a winner, and for every winner a trophy. It is a tradition that goes right back to the beginning of motor sport.

Generally speaking, the bigger the race, the bigger the winner's trophy. Even success in a tiny speed event would earn the winner a small cup, but for major events only a substantial trophy would do. The epic Borg-Warner Trophy, for instance, awarded to winners of the Indianapolis 500 in the US, stands at over 5 ft (150 cm) tall.

Traditionally, racing trophies were silver or gold cups that would be engraved with the name of the winner each time they were presented – or in the case of the Borg-Warner Trophy, the winner's face. Shields, figurines, and stylized steering wheels were also favourite designs.

Modern trophies come in a wider range of shapes, and often blend metal with polished wood and glass. These are more commonly given to drivers at recently introduced events, as long-established races often retain their valuable original trophies.

Sometimes there is an inextricable link between a trophy and a race, as with the Oulton Park Gold Cup or the Royal Automobile Club (RAC) Tourist Trophy races, or between the trophy and a championship, as with the NASCAR Winston Cup. Winners may also be presented with a victory laurel wreath and a celebratory drink – often champagne, although the winner of the Indy 500 is given a glass of milk.

ROYAL AUTOMOBILE CLUB (RAC) INTERNATIONAL TOURIST TROPHY, 1905

Gordon Bennett Cup was one of the first high-profile trophies

GORDON BENNETT CUP TROPHY, 1904

RAC DEWAR TROPHY, 1906

Segrave Trophy is awarded for achievement on land, sea, in air, or on water

RAC SEGRAVE TROPHY, 1930

KEY DEVELOPMENT

Champagne celebration

Dan Gurney was handed a celebratory bottle of Moët et Chandon champagne after winning the 24 Hours of Le Mans in 1967, and after flipping off the cork he took aim at journalists who had said he and fellow American A.J. Foyt would never win the race. Press photographers, Henry Ford II, and team manager Carroll Shelby were soaked, too. It was something nobody had ever done before, and the spraying of champagne quickly became a tradition.

LEWIS HAMILTON SHOWERS VALTTERI BOTTAS **WITH CHAMPAGNE AT THE 2017 MELBOURNE GRAND PRIX.**

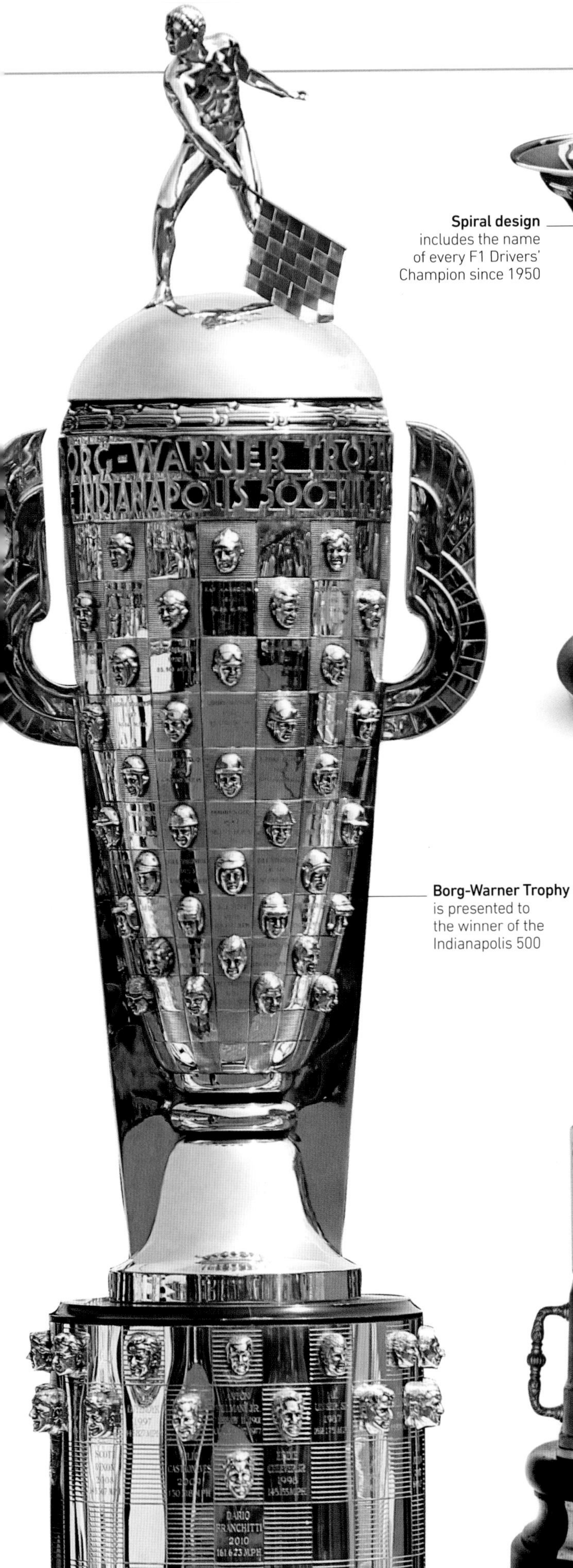

Borg-Warner Trophy is presented to the winner of the Indianapolis 500

BORG-WARNER TROPHY, 1936

Spiral design includes the name of every F1 Drivers' Champion since 1950

FORMULA 1 WORLD DRIVERS' CHAMPIONSHIP TROPHY, 1950

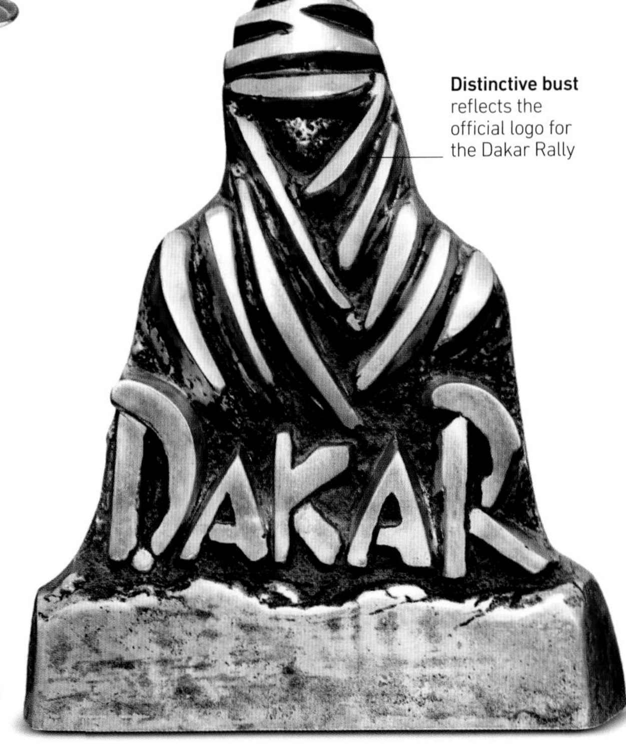

Distinctive bust reflects the official logo for the Dakar Rally

DAKAR RALLY INDIVIDUAL TROPHY, 1978–2011

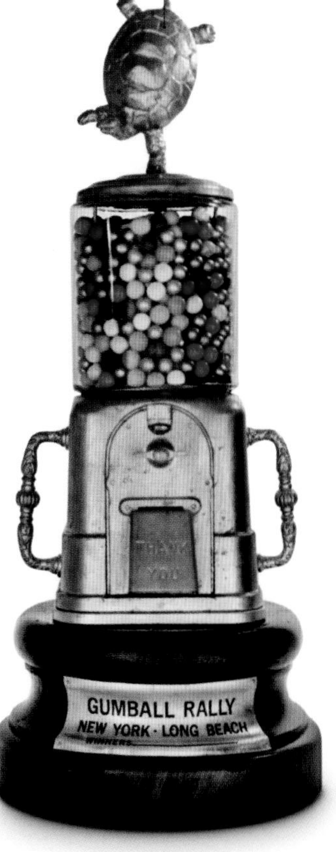

Dancing turtle atop a gumball machine featured in the 1976 film *The Gumball Rally*

GUMBALL RALLY TROPHY, 1976

24 HOURS OF LE MANS TROPHY, 2015

This trophy was awarded for the Fitzgerald Glider Kits 300 NASCAR Xfinity Series race on 22 April 2017

NASCAR XFINITY SERIES TROPHY, 2017

A home for your car

Keeping your car safe and dry has been a concern for drivers since the dawn of motoring. Suburban homes with purpose-built garages were ideal – but there were other solutions.

As the preserve of the wealthy, many early cars had their own accommodation. Some owners converted stables and coach houses into designated shelters, while others had "motor houses" constructed, which might even include cramped accommodation for a chauffeur. The word "garage" comes from the French verb "garer", meaning "to shelter", and it was first used in English in about 1902. By the early 1930s, new family houses were routinely built with an integral or matching free-standing garage as part of the package.

If you lacked the space or the money to build your own garage, then a foldable "tent" garage was a cheap and effective solution. These started to be advertised in car magazines as far back as the 1940s, and even today they remain a good option to protect paintwork from grime and the elements. Do-it-yourself garage kits appeared in the 1930s, and these were usually made of wood, with a simple corrugated iron sheet for the roof and sometimes the outer walls. By the early 1950s, the new wonder material of sheet asbestos was also in widespread use. This was much lighter than wood and had excellent insulation properties – although it was a health hazard, no-one knew it at the time.

In 1952, British tile manufacturer Marley is believed to have pioneered the concept of the sectional concrete home-build garage kit. Its pre-cast uprights were designed so that matching panels, modular glazing, and ventilation grilles could be slid into place between them – an arrangement that was soon supplemented by steel up-and-over doors. The kit was designed so that a couple of strong, practical people – perhaps neighbours – could build a garage in a single weekend, spurred on by the fact that planning permission was not required. As car ownership snowballed throughout the 1960s, this new "pop-up" mentality proved enormously popular – even if the results could rarely be described as pretty.

▷ **Folding garage, 1956**
Consisting of a canvas-covered metal frame, folding garages enabled motorists to protect their cars from the elements without the need of a permanent structure. When not in use, they could be folded up and put in storage.

HOY 863

The US import market

The post-war car boom in the US began with American manufacturers struggling to meet demand. This resulted in a growing market for imported cars from Europe, which was dominated by the UK and West Germany.

On 3 July 1945, a white 1946 Super DeLuxe rolled down the assembly line at the Ford Motor Company's plant in Dearborn, Michigan, restarting production in an industry that had been put on hold for over three years by the recent war. American car manufacturers built only 500,000 cars that year – all of them lightly modified prewar models – but six years later annual car sales in the US exceeded 7 million.

Several popular brands did not return after the war, while the Kaiser-Frazer and Tucker marques were two newly established ones, aiming to capitalize on consumer demand. Before 1939, foreign cars in the US were rare, usually very expensive, and available only in such places as Hollywood, New York, and Miami, where the rich lived, worked, or vacationed, but this was about to change. During the war, more than 3.4 million cars were scrapped. Another 7 million were still in use but in poor condition. The scarcity of consumer goods meant that Americans left their money in the bank, but once the war ended, they were ready to buy.

In 1946, imports began to trickle in, led by MG and Jaguar sports cars. Three years later, Volkswagen entered the market but sold only two examples. All import sales were initially slow as the early importers worked to establish nationwide availability of parts and service.

Kjell Qvale, an MG distributor in California, assembled kits of spares for his customers to take on trips. In New York, Max Hoffman, the country's first importer of Jaguar, Porsche, Mercedes-Benz, and Volkswagen, appointed dealers on the east coast. By 1950, most of the first wave of imports were selling well, as the appetite for new cars could not be satisfied by Detroit, the centre of US car production, and MG had sold 10,000 TCs, all right-hand drives.

△ **Renault Dauphine poster, 1958**
The poster for the US launch of Renault's Dauphine in 1958. The strapline, "the frisky, thrifty family car", highlighted that the Dauphine was unlike any car made in Detroit at the time.

▷ **Car envy**
The showroom of Fergus Motors in New York, taken in 1957. The dealership imported and sold mainly European cars. Car production in Europe had recovered from the war by this time, and Americans were enthusiastic consumers. A German Borgward saloon and a British Morgan are on display here.

Size matters

European cars were different in every way from the typical Detroit car. A 1948 Chevrolet saloon weighed 3,100 lb (1,406 kg) – 1,200 lb (545 kg) more than the 1948 MG TC. Although the MG's 54-hp engine was tiny by US standards, the car felt light, quick, and nimble, unlike American cars of the time.

By 1955, Europe's war-shattered industries were recovering and the second wave of European cars, including Volvo, Saab, Fiat, and Alfa Romeo, arrived in the US. Of these newer brands, Renault saw some early success with the Dauphine: in 1960 Americans bought 102,000 of them, making Renault the top import brand.

American drivers were willing to give almost any imported car a try as long as it suited US roads. Sales of the Saab 93 were slow, in part because the car's smoky two-stroke engine required two

pints (1 litre) of oil to be added to the petrol tank at each fill-up, something most Americans were unwilling to do. Sales of the first imported Volvo, the PV444, began slowly because it was initially available only in California and Texas, and its dated, prewar, Plymouth-like styling did not help matters.

In 1958, Toyota became the first Japanese company to export to the US (see pp.206–207), but the Toyopet Crown (originally designed as a taxi) fared poorly. Lingering anger over the war and Japan's reputation for producing cheap goods, combined with the car's high price and lack of power, resulted in only 288 being sold in two years. In 1966, Toyota tried again, this time very successfully, with the Corona, a sleek-looking, small family car with optional automatic transmission and air-conditioning.

BIOGRAPHY
Heinz Nordhoff

In 1949, its first year exporting to the US, Volkswagen sold just two cars. Six years later, when sales grew to 25,000, the company surpassed MG to become the top-selling import brand in the US. Initially, however, the Beetle, like the first Austin, Hillman, Morris, and MG saloons sold in the US, was not well suited to the American market.

While British manufacturers refused, or were unable, to tailor their vehicles accordingly, Volkswagen's determined chairman, Heinz Nordhoff, listened to American customers' criticism of the car and ordered engineering changes to improve the car's steering, braking, and performance. Sales slowly but surely improved: 328 were sold in 1950, 417 in 1951, and 980 in 1952. Then, when the company appointed dealers east of the Mississippi, Beetle sales took off, soaring from 1,214 in 1953 to 8,895 in 1954.

Except for being outsold by the Renault Dauphine in 1960, the Beetle was America's top imported car through the early 1970s, with sales peaking at 420,000 in 1968.

HEINZ NORDHOFF, **VOLKSWAGEN'S CHAIRMAN, ENSURED THE SUCCESS OF THE MANUFACTURER IN THE US.**

△ **All aboard**
The popularity of British cars in the US continued into the 1960s. Here, Minis, Hillmans, Metropolitans, and MGs are among the cars lined up ready for export at Birkenhead Docks, near Liverpool, in June 1960.

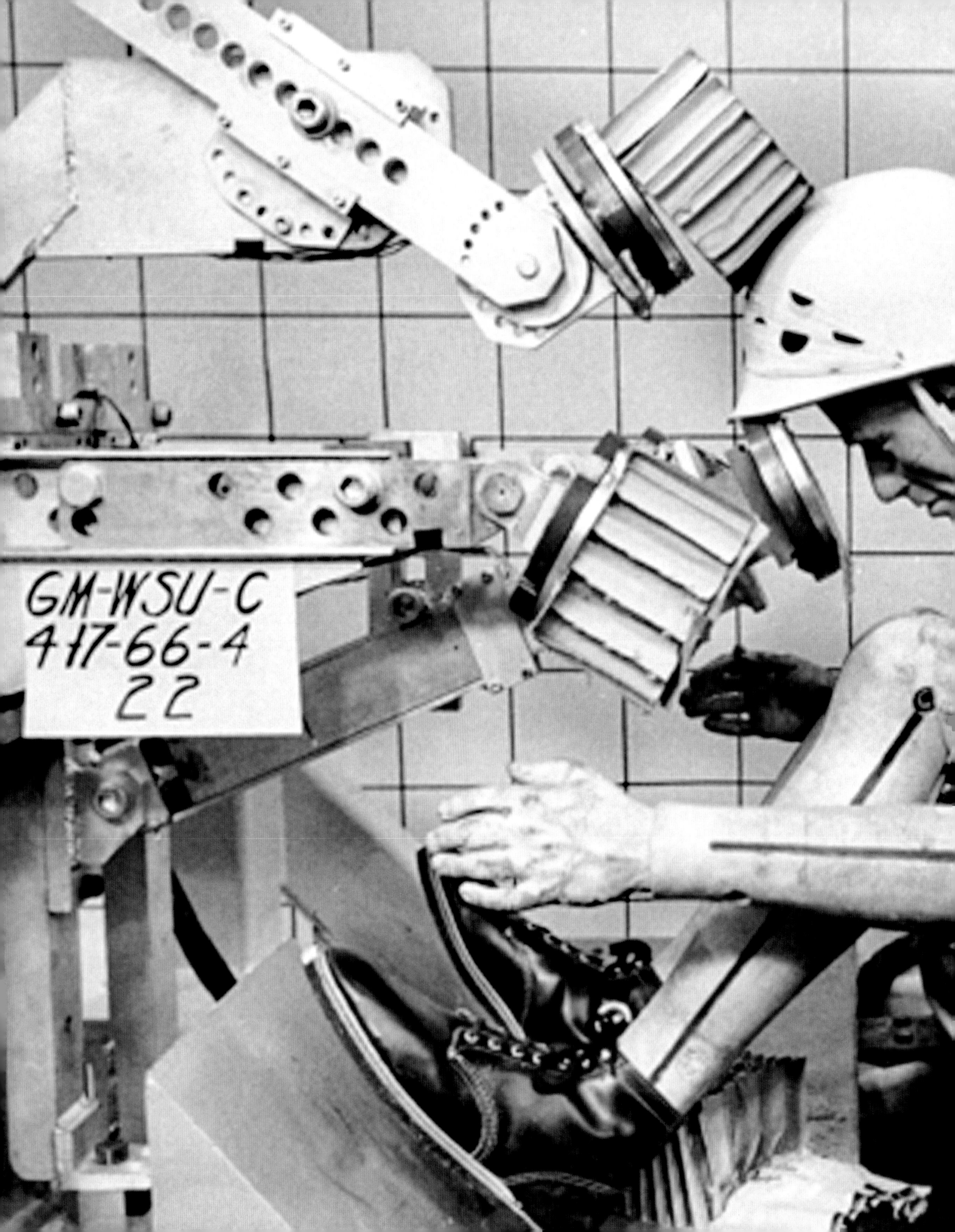
GM-WSU-C
4-17-66-4
22

1961–1980

TECHNOLOGY AND SAFETY

1961–1980

Technology and safety

The 1960s was the golden era of the classic car, and many game-changing models hail from this period. Among small cars, there emerged the revolutionary Mini and Renault 4, while stylish saloons included the Ford Falcon, Ford Cortina, BMW 1500, and Renault 16. To get the blood racing, there were now sports cars such as the MGB, Porsche 911, Alfa Romeo Duetto, Jaguar E-type, and Lotus Elan – and, even more thrilling, the first "supercars" such as the Lamborghini Miura, Ford GT40, and Ferrari Daytona.

Fun, flair, and freedom

The reasons for this explosion of innovative design were many. There were forward strides in materials, technology, electronics, motor sport thinking, mass manufacturing techniques, and Italian styling influences. New power units, including rotary engines and gas turbines, had caused a stir, but the most significant area of car design, from its birth in the 1960s to its flourishing in the 1970s, was in packaging – in particular mixing front-wheel drive with versatile hatchback body styles. This was something that the new Car of the Year awards in various countries saluted, as output from all the major manufacturers snowballed.

There was a definite shift in focus towards cars offering sporty enjoyment, with the traditional open two-seater giving way to slick coupés that were fast and reliable. Compact new hatchbacks also featured in the fashionable motor sport of the era – rallying – and retail versions, the so-called "hot hatches", were a smash hit. Motoring for sheer fun encompassed everything from dune buggies to custom cars.

THE CHEVROLET CORVETTE STINGRAY BECOMES AN INSTANT CLASSIC

MANUFACTURERS COME UNDER PRESSURE TO IMPROVE CAR SAFETY

"At the **change of the decade**, society's concerns also extended to **pollution**."

On another tack, successful businessmen were now eschewing chauffeurs and driving themselves in some of Europe's executive sports saloons.

Safety and the environment

A more sombre but vital consideration emerging in the 1960s was how seriously safety was being taken. In the US, campaigner Ralph Nader had made the public aware that they did not have to accept cars that could physically harm people in the course of normal use.

In response to safety concerns, the US government established mandatory safety standards for vehicles, and the industry started to incorporate active and passive safety measures, such as seatbelts. At the change of the decade, society's concerns also extended to pollution, and the role that cars played in generating it.

Fuel-burning muscle cars – exciting though they were and looked – fell rapidly from favour as cleaning up exhaust emissions became a top priority. American cars shed metal, length, and power to comply, but Detroit had a fight on its hands as Japan's smaller, cleaner, more fuel-efficient cars were already finding millions of satisfied new customers.

Traffic and parking pressures were steadily mounting on drivers around the world. No wonder that in-car entertainment – FM radios and cassette players – was a big growth area. Nonetheless, the road environment was being made more intuitive with, for example, careful research into road signs to create consistency and boost safety. And ever more nations were embracing the car, from Iran to South Korea, as a way to mobilize their societies and economies.

PARKING METERS BECOME A FAMILIAR SIGHT IN CITIES AROUND THE WORLD

CARS REMAIN A SOURCE OF PLEASURE DESPITE NEW RESTRICTIONS

Japanese cars go global

The shape of the world car industry began to change as Japan arrived on the scene, sparking huge growth and a shift in customer expectations. In Europe, carmakers began to consolidate, in part to meet the new challenge.

△ **The start of their journey**
A pier in Yokohama is filled with Japanese-made Datsun Cherry 100A and Sunny 120Y cars awaiting shipping for export to foreign markets in 1975. By the 1970s, Japan's auto exports were a huge industry.

Japan's ascendancy on the world's motoring stage began in earnest in the 1960s with the likes of the Toyota Corona and Nissan Bluebird – sturdy, unglamorous, family saloons that capably did the job. Alongside these, enthusiasts loved the Datsun 240Z, Mazda Cosmo 110S, and Honda S800 sports cars, and still do to this day.

During the 1970s, Japan's car production increased significantly. The first, pivotal Honda Civics and Accords established Honda as a maker to watch, especially in the US. The fuel crisis of 1973 saw a huge demand for affordable, fuel-efficient compacts, and Japan was perfectly positioned to meet that call with a growing range of models.

This was the age when the industrial might of Toyota, Nissan, Honda, Mazda, Mitsubishi, Subaru, Suzuki, Isuzu, and Daihatsu began to make itself felt. This was the era of the Datsun Cherry, for example: a simple if garish small hatch, well equipped for the price, convenient, economical, and reliable. These are qualities that won the car many satisfied owners.

Engagement with Europe

Datsun became a serious player at a time when Europe's automotive industry was beginning to restructure. In the UK, the ill-fated British Leyland conglomerate, created in 1968, was partly nationalized in 1975. On paper, British Leyland might have worked, but the company was beset with managerial issues, industrial unrest, and cars that rarely fulfilled their promise. By the late 1970s, British Leyland had partnered with Honda to create a new generation of cars, starting with the Triumph Acclaim, and Japan's influence within the industry continued to grow.

In Italy, Lancia and Ferrari came under Fiat's wing. In France, Peugeot took over an ailing Citroën in 1975 and then absorbed Chrysler Europe in 1978, resurrecting the Talbot marque, while Hillman, Singer, and others disappeared. For the Japanese, it was all go – except that their booming car production led to years of trade friction and quotas in both the US and Europe.

◁ **The first US import**
A Toyota Toyopet is hauled onto a ship for export to the US in 1957. This was the first Japanese car to be imported into the US, and it led the way for a huge demand for Japanese cars from the 1960s onwards.

KEY EVENTS

- **1957** The first model of the Toyota Corona is released.
- **1963** Honda launches its first production consumer car, the S500.
- **1966** The Toyota Corolla is launched. It becomes the best-selling model of all time.
- **1960s** Japan's motor industry begins producing a large number of *kei* cars – "light automobiles" (see pp.262–63).
- **1967** The Japan Automobile Manufacturers Association (JAMA) forms.
- **1972** The Honda Civic is released.
- **1973** The Organization of Arab Petroleum Exporting Countries (OAPEC) declares an oil embargo, causing a rise in oil prices. Affordable cars manufactured in Japan and other countries see a rise in popularity as a result.
- **1970s** Nissan's Datsun-badged range of cars becomes popular in export markets.
- **1980** Japan overtakes the US to become the world's leading car manufacturer.

***A HONDA 1300X* ON DISPLAY AT THE TOKYO MOTOR SHOW IN 1970.**

△ **Industry guests** view new models at the 10th All Japan Motor Show in 1963 in Tokyo, Japan. This is the display of Prince Motors, which was taken over by Nissan.

International rallying

In the 1960s rallying became a spectacular sport, pitting cars and crews against each other over challenging terrain that ranged from forest tracks in Europe to rutted dirt roads in Africa.

Good navigation and driving consistency were the key features of rallying up until the 1960s, although there were often a few timed tests as well. The RAC Rally featured "driving tests" in which competitors raced their cars around a tightly cornered temporary course and reversed them into "garages" laid out with cones – all against the clock. Meanwhile, the Monte Carlo Rally included a race around the Monaco Grand Prix circuit.

International rallies began to incorporate timed "special stages" until top-level rallying (such as the Group B events introduced in 1982) became a fundamentally different form of motor sport. As the events developed, so did the ideal rally car. Big, comfortable saloons with plenty of space for a crew of two, even three, were no longer competitive. Rallies were now won by high-performance cars with good handling. BMC's Mini Coopers came to the fore with the "Flying Finns" Timo Makinen and Rauno Aaltonen, and Northern Irishman Paddy Hopkirk. As rally cars became more powerful, so traction became all-important, and rear-engined, rear-wheel-drive cars like the Porsche 911 and Alpine A110 excelled. Into the 1970s Ford's nimble RS Escort became the car of choice for a huge number of rally competitors in club events as well as internationals, not least because Ford made it easy to buy the parts needed to turn a standard Escort into a rally-winner. But there was stiff competition at top-level events from the Ferrari-engined Lancia Stratos and the well-prepared Abarth Fiat 131s.

Alongside the relatively short European rallies, there were marathon events such as the transcontinental World Cup rallies, the East African Safari Rally in Kenya, and the Bandama Rally in the Ivory Coast. These covered thousands of miles and favoured tough, reliable cars and experienced drivers: Ugandan Shekhar Mehta won a record five Safaris between 1973–82 driving Nissans.

▷ **An uphill climb**
Local people seem unconcerned as an Austin 1800 rally car speeds past, kicking up the dust on a narrow road along the Great Rift Valley, Kenya, in 1969. Taking place over unmade roads, African rallies were gruelling, for both men and machines, and encounters with wild animals were inevitable.

"One stage was over 900 km long. I would start to **fall asleep while driving** and Gunnar Palm used to **hit me with the pace-note book** to keep me awake."

HANNU MIKKOLA ON THE WORLD CUP RALLY

Powerful 3.8-litre, double-overhead camshaft, straight-six engine produces 265 bhp

Side-opening tailgate situated above sophisticated independent rear suspension

△ **Jaguar E-type, 1961**
Derived directly from the Le Mans-winning D-type racing car, the E-type's dart-like profile and sleek chrome detailing had an immediate impact. It offered 150 mph (240 km/h) performance at a third of the cost of a Ferrari.

Optional aluminium bonnet reduces weight

Steel wheel spinners with two-winged centrepieces

△ **MGB, 1962**
Nimble and quick off the mark, the well-proportioned MGB was a capable, long-legged cruiser. Powered by a 1.8-litre, straight-four engine, it had a top speed of 103 mph (166 km/h), and was popular among driving enthusiasts.

"I spent **a lot of money** on booze, birds and **fast cars**. The rest I just **squandered**."

GEORGE BEST, BRITISH FOOTBALLER

△ **Datsun Fairlady, 1965**
The Japanese manufacturer took particular note of MG's success in the US and produced its own two-seater sports car in a similar idiom, using the ever-reliable 1.6-litre engine from the Datsun Bluebird saloon.

△ **Alfa Romeo Spider, 1966**
This stylish, sporty car from Italy combined classic looks with a five-speed gearbox and all-round disc brakes as standard equipment. Versions of this original Spider were still on sale in the 1990s.

The golden age of sports cars

▽ **Chevrolet Corvette**
A classic American sports car, the Corvette of the 1960s looked mean and had speed to match thanks to its 5.3-litre V8 engine. This second-generation car drew heavily on the Corvette's widespread successes in sports-car racing.

Sports cars built for all-out fun – mainly two-seater roadsters – had been around since the 1920s. However, they reached their pinnacle in the '60s when they became desirable status symbols, projecting an image of financial success and carefree living.

Many '60s sports cars were also technically accomplished and stylistically cutting-edge as well as gorgeous to look at. Some drew directly from racing car designs, while Italian manufacturers built cars that offered a scintillating driving experience thanks to responsive engines and expertly balanced suspension systems. the UK's mass-produced MGs turned the sports car into a popular commodity. Many sports car marques, including the UK's Triumph and Germany's Porsche, tapped into strong demand from California and the US northeast, centred around New York.

△ **Do-it-yourself sports car kits**
First introduced in the 1950s, sports car kits – such as for this Ginetta G4 – were ideal for drivers who could not afford a brand new factory-made one. They could be assembled cheaply by a competent mechanic, offering the owner a large tax saving.

The road sign revolution

△ **"Slippery road" warning**
According to Kinneir and Calvert's system, the triangular shape of this sign signifies "warning". That and the pictogram of the skidding car convey all the information in the simplest possible way.

The Anderson Committee of 1957 was set up to create a new, unified road signage system for the UK's motorways. It employed two graphic designers whose work is still in use today and whose influence was felt worldwide.

▽ **Sign language**
Jock Kinneir (right) watches one of his road signs come to life. Rectangles were reserved for destination information only.

The advent of the motorways meant greater speeds – and with far higher levels of traffic than ever before, it was time to do something about the standard of the UK's road signs. Until the 1950s, these had been a jumbled mass of typefaces, sizes, and styles – which had served well enough during the early days of motoring, but the increased popularity and speed of cars meant that the system needed to be updated. In addition, the UK had been flouting the 1931 Geneva Convention concerning the Unification of Road Signals, which stressed that visual keys were more important than written signs, lest foreign motorists become confused.

In 1957, the British government set up a committee chaired by Colin Anderson, CEO of the P&O-Orient shipping line, to review the nation's system of road signs. Anderson had already employed graphic designer Jock Kinneir to produce baggage labels for P&O; Kinneir, and his former pupil Margaret Calvert, were given the task of redesigning road signage.

Kinneir and Calvert had worked together before on the artwork and signage for Gatwick Airport. Their work

▷ **Motorway sign, 1958**
The signs devised by Kinneir and Calvert were first tested in London's Hyde Park and a nearby underground car park. The first signs to see actual service were installed on the new Preston bypass motorway in 1958.

sought to reduce the signs to the bare essence of information. For motorways, they created a new typeface (called "Transport") and applied it to signs that offered basic bird's-eye views of junctions. Unlike earlier British road signs, the new signs had upper- and lower-case characters to ensure clear legibility at speed. All signs had a blue background and white lettering made of a reflective material that could be lit up by car headlamps at night. The scale of the lettering, the width of the borders, and the thickness and shape of directional lines were all standardized.

Design for the nation

However, while work was progressing on the motorway signs, those of the UK's other roads also needed attention. In 1961, designer Herbert Spencer published two articles in *Typography* magazine detailing the discord between signs that he had seen on a single journey between central London and Heathrow airport.

The Worboys Committee of 1963 commissioned Kinneir and Calvert again to review the signs across the rest of the British road network. The committee aimed to bring British road signs into alignment with the 1949 Geneva Protocol: using pictograms to convey essential information, with triangular sign shapes for warnings, circular for commands, and rectangular for advice on destinations.

Kinneir and Calvert retained the bird's-eye diagrams and the Transport font for signs giving route distances and directions. For primary routes, they adopted white lettering on a mid-green background, with road numbers in yellow. For secondary routes, they used black lettering on a white background.

Margaret Calvert designed many of the pictograms. For the "Children Crossing" sign with the figure of a girl leading a boy, the girl was modelled on herself as a child; the "Cow" sign for farm animals was modelled on a cow that she had known. The Kinneir-and-Calvert road signs have seen minor updates, but have not fundamentally changed since their introduction in 1965. Their rigorous design has been copied around the world.

BIOGRAPHY
Margaret Calvert

Margaret Calvert OBE was born in 1936 in South Africa. She moved to the UK in 1950, where she attended the Chelsea College of Art. There, she met tutor Jock Kinneir. Kinneir asked Calvert for her help in creating signs for the then-new Gatwick Airport, and Calvert was instrumental in choosing the black-on-yellow colour scheme. Having then worked together on the road signs for the British motorway network and for British Rail, Jock Kinneir made her his business partner in 1964 and renamed his company Kinneir Calvert Associates. Later projects included several unique typefaces, including the "Calvert" typeface used on the Tyne and Wear Metro system. Calvert was made Officer of the Order of the British Empire (OBE) in 2016 for services to typography and road safety.

***MARGARET CALVERT* SURROUNDED BY HER DESIGNS, LONDON, 2015.**

"**Direction signs** and street names... are as vital as a drop of **oil** in an **engine**."

JOCK KINNEIR, 1965

"Here is **full power** potential for **instant acceleration** and **outstanding** climbing ability."

CHRYSLER PROMOTIONAL FILM, 1963

Space-age motoring

In the decade that saw a man walk on the Moon, Chrysler came close to launching the first jet-powered car – the Chrysler Turbine.

Although Chrysler was not the only manufacturer testing jet engines, the Detroit corporation came the closest to putting a quiet, smooth-running gas-turbine engine into a production car. From 1962–64, Chrysler publicly tested a fleet of five prototypes and 50 production cars that it had built in Italy with Ghia. These cars, designed by former Ford stylist Elwood Engel, looked similar to the 1961 Ford Thunderbird, but were bespoke in every way.

The jet-age styling carried through the entire car, from the finned headlight surrounds to the interior, which had four leather bucket seats, an aeroplane-inspired instrument cluster, and a centre console resembling the internal shaft of the turbine engine. The highly distinctive tail-lights looked like rocket thrusters.

More than 30,000 people applied to road test the car, and Chrysler selected 203 families, who logged more than 1 million miles. During the three-month trials, each driver kept a log book and described the car's fuel economy, reliability, and performance in various traffic conditions.

Mechanical failures were rare, and most of these were due to drivers using the wrong fuel. After the test, Chrysler had tough decisions to make. Drivers loved the cars, but fuel economy and performance from the 130-hp jet engine were no better than a Chrysler vehicle with a mid-sized V8. The engine was also easily 10 times more expensive to produce than a conventional V8. With emissions regulations tightening, Chrysler chose not to build the production jet car, which would have debuted in 1966.

After the test, 46 of the Turbine cars were crushed in a Detroit scrapyard. The other nine cars were deactivated and sent to museums. Chrysler continued research into turbine-powered cars until 1977. It finally ended when engineers still could not reduce fuel consumption and emissions.

◁ **Height of fashion**
With its cutting-edge design and promise of space-age technology, the Chrysler Turbine Car made the ideal prop for this photoshoot for *Vogue* magazine in 1963. However, the engines were too expensive for the cars to enter production.

Safer car design

The explosion in car use inevitably led to a corresponding increase in the number of road accidents. Safety really started to come under the spotlight in the 1960s, propelled by vocal campaigners and advances in technology.

The unsafe nature of cars was clear from the dawn of motoring. The first recorded death by motor vehicle was in 1896, when pedestrian Bridget Driscoll was struck by a car in London. By 1955, over one million people had died on the world's roads; by the end of the century, the same number of people were being killed worldwide each year.

Early car designers gave little or no thought to safety: in a crash, steering columns were like lances aimed at the driver's chest, while an array of protruding levers presented similar hazards. The first crash barrier test was performed by General Motors as early as 1934. Monitoring the effects of accidents on human occupants started, rather gruesomely, using dead bodies. Live human volunteers followed, and these were replaced by crash test dummies in 1949.

Since the bodywork of early cars was not designed to "deform", the full force of any impact was transmitted directly to the occupants. Mercedes-Benz was the first company to address this, engineering "crumple zones" into its vehicles in 1959. These served to absorb at least some of the impact forces.

▷ **Prompting development**
Ralph Nader's 1965 book *Unsafe at Any Speed* changed attitudes towards safety forever – but only after a concerted industry campaign to discredit him.

Belt up

The single biggest development in the history of road safety was the invention of the seatbelt. What began life as a lap seatbelt introduced

"... the automobile has **brought death**, injury... **to millions**."

RALPH NADER, PREFACE TO *UNSAFE AT ANY SPEED*

by American carmaker Nash to its Rambler model in 1950, truly came of age in 1958, when Volvo's Nils Bohlin of Sweden invented the three-point lap-shoulder belt. Knowing how important his invention was, Bohlin convinced Volvo to waive its patent rights so that any automobile manufacturer could use it.

Soon, lobbysists in many countries were arguing that seatbelts should be fitted in all new cars. In the US, a law to that effect was passed in 1968, but it took longer for other countries to be convinced. In the UK, wearing seatbelts only became mandatory in 1983.

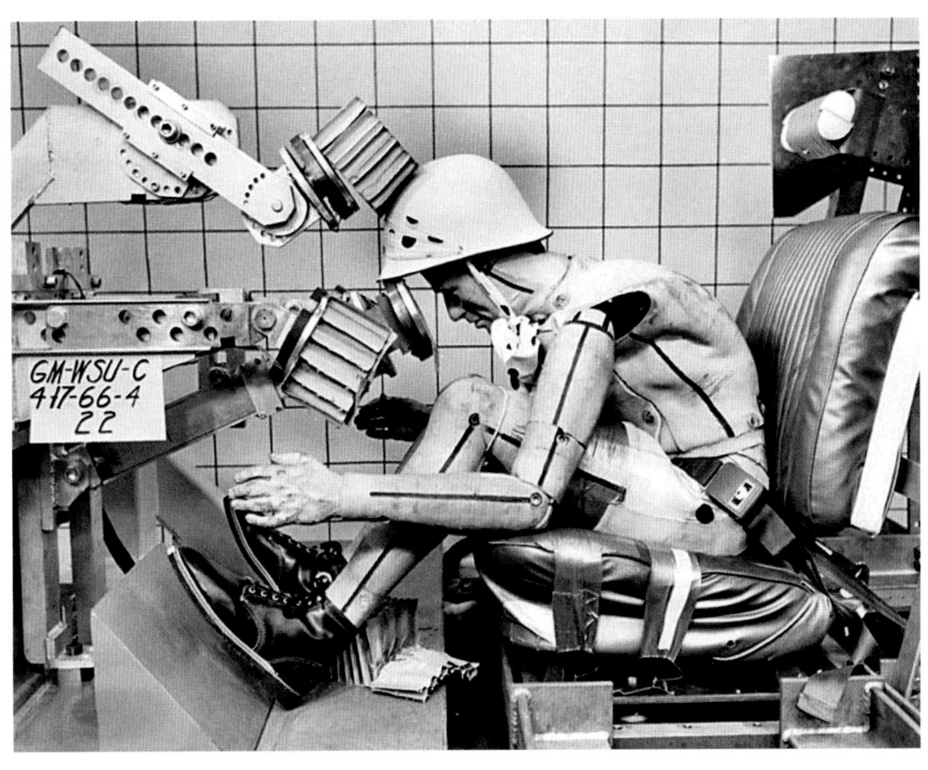

◁ **Simulating injury**
Car manufacturers use mannequins and crash sleds to assess the safety of their vehicles. Safety has always been key to a brand's success.

Unsafe at Any Speed

One of the loudest voices in the road safety lobby was Ralph Nader. A lawyer, he wrote the bestselling book *Unsafe at Any Speed* (1965), which criticized General Motors' Chevrolet Corvair. Nader claimed that accidents were being caused by cost-cutting measures in the Corvair's suspension design. The book forced GM to change its designs, and precipitated the 1966 National Traffic and Motor Vehicle Safety Act, which set strict new safety standards for cars.

New safety tech

The 1960s also propelled a rapid progression in safety technology. Jensen's 1966 FF was the first production car with anti-lock brakes to mitigate skidding, while Volvo introduced child-proof door locks in 1972. GM invented the very first airbag (protecting the driver in a front-on collision) in 1973, but it would take until 1981 for Mercedes-Benz to launch the first production airbag in its S-Class.

▽ **Safe testing**
The crash test dummy – an idea borrowed from the aircraft industry – quickly spared live volunteers their bruises.

BIOGRAPHY

Ralph Nader

The world's most celebrated safety campaigner, Ralph Nader was born to Lebanese parents. A Harvard Law School graduate, he served as a cook in the army before becoming a lawyer. His 1965 book *Unsafe at Any Speed*, was a publishing sensation – so much so that General Motors hired private detectives and attempted to discredit him. Nader took GM to court and his settlement of $425,000 paid for him to set up an institute for legal activism. When new car safety laws were passed in the US in 1966, the Speaker of the House of Representatives attributed them to Nader's "crusading spirit". The anti-establishment figure even ran – unsuccessfully – for US president four times.

***RALPH NADER'S* FAMOUS BOOK GENERATED A SENATE HEARING IN THE US IN 1966 TO DISCUSS THE LACK OF SAFETY FEATURES IN CARS.**

WOODEN-SPOKED, ROLLS-ROYCE, 1906–25

WIRE-SPOKED, AUBURN, 1910

Alloy wheels first used by Bugatti in the 1920s, on the Type 35

PRESSED-STEEL, HILLMAN MINX, 1936

METAL WHEEL DISC, ALFA ROMEO 8C COUPÉ, 1938

CHROME WHEEL TRIM, CADILLAC SERIES 62, 1959

Wheels of fashion

Continual evolution in materials and construction has made wheels stronger and lighter. They have become a key part of a car's styling, differentiating one make and model from the next.

The first cars used wire wheels with solid rubber tyres, or heavy wooden or steel "artillery" wheels. These were slowly replaced by wire-spoke wheels in the 1910s and '20s. Ettore Bugatti followed a different route, however, and fitted some of his cars with wheels cast from aluminium alloy, initially with integral brake drums and removable rims. Few other showroom cars featured alloy wheels until the 1960s.

By then, everyday cars were being fitted with pressed-steel wheels, which were relatively light, very strong, and cheap to manufacture in mass quantities. Chrome-plated hubcaps were fitted to deluxe models.

Magnesium or aluminium alloy wheels, fitted to sports cars from the 1960s, became common in the 1980s. Today, the cutting edge of wheel design is the use of carbon fibre composite materials for even lighter weight.

CAST ALLOY, LAMBORGHINI COUNTACH, 1990

Cast eight-spoke wheels are lighter and stiffer than wooden or wire wheels

BUGATTI TYPE 35B, 1927–30

CHROMED WIRE-SPOKED, JAGUAR E-TYPE, 1964

CAST ALLOY, FERRARI DAYTONA, 1968

ROSTYLE FAKE ALLOY. MG MIDGET, 1969

COMPOSITE, MERCEDES-BENZ CONCEPT CAR, 2017

KEY DEVELOPMENT

Aftermarket alloys

Swapping wheels has always been a favourite way for enthusiasts to update a car's appearance and, possibly, improve its road manners. From the early 1960s, racing cars and high-performance road cars were available with alloy wheels from brands such as Campagnolo, Cromodora, Halibrand, and Fuchs, and there was a growing market for "aftermarket" wheels that car owners could fit themselves. The earliest were made from magnesium alloy and known as "mags". This term was later used for any alloy wheels; most road wheels were actually aluminium. Keystone and Cragar in the US, Minilite and Wolfrace in the UK, and ATS, Speedline, and BBS in Europe became synonymous with a growing trend for customizing cars that reached its height in the 1970s and then the tuning boom of the 1980s.

***THE MONKEEMOBILE*, BUILT IN 1968 FOR THE US POP GROUP THE MONKEES, FEATURED CRAGAR ALLOY WHEELS.**

British cars set the pace

In the 1960s, the UK motor industry led the world in innovative design, producing cars that had both sophistication and style. Even now, some of the UK's most famous marques trade on the success of their 1960s models.

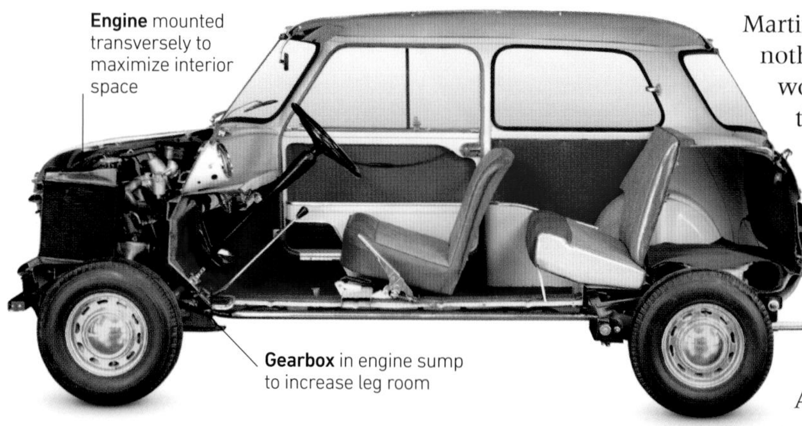

△ **BMC Mini**
The best-selling Mini struck a chord in 1960s Britain. Innovative, small, and cheeky, it epitomized the plucky British spirit, and put an end to the European microcar.

The Mini, one of the UK's biggest automotive successes, started life in 1959. It proved perplexing at first, owing to its size and front-wheel drive, although it soon rendered the European microcar redundant.

However, the UK was not just about miniatures. It also created some sublime sports cars and the grandest of grand tourers. In 1961, an undisputed game-changer was launched: the Jaguar E-type. This svelte sports car offered a genuine 145-mph (233-km/h) performance in an eye-catching body – and for half the price of an Aston Martin. For the money, there was nothing to match it anywhere in the world, and it is one of the few cars that everyone agrees can be called an icon of motoring.

The Jensen FF

The Jensen FF joined the pantheon from the day it was launched in 1966. This grand tourer followed a winning formula for British GTs: an American V8 engine matched to an Italian body, in this case one that was near-identical to Jensen's own Interceptor. But the FF took the concept of the 6.3-litre Interceptor and rammed it with the latest technology.

The FF was so named because it used the patented Ferguson Formula four-wheel drive (4WD) system. The tractor manufacturer, which had developed the 4WD technology and trialled it in its R4 and R5 prototype cars, licensed it to Jensen, who in turn produced the world's first high-speed, 4WD GT car. Dunlop Maxaret anti-lock brakes (derived from aircraft) were also fitted, making the FF arguably the safest car in the world when it was launched. It remained in production until 1971.

KEY EVENTS

- **1959** The Austin Mini Seven and Morris Mini-Minor go on sale. A total of 5,387,862 Minis are produced up to 2000.
- **1961** The Jaguar E-type begins its 14-year life. Enzo Ferrari calls it "the most beautiful car ever made".
- **1962** Ford introduces the Cortina, a rear-wheel-drive saloon that was built to cruise comfortably all day.
- **1963** The Lotus Elan brings sublime new levels of sports car responsiveness, and a novel backbone chassis.
- **1963** Rover's futuristic 2000 pioneers the compact executive saloon.
- **1966** The Jensen FF updates the grand tourer with its four-wheel drive and anti-lock brakes.
- **1968** The Ford Escort replaces the Anglia, setting the benchmark for small saloon cars.

***THE JENSEN FF*, ONE OF THE MOST TECHNICALLY AUDACIOUS CARS OF ITS TIME.**

◁ **Jaguar E-type**
The Jaguar E-type remains an arresting sports car several decades on. It proved that performance and style are easily combined.

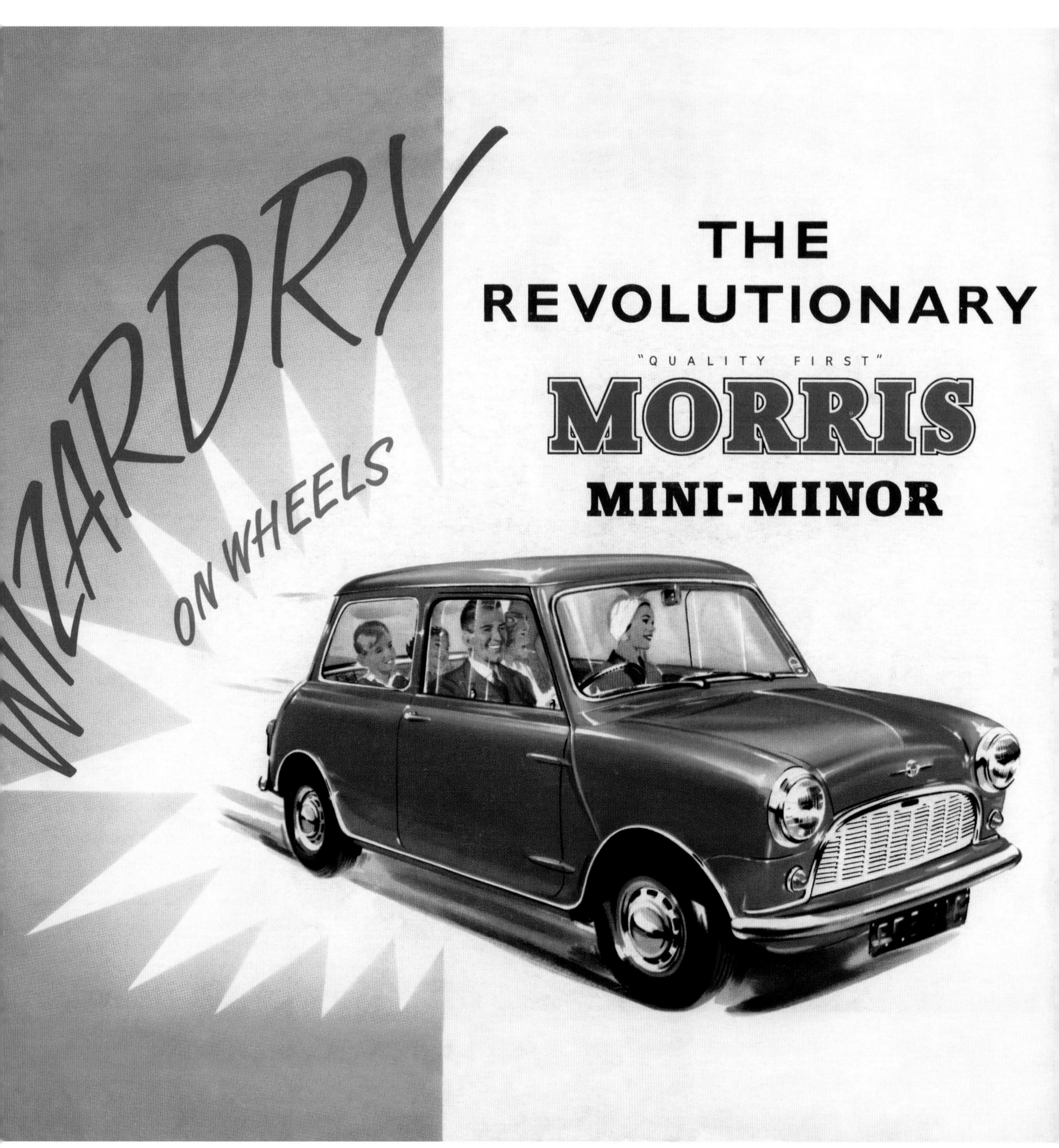

△ **The Morris Mini-Minor**, launched in 1959, was a hit among conservative car owners. In 1969, the Mini became a marque of its own.

▽ Meyers Manx Dune Buggy, 1968
An ultra-light plastic body, soft, fat tyres, and a super-reliable air-cooled engine were the material assets of the dune buggy. This Meyers Manx with its Chevrolet Corvair flat-six engine was personally owned by Steve McQueen – seen here at the wheel with Faye Dunaway, in a scene from their 1968 film *The Thomas Crown Affair.*

△ **Fiat 128, 1972**
The manufacturer's first model with a transverse-mounted engine and front-wheel drive. It was available as two- and four-door saloons, and also formed the basis of the 127 hatchback.

△ **Toyota Corolla, 1966**
The Corolla was designed to out-class its rivals from the beginning, and featured a specially developed engine and all-new components. Drivers could enjoy bucket front seats, a heater, and a radio.

Popular small cars

Every so often, a new model is launched that can be described as game-changing within the car industry. Whether it does this by inventing a whole new genre, by introducing a new technology or design, or by mobilizing the masses in a way that few have done before, such a car leaves its mark on motoring for decades to come. These cars are not necessarily pinnacles of technology or style. Instead, they are motors that bring new levels of practicality, reliability, and handling at a price within reach of the typical motorist. In the 1960s, such cars included the Mini, which reinvented small-car design, and the Toyota Corolla, which offered a level of reliability that served as a wake-up call to other manufacturers.

Leading contenders of the 1970s include the Ford Pinto, which was among the first small hatchback cars developed for the US market, although concerns over crash safety tarnished its reputation there. However, perhaps the most influential car of the period was the Volkswagen Golf. With its mix of good build quality, practicality, safe handling, and style, it was copied but never beaten by its rivals.

▷ **Small is beautiful**
Iconic small cars the Mini, the Citroën 2CV, and the Renault 5 (front to back) are pictured driving together in a French publicity photograph taken in 1979.

△ **Brochure for the Renault 5 TX, 1981**
Even before the Golf came along, Renault had created the three-door "supermini" class with its much-loved 5; this TX is one of the last Mk1s.

△ **Ford Pinto, 1971**
The Pinto was designed to compete in the US with cars imported from Japan and Europe. Despite early controversy about the risk of fire in the event of a crash, it proved popular, with more than 3 million built.

△ **Volkswagen Golf, 1974**
Volkswagen replaced its long-established Beetle with the Golf, which had a front-mounted, water-cooled engine, and a boxy, roomy body. It instantly became the benchmark family hatchback; the GTi is shown here.

“...we have to **develop a car** that our customers **are proud** to own.”

TATSUO HASEGAWA, CHIEF ENGINEER OF TOYOTA COROLLA

Crossing the Darién gap

What began as a publicity stunt to showcase the new Range Rover of 1971 unfolded as a gruelling expedition across 100 miles (180 km) of the most hostile, swamp-ridden jungle on Earth.

The Pan-American Highway stretches some 18,000 miles (29,000 km) from Alaska in the north to Tierra del Fuego in the south – except for a 100-mile (180-km) section in Panama, where the terrain was so difficult that no vehicle had crossed it before 1972. In 1962, Chevrolet had sent three Corvairs to traverse this section – called the Darién Gap – and ended up abandoning them in the jungle. A decade later, Land Rover took on the challenge, hoping to generate press coverage for its newly launched Range Rover.

Heading the expedition was Colonel John Blashford-Snell of the British Armed Forces' Royal Engineers. He hand-picked the team of 64, many of whom came from his own regiment. The rest were scientists, aircrew, and Range Rover-trained mechanics from the 17th/21st Lancers. To provide some of the hard labour required to get the vehicles through the most difficult parts of the Gap, the Colonel reportedly exchanged a case of Johnnie Walker Black Label whisky for 12 murderers who were doing time in a Panamanian prison, and who were promised their freedom at the end of the expedition.

Basing their preparations on the account of Brendon O'Brien, an Irishman who had walked the Gap in three months and survived, the team took 28 horses for transporting supplies, and motorized wheelbarrows known as "Hillbillies" for carving out the track for the Range Rovers to follow. A support aircraft loaned by the British Army

△ **Survivor of the Darién Gap**
One of the two Range Rovers used by the British Trans-Americas Expedition of 1971–72. Today, it resides at the British Motor Museum in Gaydon.

▽ **River crossing**
A Range Rover survives one of the many river crossings required by the traverse of the Darién Gap. Later, one of the vehicles overturned in the Tuira River, but was winched out, dried, and repaired.

provided critical reconnaissance information and dropped crucial supplies along the route.

Crossing the Gap

Problems emerged early on. Broken differentials on both cars brought the expedition to a halt within the first few days in the Gap. When the team mechanics had done all they could, Land Rover flew in redesigned differentials to get them on their way.

The differentials took a week to fit, but almost as soon as they were underway again, the Hillbillies, which had only been tested in the UK, got their track treads jammed with mud and had to be abandoned. As a substitute, the team sourced an old Series IIA Land Rover, made it as light as possible, and kitted it with the cables, saws, and chains that the engineers needed to make a track.

▷ **Delivering supplies, 1972**
A "Huey" helicopter lands in the jungle to deliver new differentials for the Darién expedition's damaged Range Rovers, with team member Gavin Thompson in the foreground.

The journey was fraught with other challenges. It had been timed for the dry season, but the rainy season ran late, turning the swamps into thick mud. The engineers carving out the track suffered from trench foot, others fell ill with malaria, and scorpions, snakes, and biting ants were daily companions. The conditions also rendered the Rovers' swamp tyres a liability, building up so much inertia in the mud that they started to damage the vehicles. Standard off-road tyres had to be flown in to replace them.

Although the Range Rovers were driven across as much of the terrain as possible, two Avon rafts were used for crossing numerous rivers. During one perilous crossing over the Tuira River, a rapid overturned one of the rafts, submersing its Range Rover. Although the vehicle was winched out, water had seeped into every crevice. It took the mechanics 36 hours to pull the motor apart and get it going again. Incredibly, both cars survived the 100-day ordeal, and all of the British team members lived to tell the tale.

"We were told it was **swamp**... it **wasn't**... it was **mud**."

GAVIN THOMPSON, ON THE 1972 DARIÉN GAP EXPEDITION

LIFE BEHIND THE WHEEL

The Corvair expedition

In 1959, Chevrolet launched its family-sized Corvair, touting it as a car that was able to handle any conditions. To generate publicity, a Chicago Chevrolet dealer sponsored the first vehicular crossing of the Darién Gap in 1962, with an entourage of three red Corvairs, accompanied by several Chevrolet pickup trucks equipped with power winches.

The air-cooled aluminium engines provided enough power to get the cars through the first valleys, but as they hit increasingly steeper gullies the team resorted to making bridges out of logs. Conditions proved too much for one of the cars, which was left behind in the jungle. The other two made it to the Colombian border, but they, too, were abandoned. The Land Rover team found their remains a decade later.

THIS ABANDONED CHEVROLET CORVAIR **COULD STILL BE FOUND IN THE JUNGLE OF THE DARIÉN GAP MANY YEARS LATER.**

Alternative car culture

Even before World War II, a new type of car began to appear in the US, particularly on Californian roads: highly modified vehicles with flathead Ford V8 engines, stripped-down bodywork, and wild paint jobs.

△ **Road Agent, 1965**
Painted in vivid pink with an orange-tinted bubble roof, the Road Agent was one of many cars made by artist and custom car builder Ed "Big Daddy" Roth – a leading figure in the 1960s US custom car scene.

The earliest custom cars, built during the 1930s and '40s, were commonly based on Model Ts, 1920s Chevrolet roadsters, and 1932 Fords. They soon gave rise to an alternative motoring scene with a language of its own, using terms such as "lakester", "streamliner", and "hot rod" to describe the different styles of modification.

The scene expanded rapidly following the end of World War II, when former military engineers and mechanics sought to use their technical skills as they rejoined civilian life. For them, tinkering with cars quickly grew from a hobby to a new industry.

At the same time, car manufacturers resumed production following the war, which created a ready supply of used models, as owners could finally trade up. Wartime surplus also provided a source of cheap materials, including the external fuel tanks taken from fighter planes, such as the P-38 and P-51. The teardrop-shapes of these tanks made them extremely aerodynamic. They were also sturdy and could be adapted to accommodate auto components, giving rise to some of the wildest custom cars that came out of California.

The custom business

As the scene grew, it also became professional. In southern California, workshops opened that specialized in building custom cars. Engineers and businessmen started mail-order companies to supply parts, such as camshafts, carburettors, and intake and exhaust manifolds, which greatly improved the performance of standard engines. While the original custom car scene focused on US-built models, a new wave developed with the arrival of the first imports – especially Volkswagen, whose Beetle and Type 2 van or "bus" became the hippie generation's favourite modes of transport in the 1960s. Baby boomers coming of driving age in the early '60s rebelled against their parents' values and chose to customize VWs in the hippie ethos of peace, love, and fun. Several manufacturers saw this trend and tried to cash in. In 1964, the British

▷ **Customized Mini Moke, 1966**
Originally designed as a lightweight alternative to the Land Rover, the Moke's low ground clearance proved a problem off-road. A seen here in a promotional photo for pop group the Beach Boys, it later found a cult following as a beach car.

"... when you **had an idea**, the key was **'different and fun!'**"

DAN WOODS, CUSTOM CAR BUILDER

Motor Corporation launched the Mini Moke, a small open-top four-seater based on the ground-breaking Austin Mini. More than 50,000 Mokes were built in three countries before production ended in 1993. The Moke was conceived as a light military vehicle modelled on the Jeep, but in the spirit of the late 1960s, it too became a hippie emblem. It was also used as a beach car for open-top fun in holiday hotspots – a role it shared with the plastic-bodied Citroën Méhari, which was derived from the 2CV. Yet another holiday fun car was the Ghia Jolly, a cut-down Fiat 600 often seen with a fringed fabric "Surrey top" to protect passengers from sunburn. This car was a favourite with wealthy yacht owners, who would keep one onboard for land-based excursions.

In addition to the Moke, the Jeep also inspired manufacturers in Japan, such as Mitsubishi and Suzuki, to build their own variations. Meanwhile, in the Philippines, surplus war Jeeps were turned into richly decorated Jeepneys.

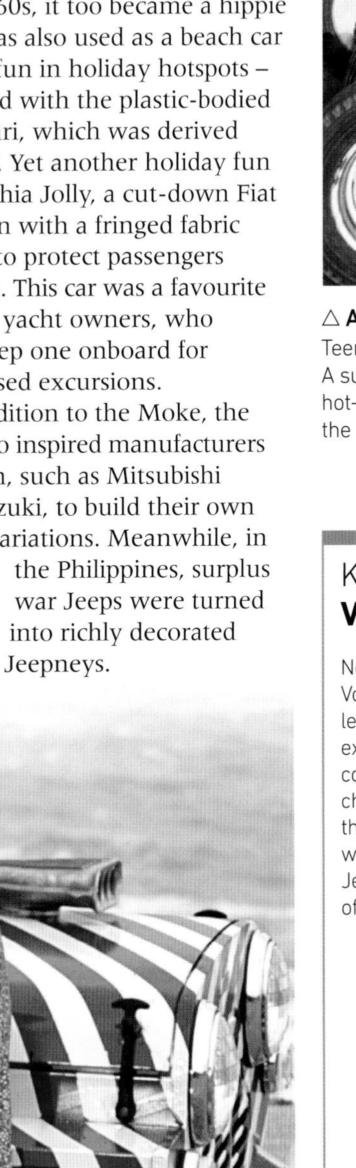

△ **American hot rods**
Teenage boys work on their hot-rodded cars. A supply of inexpensive vehicles helped make hot-rodding a popular hobby in the US from the 1950s onwards.

KEY DEVELOPMENT
Volkswagen Type 2 Kombi

No vehicle defines the '60s hippie generation better than the classic Volkswagen Kombi/camper van. It was solid, roomy, and reliable, and lent itself to customization. The giant chrome VW logo on the nose, for example, often gave way to custom-made peace signs. The VW camper could be slept in, camped in, and even lived in. It was not fast, but it was cheap to run and cheap to fix. Some of these vehicles today are among the hottest collectibles, with the most coveted being mid-'60s models with 21 windows and flip-open windscreens. When Grateful Dead leader Jerry Garcia passed away in 1995, Volkswagen ran an ad with a drawing of a '60s van showing a teardrop falling from its headlight.

BRIGHTLY PAINTED VW CAMPER VANS, **SUCH AS THIS '70s MODEL, STILL EVOKE THE SPIRIT OF THE HIPPIE MOVEMENT LONG AFTER IT DIED OUT.**

The Dodge Deora

Some of the most famous cars of the '50s and '60s were the one-off customs made by the likes of George Barris and Dean Jeffries in California. However, Detroit brothers Mike and Larry Alexander – the A-Brothers – created one of the most sensational of all: the Dodge Deora.

After World War II the custom craze in the US flourished in California, with speciality workshops creating unique one-offs for film stars and other wealthy customers. Typically, these cars were extremely powerful, chrome-trimmed, and decorated with wild paint schemes and flashy colours.

Two brothers based in Detroit, with no formal training in automotive design, created one of the world's most famous concept vehicles: the Dodge Deora. Mike and Larry Alexander began working on custom cars in Detroit in the 1950s, specializing mostly in unique paint jobs and light customization work. The brothers' obvious talent caught the attention of General Motors designer Harry Bentley Bradley, who drew custom cars for magazines in his spare time. In 1964, Bradley commissioned the Alexander brothers to create the Deora from a Dodge A100 truck.

The Alexanders moved the engine and gearbox from under the bonnet to the rear of the vehicle. They fitted the rear tailgate of a 1960 Ford Country Sedan station wagon to the front of the car, installed a modified Oldsmobile Toronado steering wheel, plus a full array of Stewart-Warner gauges to the dashboard. To enter the Deora, the driver lifted the windscreen, swivelled the rear tailgate, and swung the steering column out of the way. Deora was such a sensation at its debut in Detroit in 1967, that Bradley, then working for toy maker Mattel, included it in the first batch of Hot Wheels® toy cars in 1968. It was the world's first concept pickup truck when displayed by Chrysler in a two-year deal. The Alexander brothers lived to see the restored Deora sell at auction to a private buyer in 2009 for $324,500.

▷ **Design on display**
First exhibited by the Alexander Brothers at the Autorama show in Detroit in 1967, the Dodge Deora was leased by Chrysler for two years to form part of their display of concept cars. The vehicle was then kept in storage until the 1990s.

> “**Wouldn’t you** like to own a car that is **totally different** from that of **Mr Average Public**?”

GEORGE BARRIS, *CUSTOM CAR CHRONICLE*, 1953

△ **Opel Manta GT/E, 1970**
Based on the Opel Ascona, the Manta was a stylish and sporty-looking GT that became a key rival to the Ford Capri. It had distinctive round tail-lights, and took its name from the Manta Ray concept car of 1961.

△ **Alfa Romeo Alfasud Sprint, 1976**
The pretty Alfasud Sprint used flat-four engines in a body reminiscent of the larger Alfetta GTV. Although front-wheel-drive, the excellent chassis gave the car great road handling.

"**Ford Capri**: the car you **always promised yourself**."

FORD ADVERTISING SLOGAN, 1969

△ **Ford Mustang hardtop, 1965**
The Mustang formula of a simple saloon floorpan with a sharp body and multiple power choices worked a dream – as did the vast options list, which meant that it was rare for two Mustangs to be alike.

△ **Porsche 924, 1976**
This 2+2 seater coupé was intended to broaden Porsche's appeal. With a water-cooled, Audi-supplied engine at the front, it was a big departure from Porsche's traditions, and it proved very popular. Volkswagen helped design the car.

△ **Toyota Celica MkII, 1977**
Toyota saw the success of cars such as the Ford Mustang and Capri, and decided to enter the market. For both the first- and second-generation series, there were versions with a "Liftback" third door.

A new kind of style

Combining style and sporty performance in a hardtop capable of carrying four people in a 2+2 configuration that ensured all-weather driving, grand tourers, or "GTs", were originally luxury cars for the wealthy. However, after World War II, a new generation wanted cars that were different from the sensible but dull models that their parents drove. Something new was needed, something with a little glamour, but still sensible enough to be viable – in short, a GT car on a budget. In the US, Lee Iacocca's dream of the Ford Mustang as a sporty "personal car" proved a huge success, and other manufacturers from around the world rushed to introduce their own contenders. The two most popular in Europe were the Ford Capri and the Opel Manta, with Japan's Celica in hot pursuit.

▽ **Ford Capri, 1969**
Designed as a Mustang for Europe, the Ford Capri shared components with other Ford models – including the Cortina and Escort – making it affordable to buy and maintain.

Feeding the meter

As car use soared in the second half of the 20th century, the demand for places to park increased dramatically. Parking meters, and "meter maids" to patrol them, restored order to crowded city centres.

To keep parking chaos at bay, motorists started to be charged to park their cars. Multi-storey car parks had been in existence since the early part of the 20th century, but coin-operated parking meters by the side of the road did not arrive in the US until 1935, and in the UK until 1958. By the early 1940s, more than 140,000 parking meters were operating in the US.

Yellow lines and meter maids

In 1960, yellow lines started to be painted on the edges of British roads, restricting parking to certain times – or no time at all if they were double lines. Staff needed to be hired to patrol the streets, and in September 1960, the very first traffic wardens appeared in the London Borough of Westminster. There were 40 in all, dressed in semi-military style with distinctive yellow-banded caps, handing out fines of £2. Traffic wardens have been instilling fear and loathing among motorists ever since – and even, on occasion, love. The phrase "meter maid" was coined in the US in the late 1950s for such wardens, and popularized in The Beatles song lyric "Lovely Rita meter maid".

▷ **Checking a meter, 1964**
Parking Enforcement Officer, Willa Chandler checks a meter in Pittsburgh, Pennsylvania.

Alternative solutions

Accommodating idle cars in built-up areas has perplexed city planners for over a century. As early as 1905, a semi-automated car park was opened in Paris. Lifts carried cars to its upper levels, where an attendant would then park them.

The Auto Stacker in Woolwich, south-east London went further, incorporating a totally automated system of conveyor belts, lifts and dollies to carry cars to its 256 spaces on eight storeys. It opened in 1961 but was so complicated that it malfunctioned on its opening day, and closed within months.

Parking meters and an army of traffic wardens, it seemed, were the best way to maintain order and raise revenue. In Australia, though, a very different type of parking attendant arrived in 1965.

Drivers arriving at Queensland's Surfers' Paradise holiday resort were met with parking charges, but any who over-stayed avoided a fine by having their meters "fed" by unofficial Meter Maids clad in gold bikinis. It was the controversial idea of developer Bernie Elsey, and despite calls for political correctness, the maids remain a feature at Surfers' Paradise to this day.

◁ **Parking in Germany, 1982**
Ranks of cars stand at parking meters in Hanover, Germany. Parking for hours on end required frequent trips to the meter.

KEY EVENTS

- **1933** The first working coin-operated parking meter, The Black Maria, is invented in the US by Holger George Thuesen and Gerald A. Hale.
- **1935** The world's first parking meter, the Park-O-Meter No.1, is installed in Oklahoma City in the US.
- **1954** The first automated car park opens in the US. For a monthly fee, a magnetic key card allows entry and exit via a barrier.
- **1954** Australia's first parking meter is installed in Hobart, Tasmania.
- **1958** The UK's first parking meters start operating in London. The price of one hour of parking is 6 pence.
- **1960** UK traffic wardens issue their first ticket to Dr Thomas Creighton, who parks to treat a heart attack victim. His fine is dropped after a public outcry.
- **1974** The UK's first automatic "pay on foot" car park is opened in Oxford. Drivers collect a ticket on entry, then pay to raise the exit barrier.

***PARK-O-METER* PARKING METER ON CITY STREET, OKLAHOMA CITY, 1939.**

△ **Casino dancer Audrey Crane** feeds a parking meter in London. Only one type of coin was accepted, so it was important to have the right change.

The first air ferry

Before ferry boats became the default way to cross the English Channel by car, drivers could take their vehicles overseas by plane on the world's first air ferry service.

Silver City Airways operated the first air ferry for cars in 1948 using a Bristol Freighter, originally designed to transport military vehicles. The hop over the Channel took just 19 minutes, the aircraft cruising at just 1,000 ft (330 m). The service ran each summer, from July to September, initially from a grass airfield at Lympne in Kent, UK, to Le Touquet on the north coast of France. A one-way trip for a family car and four passengers cost £32.

The service gained in popularity, moving to a new, purpose-built airport, Ferryfield, at Lydd on the Kent coast. Routes were added from Lydd, Lympne, Gatwick, and Southampton in the UK to Le Touquet, Calais, and Cherbourg in France, and Ostend in Belgium. In 1955, new routes linked Stranraer in Scotland with Belfast in Northern Ireland, and Birmingham with Le Touquet.

By then a competitor, Channel Air Bridge, was offering services from Southend-on-Sea in Essex to Calais, Ostend, and in the Netherlands, Rotterdam. The two companies merged in 1963 to form British United Air Ferries. Further routes were added, including long-distance flights to the Swiss cities of Basel, Strasbourg, and Geneva – the latter made famous in the contemporary James Bond film *Goldfinger*. The ageing Bristol Freighters were replaced by the four-engined Carvair, a conversion of the Douglas DC-4 created by Freddie Laker, which could carry five cars.

Aer Lingus adopted Carvairs for car ferry flights to Ireland in 1963, but by then the popularity of air ferries was waning. The existing roll on/roll off ferry boat services took over (see pp.186–87), being vastly cheaper, and the age of air travel for cars ended in 1977.

▷ **Silver City Airways Bristol Freighter**
This twin-engined aircraft had a pair of large outward-opening doors in the nose, giving access to a load bay that could take cars, motorcycles, bicycles, and passengers.

SILVER CITY
K
SILVER CITY
303 AMG
SILVER
CITY

Adapted designs

Developing countries lacked high-quality road networks, and their locally-built cars, often based on Western designs, had to cope with arduous conditions.

△ **The Hindustan Ambassador**
The Ambassador is still a feature in many Indian cities, where it serves as a reliable taxi.

Iran had no motor industry of its own until 1966. Then factories began producing a version of UK auto manufacturer the Rootes Group's Hillman Hunter model, using "completely knocked-down" (CKD) kits imported from the UK – essentially, mass-produced kit cars. This car was called the Paykan, which is Persian for "arrow" (the Hunter's development codename was Arrow) – or less formally the "Persian chariot".

In the 1970s, the Hunter was discontinued in the UK, and the production equipment was sold to Iran, so all the parts needed to build the Paykan could be made locally. After the problems Rootes had with the advanced Hillman Imp, it had designed the Hunter with relatively conservative engineering – and that made it a tough and reliable machine, which was ideally suited to the rough roads and patchy maintenance it faced in Iran.

Iranian icon

The "Persian chariot" proved to be very popular, with up to 120,000 built every year at the factory in Tehran. Design updates freshened the exterior and interior, a pickup variant was developed, and the original 1.7-litre Rootes engine was replaced by a more modern 1.6-litre Peugeot unit. The Hunter was phased out in 1979, but the Paykan continued in production far longer. Its low price, simplicity, and toughness made it a favourite with Iran's private and business drivers, and almost half the cars in Iran were Paykans. The saloon staggered on until 2005, when it was replaced by the Peugeot 405-based

▽ **Ladas in 1972**
Manufactured at the Volga Automobile Plant (VAZ), the Lada's simple design, ease of repair, and capability in challenging driving conditions, such as snow, made it a favourite with export markets.

"... despicably **clumsy** and **bodger-built**."

L.J.K. SETRIGHT, DESCRIBING THE LADA IN *CAR MAGAZINE*, 1992

◁ **Paykan cars in modern Tehran, Iran**
The Paykan has achieved astonishing longevity in Iran, and can still be seen on the streets of Tehran today.

Samand, but the pickup continued in production until 2015. Third-party Iranian manufacturers were still making spare parts for the Paykan as recently as 2017.

Existing designs

Adopting Western car designs was a cost-effective way of introducing car manufacture to many countries. In India, the Hindustan Ambassador (see pp.286–87) was a Kolkata-built version of the British Morris Oxford Series III, produced from 1957 to 2014. The solidity of the 1950s Morris design was ideal in India, where cars were used on rough dirt roads and regularly overloaded with people and cargo. In the 1980s, Hindustan built the Contessa, based on the 1970s Vauxhall Victor FE, and there were Indian versions of other European cars including the Rover SD1 and Fiat 124. The latter – a compact four-door saloon with boxy styling that was awarded Car of the Year in 1967 – was also adapted to Russian conditions and built as the VAZ 2101. During the Soviet era, private car use was rare. Trains were the main method of moving people and goods around the country, so there was little investment in the road network, resulting in very poor roads in many areas. Consequently, the ride height of the car was increased to cope with rough surfaces, the bodywork was built from thicker steel, and the rear disc brakes were replaced by drums. To cope with the unfavourable conditions of the harsh Siberian winters, a starting handle was provided and there was also a manual auxiliary fuel pump.

After the Fiat 124 was replaced by the 131, the Russian version was sold in Europe as an economy car – the Lada. A revamp in 1979 introduced revised styling and new engines, and it became a good source of income for the Soviet government, including being used in barter arrangements, most notably in exchange for large quantities of Coca Cola. The Lada was exported worldwide to a wide range of countries including Brazil, New Zealand, Canada, Finland, and Sweden. It continued in production well into the new millennium – the final ones were made in 2012, but even then it lived on in Egypt, where it was a popular taxi.

KEY DEVELOPMENT

The rise and fall of China's bicycles

For much of the 20th century, most of the population of China travelled by bicycle. More than 25 million Flying Pigeon bicycles were built every year and it became the most popular vehicle on the planet, costing the equivalent of two months' average salary and with a waiting list stretching into years. However, the reform of China's economy came with an increasing demand for cars and a sharp fall in the use of bicycles – in Beijing, 63 per cent of commuters cycled in 1980, but by 2017 the number was reduced to less than 12 per cent.

***COMMUTERS IN GUANGZHOU*, CHINA, PUSH THEIR BICYCLES THROUGH THE STREETS IN 1979.**

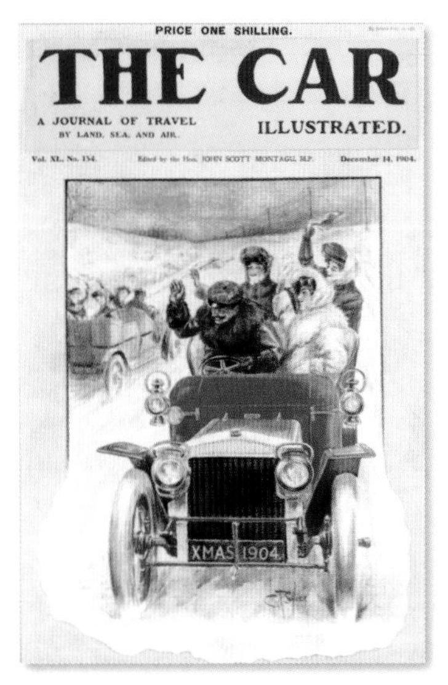

THE CAR ILLUSTRATED, 1904

L'ILLUSTRATION, 1928

THE AUTOCAR, 1928

OMNIA-SALON, 1930

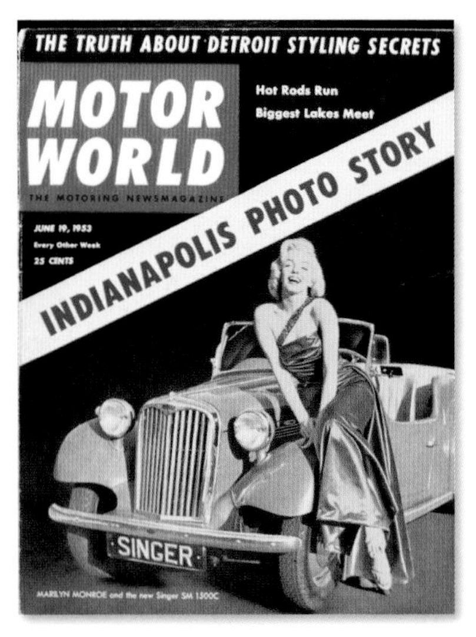

MOTOR WORLD, 1953

THE MOTOR, 1955

DAS STERNCHEN, 1955

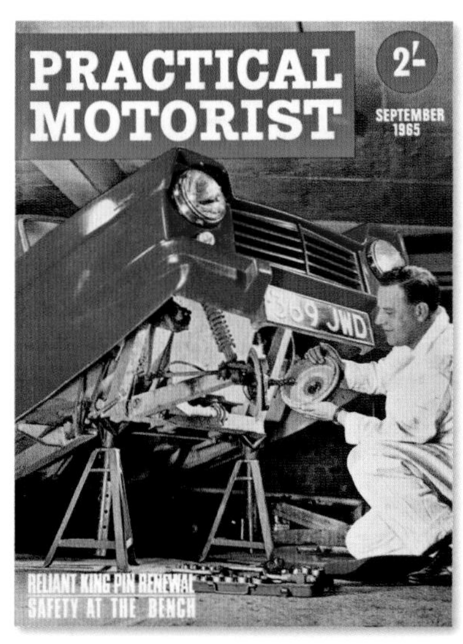

PRACTICAL MOTORIST, 1965

Speed reads

Most newsagents or station bookstalls have a section devoted to car-related titles, and thousands have been published across the world.

November 1895 saw the publication of the first two magazines on cars and motoring. In the US, *The Horseless Age* showcased the transition from horse-drawn vehicles to the internal combustion engine. In the UK, *The Autocar* focused solely on "the mechanically propelled road carriage". Both magazines – under their current names, *Automotive Industries* and *Autocar* – are still going strong.

Indeed, almost every nation with a car industry has produced at least one magazine that is highly respected by readers and the global car industry, including *Auto Motor Und Sport* in Germany; *Car and Driver* in the US; and *Car Graphic* in Japan. Alongside the latest car and industry news, most of these publications conduct road tests in forensic detail – to verify manufacturers' performance claims – and report on motor racing. Some titles specialize in car design, others in classic cars, car maintenance, buying advice, or modification. Even in the age of the Internet, car magazines continue to prove popular with readers, with new titles constantly being launched.

L'AUTO-JOURNAL, 1983

MOTOR, 1935

SPEED, 1936

MOTOR SCHAU, 1938

MOTOR UND SPORT, 1939

QUATTRORUOTE, 1967

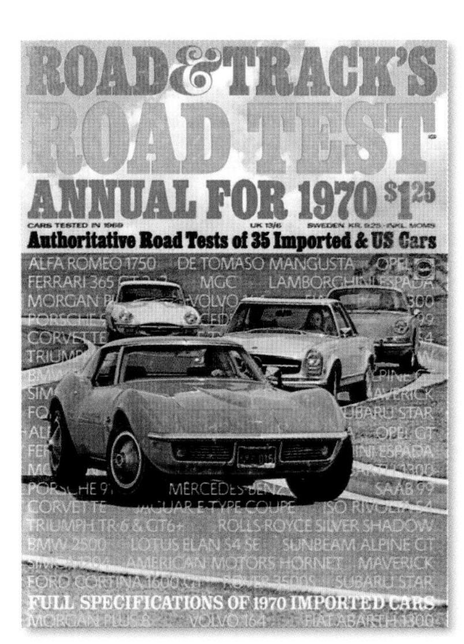

ROAD & TRACK, 1969

DAILY EXPRESS MOTOR SHOW REVIEW, 1970

KEY DEVELOPMENT
Haynes repair manuals

The official workshop manual that carmakers produce for each model is intended for professional mechanics. In 1965, however, independent publisher John Haynes devised one that ordinary owners could follow when working in their garage at home. He took the car apart (the first was an Austin-Healey Sprite) and then rebuilt it, photographing and noting what was required so it was easy to understand. Haynes Manuals became a huge success in Europe and the US in the 1970s and '80s, at a time when DIY repairs were still relatively easy. They are still on sale today.

***HAYNES MANUALS* OFFER CLEAR ADVICE FOR THE AMATEUR MECHANIC.**

Viva Fiesta!

The Fiesta has been called one of the motoring industry's greatest success stories: a Spanish-made supermini that achieved cult status during the 1980s and became the bestseller in its class in the UK and Germany.

The car that would become a household name began as Project Bobcat in the boardroom of Ford of Europe in 1972. The idea for a compact, front-wheel-drive hatchback had been floating around for a few years – it was conceived as a "supermini", adding versatility and comfort to the safe, enjoyable handling of the famous Mini itself. Now it was firmly on the drawing board, with the support of chairman Henry Ford II. In 1973, the oil crisis added urgency to the production of the new, fuel-efficient model, and by 1974 it was ready to go into production in Spain, where Mr Ford had been cultivating ties with government and industry. It seemed appropriate to give the new model a Spanish name; the European directors offered up Bravo, but the chairman chose Fiesta.

Fresh off the production line at the purpose-built Ford plant in Valencia, the one-litre Fiesta was launched in Germany and France in 1976, and in the UK in 1977. It quickly made an impact, but it was not until 1981, with the release of the sporty 1.6-litre, 100-mph (161-km/h) XR2, that the Fiesta developed a sporty persona. The transition to a more fuel-efficient version in 1983 with the Mk2 strengthened the Fiesta's appeal, especially after the introduction of the popular 1.6-litre diesel model.

The Fiesta had been withdrawn from the US as early as 1980 due to poor sales (although it was later reintroduced in 2010), but its European fan base continued to grow throughout the 1990s and into the 21st century as one new generation slowly followed another. Why has the Fiesta been so enduringly popular? The secret, if there is one, is in its mix of value, lively performance, thrift, and most of all its versatility. Back in 1976, it was the first car of its kind to offer front-wheel drive, a transverse engine, folding seats, and a hatchback that opened down to sill level, making it very easy to load cargo. In setting the template for the small hatchback it also became the long-term class leader.

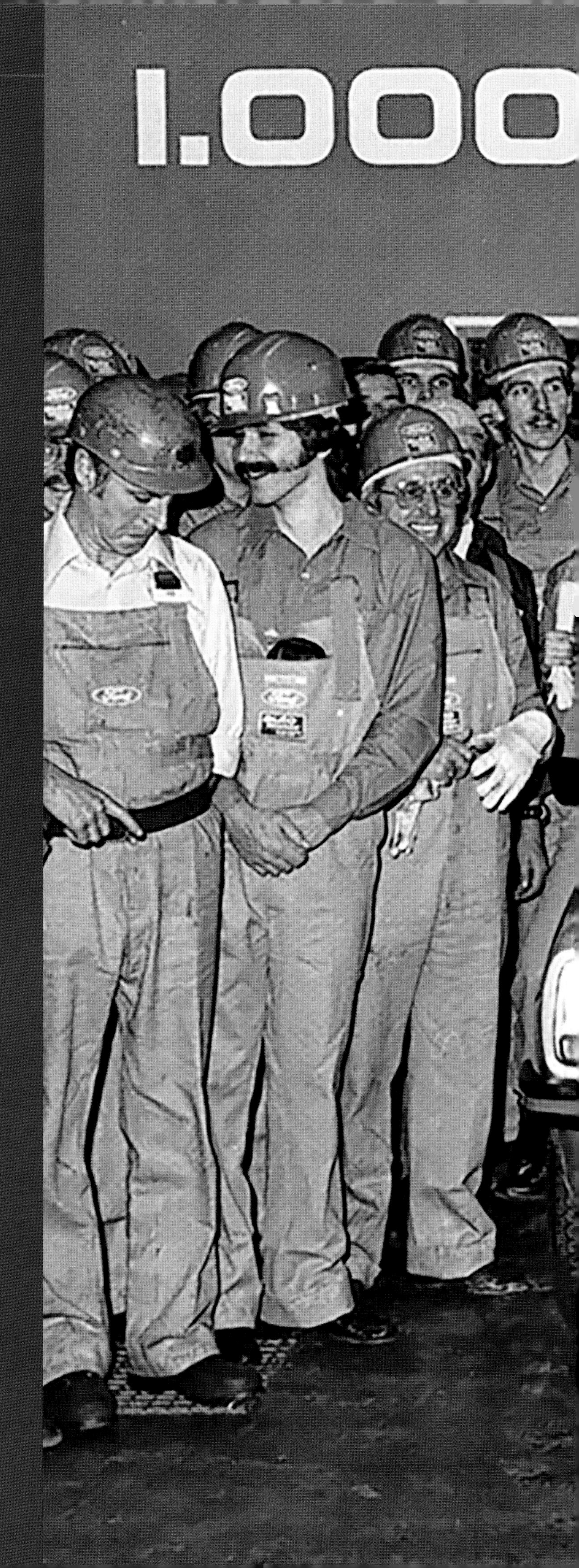

▷ **Ford Fiesta milestone, 1979**
Factory workers celebrate the construction of the one-millionth Fiesta Mk1 at the Ford plant in Valencia, Spain. Now in its seventh generation (Mk8 in the UK), a further 15 million models have since been sold worldwide.

000 FIESTA
Ford
FIESTA

Fuel crisis drives change

Changes to the environment forced carmakers to think much harder about fuel consumption. Smaller, more efficient cars soared in demand as governments legislated to reduce pollution.

△ **Fiat 500D, 1960**
The 500 series was Fiat's response to the need for smaller, more efficient, less fuel-reliant cars. It was the ideal vehicle for space-pressed city-dwellers, and more than 4 million were produced between 1957 and 1974.

Miles per gallon? As priorities go, fuel economy did not feature highly on most consumers' lists for most of the 20th century. Oil was gushing from the ground and fuel prices were low. Hardly any car companies bothered to advertise fuel economy figures. Only the occasional oil company would organise "economy runs" to boast how thrifty a car could be using its fuel. But then in 1956, the Suez oil crisis hit and fuel supplies to Europe were suddenly strangled. Car companies rushed to build frugal new cars, resulting in a boom for "bubble" cars (see pp.180–181) and the development of ultra-compact cars, such as the Austin Mini.

But when the oil taps turned back on, fuel consumption eased back out of the public's mind. It would take the 1973 oil crisis to make more permanent changes, and this time the impact was felt as much in the US as in Europe. An oil embargo was imposed by the oil-rich countries of the Middle East in response to the US's support for Israel during the Yom Kippur War. The embargo only lasted five months, but during that time the price of a barrel of oil soared four-fold. Suddenly, how far you could go on a single tank was vitally important.

▽ **Panic buying**
Motorists rush to fill up at a petrol station in Berlin, Germany, at the height of the oil crisis of 1973. Shortages were caused by an embargo imposed by various Middle Eastern oil producers.

◁ **Smog engulfs Los Angeles, 1975**
Due to its geography and enormous number of cars, Los Angeles was beset by smog in the 1970s. The phenomenon forced the car industry to produce cleaner, less air-polluting engines.

In the US, the enthusiastic pursuit of horsepower during the 1960s led to a generation of so-called "muscle cars". These big, heavy coupés with immense V8 engines had power outputs up to and beyond 500 bhp. It was common to see single-figure fuel consumption figures; indeed, in 1973 the average American car's fuel economy was just 16.1 miles per gallon (5.5 km per litre).

Faced by the 1973 oil crisis and soaring prices, consumers suddenly demanded fuel-efficient cars. American manufacturers with their gas guzzlers were severely caught out, and economical imports from Europe and Japan cleaned up – not only the competition, but also their engines.

The law got involved, too. In the US, the government introduced Corporate Average Fuel Economy (CAFE) regulations in 1975. These insisted that the average fuel consumption of a manufacturer's range should not exceed a stipulated figure. This figure became more stringent year on year, from 15 mpg in 1978, to 18 mpg in 1981, and by 1990 to 23 mpg, where it stayed right up until 2011.

Inefficient cars were hit by new taxes. In the US, the Energy Tax Act of 1978 – the so-called "gas guzzler tax" – penalized new vehicles whose fuel economy failed to meet certain levels. In 1980, for instance, $200 was levied on any new car averaging 11–12 mpg, and $550 if below 10 mpg.

Mission against emissions

It was not just what engines consumed that caused problems; it was what they emitted, too. California led the charge on tackling noxious fumes from vehicles because Los Angeles was being engulfed in smog on a daily basis. Not only was LA the most car-dependent city in the world, its geographical location meant that pollution did not dissipate. Huge clouds of smog clung to the city in a toxic shroud.

The first effort to control car pollution came as early as 1960, when California required all new cars sold there to be fitted with positive crankcase ventilation (PCV), which drew unburned hydrocarbons back into the engine's intake to be burned.

Next to be targeted were exhaust emissions. Again, California led the way for the 1966 model year, with the first emission test cycle limiting particulate emissions. Gradually this rolled out across the entire US. Manufacturers had to detune engines to meet the standards, which starved them of power. The answer to reducing emissions without strangling power was the exhaust catalytic converter, which was introduced for the 1975 model year in the US.

It all worked. In Los Angeles, volatile organic compounds declined by a factor of 50 between 1962 and 2012, while pollutants, such as nitrogen oxides and ozone, declined by 80 per cent during the same period. Other countries soon adopted similar regimes; Japan, for instance, began offering tax breaks for low-emissions cars in 1973.

> "The **smog** was **heavy**, my eyes were **weeping** from it."
>
> JACK KEROUAC, *THE DHARMA BUMS*

DRIVING TECHNOLOGY

Catalyst for change

After an engine has burnt its fuel, any residue, including pollutants, is emitted via the exhaust. French inventor Eugene Houdry discovered that using a catalyst in the exhaust system could change these pollutants into less harmful ones by chemical reaction, converting carbon monoxide (CO) and unburned hydrocarbons (HC) into the less harmful compounds, carbon dioxide (CO_2) and water (H_2O).

Houdry patented a device for car exhausts in the 1950s, but it was not until 1975 that catalytic converters (or "cats") started to be widely fitted to cars. The US led, followed by Germany and Sweden in 1985. By 1992, it was compulsory for all new cars sold in the EU to have a cat fitted.

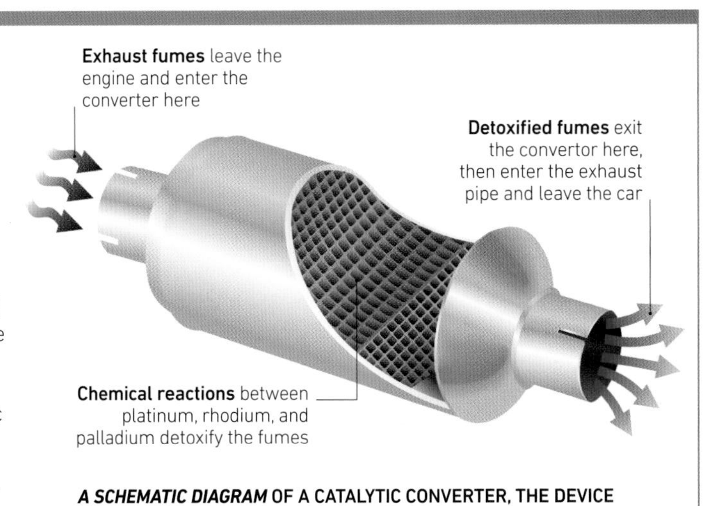

A SCHEMATIC DIAGRAM **OF A CATALYTIC CONVERTER, THE DEVICE THAT MAKES EXHAUST FUMES LESS TOXIC.**

Making a monster

The car-crushing exploits of giant-wheeled monster trucks and cars are an all-American stadium spectacle that has become popular worldwide. However, it all started by accident.

The monster truck and car craze, which has endured since the 1980s, happened thanks to 4x4 specialist Bob Chandler. After a motorcycle accident ended his construction career, Chandler opened a pickup truck repair and accessories business. He himself drove a 1974 Ford F250 pickup to 4x4 events and on camping trips. When its axles broke, Chandler fitted bigger and stronger ones, followed by ever-larger wheels and a more powerful engine. By 1979, it had a jacked-up stance and four-wheel steering, an idea inspired by the military.

As a joke, Chandler videoed the truck driving over two wrecked cars, and showed the tape in his shop. A promoter who saw it persuaded Chandler to repeat the stunt at a vehicle show in Denver, Colorado. The truck, by now christened "Bigfoot", became a massive draw, and it was soon appearing all over the US. Chandler produced more Bigfoot trucks, often fitted with 66-in (168-cm) wheels and massively powerful engines, which performed spectacular stunts to adoring crowds – in 1983, 68,000 people saw Bigfoot at one venue. Chandler eagerly took bookings, because he thought the craze could not possibly last. As it turned out, it did, and other people began making similar vehicles, which soon became known as monster trucks, along with car versions.

△ **The car-crush that started it all**
This still from the original home video shows the souped-up truck crushing cars for the very first time. Subsequent versions of Bigfoot featured even bigger tyres than these.

▷ **Bigfoot-mania**
By the 1980s, Bigfoot was a household name in the US – here it crushes three Toyotas and a Saab in 1985. Bigfoot still tours, and Chandler has built an electric version.

Ford
460
POWERED
BIGFOOT
Motorcraft
GOODYEAR

Cars of the silver screen

Cars have featured in films ever since the motor and movie industries began in the 1890s. However, the years 1960 to 1980 were a golden era of cinematic driving action.

◁ ***Bullitt* Ford Mustang GT, 1968** Starring Steve McQueen as a Mustang-driving police detective, the epic car chase scene in the film set the benchmark for high-octane action in movies for decades.

Cars have had many famous roles in films, usually as police, race, or criminals' getaway cars. But by 1960, as television became popular across the globe, the car also became a leading fixture of the small screen.

▽ ***The Italian Job* Mini Coopers, 1969** Although not suited to carrying stolen gold bullion, due to its small size, the Mini Cooper was ideal for fleeing through the confines of Turin.

Style icons

In the mid-1960s, no car was hotter – or cooler – than James Bond's Aston Martin DB5. The sleek silver GT was delivered to Sean Connery's 007 fitted with twin machine-guns, oil squirters, tyre slashers, a smoke-screen device, and an ejector passenger seat that disposed of thugs through the roof when Bond pressed a button hidden under the gear knob. The DB5, a product placement coup, may have saved Aston Martin from financial collapse. Similarly, the white Volvo P1800 driven by Roger Moore in 1962–69 television series *The Saint* not only made that car a must-have, it helped to establish Volvo in the US.

Some of the most famous cars on the small screen were not production cars. The Batmobile from the 1966–68 US television show was made from a 1955 Ford concept car, the Lincoln Futura, by Hollywood customizer George Barris. The Monkees drove a radically altered Pontiac GTO, the Monkeemobile (see p.219), a Dean Jeffries creation. Other TV cars – the Black Beauty from US series *The Green Hornet*, a 1966 Chrysler Imperial; the Lotus 7 in *The Prisoner*; and the flashy Ford Gran Torino in *Starsky and Hutch* – were nearly as important to their shows as the human stars.

Cars as stars

As both movie and automobile technology advanced into the 1970s, the role of cars in films expanded. While the DB5 and later Astons became a staple of 007 films, other cars had one-off starring roles in movies and became legends. The Alfa Romeo Duetto Spider in the 1967 film *The Graduate* was so emblematic that in 1985 Alfa Romeo produced a special low-cost version called "The Graduate". And the gymnastics pulled off by Mini Coopers in the 1969 crime caper *The Italian Job*, wherein thieves pull off a gold heist in Turin, breathed new sales life into the then 10-year-old car. A few cars even landed the lead role in movies, such as the Volkswagen Beetle in the Disney movie *The Love Bug* and the flying vintage-era car in *Chitty Chitty Bang Bang*, both released in 1968.

KEY EVENTS

- **1960** *Route 66* is the first US TV show to star a car, a Chevrolet Corvette.
- **1962** James Bond debuts in *Dr. No* driving a blue Sunbeam Alpine. In 1963, in *From Russia with Love*, Bond trades up to a Bentley Mark IV.
- **1964** In *Goldfinger*, 007's Aston Martin DB5 gains global exposure.
- **1966** The Batmobile rockets into the imaginations of children around the globe.
- **1969** Mini Coopers steal the show (and the gold) in *The Italian Job*, starring Michael Caine.
- **1960-70s** Land Rovers show their off-road ruggedness in just about every film featuring African, Australian, or South American jungles.
- **1977** Bond is back in *The Spy Who Loved Me*, driving a Lotus on the road and in the ocean.
- **1979** Mel Gibson's *Mad Max* and a screaming supercharged Ford Falcon XB GT put the Australian film industry on the global map.
- **1980** *The Blues Brothers* features a clapped-out Dodge Monaco police car, a red Jaguar E-type, Carrie Fisher, and a great soundtrack.

***THE GENERAL LEE* IN TV's *THE DUKES OF HAZZARD* WAS A CUSTOMIZED 1968 DODGE CHARGER.**

△ **The original "Bondmobile" Aston Martin DB5**, as seen in the spectacularly successful James Bond film *Goldfinger*, starring Sean Connery (inset), in 1964.

△ **Hyundai Pony, 1975**
Hyundai was not merely a new marque, it was South Korea's first independent step into the global car industry. With a Mitsubishi engine from Japan, Italdesign styling from Italy, and manufacturing expertise from the UK, the humble Pony was affordable and well built, boding well for Hyundai's massive future growth.

Rise of the hot hatch

Taking the small family hatchback and giving it plenty of power was an inevitable development, and one that appealed to drivers everywhere. The "hot hatch," as it became known, has been popular ever since.

△ **Talbot Sunbeam Lotus rally car**
Built on a shortened rear-wheel-drive Avenger platform, the Talbot Sunbeam had a powerful Lotus engine that made it ideal for rally driving.

Abarth Fiats and Mini Coopers from the British Motor Corporation (BMC) showed the world just how good a small family car can be when fitted with a potent engine. It was just a matter of time before hatchbacks, an increasingly popular genre of family car, were given more powerful engines.

Early hot hatches included the Autobianchi A112 Abarth, Renault 5 Alpine, and Simca 1100 Ti, but it was the Volkswagen Golf GTi of 1976 that truly launched the class. With its 110-bhp, fuel-injected, 1.6-litre engine, taut suspension, wide wheels, and purposeful livery with matt-black trim, the Golf GTi was seen as a game-changer, and captured the public's imagination. By 1983, 25 per cent of all Golfs in the UK were GTis.

Coming of age

British car manufacturing adopted a different approach to the hot hatch, its original offerings being rear-wheel-drives. Both the Talbot Sunbeam Lotus and the Vauxhall Chevette 2300HS acquired a reputation for being a handful, and indeed, being developed for rallying, they were cars for dedicated drivers. More conventional mid-sized sporting cars, such as the Morris Marina TC and the Vauxhall Firenza, were outclassed by foreign imports, which served only to boost the popularity of the ubiquitous Golf.

By the early 1980s, the hot hatch had found its niche as a front-wheel-drive, fuel-injected derivative of a standard family car. The Ford Escort XR3i, Peugeot 205 GTi, Vauxhall Astra GT/E, and MG Maestro EFi were typical of the genre. They were fast, practical, and easy to maintain. Such was the popularity of the hot hatch that it all but killed off the traditional open-top sports car.

KEY EVENTS

- **1971** The first hot hatch, the Autobianchi A112 Abarth, is introduced by Fiat. It is prepared by the company's motor sports division.
- **1974** The 1.3-litre Simca 1100 Ti goes on sale in France. Although not the fastest small sporting family car, it is the first true hot hatch, courtesy of its power output and fifth door.
- **1976** The VW Golf GTi, arguably the car that defined the genre, enters production in Germany. A GTi version of the Golf has remained on sale ever since.
- **1976** A high-performance Renault 5 is produced in France, where it is sold as the 5 Alpine – and as the 5 Gordini in the UK.
- **1979** The Talbot Sunbeam Lotus, the UK's idea of a hot hatch, hits the market. It is even more powerful than its rivals. Its 150-bhp Lotus engine ensures that it is quick – and it still is, even by modern standards.
- **1983** France's Peugeot 205 GTi, the car that made Peugeot into a revered manufacturer of hot hatches, is launched. Small, with a 130-bhp, 1.6-litre engine, it is one of the most sparkling performers of its age.

***THE PEUGEOT 205 GTI*, THE DEFINITIVE HOT HATCH OF THE 1980s.**

◁ **Vauxhall Chevette 2300HS, 1978**
The Chevette was intended as a mini Chevrolet. In the mid-1970s it was the UK's bestselling hatchback, and proved a successful rally car.

△ **The Volkswagen Golf MkII GTi** of 1983. The Golf is now in its seventh generation.

1981–2000

THE CHANGING WORLD

1981–2000

The changing world

In the early 1980s, the automotive industry had yet to fully enter the digital age. However, the typical car was about to undergo a technical revolution that would enable it to meet ever higher customer expectations.

Car meets computer

Car factories were becoming increasingly automated, which made "Friday afternoon" cars (those supposedly made by workers whose minds were already on the weekend, and so with extra faults) a thing of the past. Under the bonnet, computers and sophisticated electronics ensured that cars started more easily, maintained optimum performance, and used less fuel. Once linked to catalytic converters, the electronic "brain" eventually found in even the cheapest models also dramatically reduced emissions.

The downside to all this was the steep decline in home maintenance. Amateurs increasingly found the under-bonnet world simply too complex to contemplate. Did carmakers take advantage of this to dictate service intervals and costs? Many customers certainly thought so.

From racetrack to suburbia

The 1980s also saw advances in other areas of automotive design. By the end of the decade, supercars could top 200 mph (322 km/h) where allowed, their aerodynamics and turbocharging drawing heavily on developments in Formula 1, Indycar racing, and Group B rallying.

A new genie was out of the bottle in the US and Europe by 1984, as the versatile multi-passenger vehicle, or MPV, became the transport of choice for the big, busy family.

NEW TECHNOLOGY IS SET TO TRANSFORM CAR MANUFACTURING

ADVANCES IN RACING CAR DESIGN FILTER DOWN TO EVERYDAY CARS

"...there was almost **no such thing** as a **bad car** anymore."

For more adventurous spirits, the SUV and the pickup truck, with their off-road ability, made deep inroads into suburbia – an environment in which they were not strictly needed but conveyed a beefy image that many drivers found irresistible.

Global challenges

Overcrowding was also affecting car design. In Tokyo, where it was even hard to park overnight, a new breed of tiny, narrow vehicle, the Japanese *kei* car, thrived in response to the lack of space. Japan also kick-started the retro car movement in reaction to the blandness of so many mainstream models. This trend eventually spread to Europe, where re-imagined versions of the Volkswagen Beetle and Mini proved very popular. Globally, the "green car" also became a popular concept, largely due to the rise in profile of the green lobby, which sought to dispel the notion that the world simply had to accept more roads – with all the noise, filth, monster traffic jams, and even gridlock that came with them.

Meanwhile, US car manufacturers were having a torrid time, with many big-selling models still not shifting enough to stave off financial problems. Perhaps the biggest surprise of the time came when the Cold War ended in 1990, and the division between Western Europe and the Soviet bloc crumbled away. The world discovered how people had been living in the former Communist countries of Eastern Europe. Westerners saw how the lowly Trabant – which was made from indstrial waste rather than steel – had been keeping people mobile in East Germany. Such second-rate products would soon fade away, until there was almost no such thing as a bad car anymore.

FOUR-WHEEL DRIVES BECOME THE FAMILY CAR OF CHOICE

ENVIRONMENTAL CONCERNS TURN POPULAR OPINION AGAINST THE CAR

Computers take control

In the 1980s a new generation of in-car technology was born as computers and increasingly sophisticated electronics made their mark. The simple internal combustion engine was about to enter a new era of improved performance and efficiency.

▷ **Electronic Control Unit (ECU)**
Often referred to as the "car computer", the ECU is the brain of the engine management system. It regulates the fuel mixture, ignition timing, and variable cam timing, and controls the car's emissions.

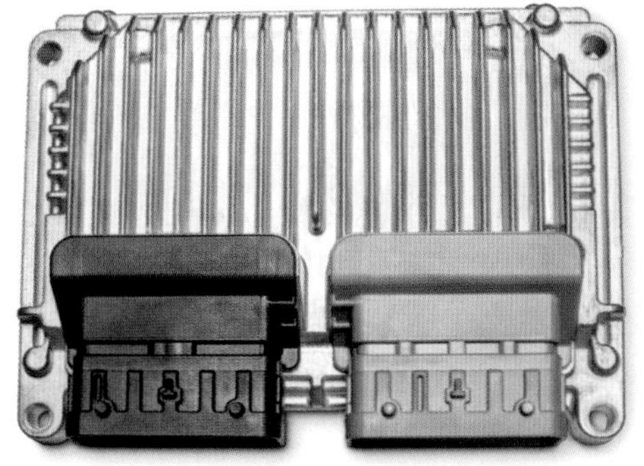

One of the most important yet seldom seen developments in popular cars has been the creeping influence of computers on the way they perform.

Electronic control units, or ECUs, have been widely fitted to cars since around 1979. At first, their main purpose was to control engine emissions in the face of ever-tougher anti-smog laws in the US (see pp.244–45). By 1981, General Motors had standardized them across all its ranges. On models such as the mostly-forgotten Buick Century Turbo Coupe they helped to create a package that offered both an energetic response for the driver and consistently less pollution.

Computers meet cars

The "solid state" revolution had begun in earnest in 1968. This was the year when the first in-car computer was fitted to the Volkswagen 1600TL. The unit was allied to the car's Bosch electronic fuel-injection, and was intended to optimize its performance for consistent power delivery.

Digital instrumentation was still something associated with on-screen sci-fi fantasy when, in 1976, Aston Martin unveiled the first LED (light-emitting diode) instrument display in its Lagonda. However, it was not until 1979 that the Lagonda went on sale, by which time Cadillac was already starting to offer the first "trip computer" in the dashboard of its Seville. This, and similar gadgets found in European cars such as the Talbot Horizon, were little more than gimmicks at the outset; digital clocks that could also calculate your fuel consumption figure and

"The **Audi Quattro**... was a **trendsetter** in **every way**."

AUDI QUATTRO, GRAHAM ROBSON

◁ **Aston Martin Lagonda, 1976**
Featuring futuristic styling, the Lagonda was the first car in the world to be equipped with an LED instrument display on its dashboard.

deliver other – mostly useless – data. However, advances were rapid, and by 1987 Oldsmobile was offering the first digital head-up display. The very first GPS satellite-navigation screen appeared in one of Mazda's Eunos cars just three years later.

Audi's game-changer

In 1980, Audi unleashed a car that was among the first that could be described as truly technology-packed – the Quattro. It featured a turbocharged five-cylinder engine and four-wheel drive, and was the safest high-performance car in the world. Just a year later, anti-lock brakes were added, and the car soon boasted a green, glowing LED dashboard display too, which ran off the still-basic electronic "brain" that regulated the car's ignition and fuel-injection. Four-wheel drive with anti-lock brakes had been seen before in a fast car, the 1966 Jensen FF, but it was the computer-driven inputs that enabled the features to function so well together in the Quattro.

As fuel injection (see box, right) became increasingly widespread – wiping out the carburettor once and for all as a means of fuel delivery to the combustion chambers of engines – more and more cars used ECUs to control their systems. Today's ECUs also oversee braking, anti-theft measures, traction control, and variable valve timing. Such is the complexity of modern car computers that engine inputs can now be processed and adapted to in real time, and ECUs can interface with other parts of the car such as the lights. However, controlling the engine remains the most processor-heavy job on the vehicle.

KEY DEVELOPMENT

Fuel injection

Fuel injection is a fuel supply system that replaces a carburettor and is now universal in new cars. Petrol is pumped electrically from the tank and sprayed straight into the engine's inlet ports, where it mixes with air before being burnt in the cylinder. On diesel and direct-injection petrol engines, fuel is injected into the cylinder rather than the inlet port. Direct injection was popularized in the 1950s in Germany by Goliath and Mercedes-Benz. Lucas developed a similar system, used in the UK by Jaguar and Triumph, among others. But it was the inception of electronic fuel injection courtesy of Bosch with its D-Jetronic that put the technology on the map.

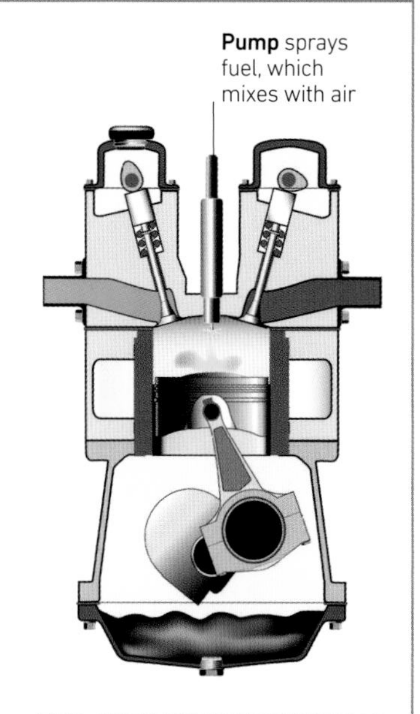

FUEL-INJECTION SYSTEMS **SUCH AS THIS ARE STANDARD IN NEW CARS AND CREATE FEWER EMISSIONS THAN CARBURETTORS.**

The 1980s was characterized by an indiscriminate technology rush throughout the industry, with turbochargers, four-wheel drive, and electronic appellations attached to even the humblest models. However, some innovations, such as electronically-controlled four-wheel steering, showed potential but later were quietly dropped as they offered negligible benefits in return for increased engineering complexity.

◁ **Audi Sport Quattro, 1980s**
With its turbo-charged five-cylinder engine, four-wheel drive, and anti-lock brakes, the Quattro was both ultra-safe as a road car, and dominant in world rallying.

A new class of car

Today's multi-passenger vehicles are the culmination of ideas that were tried by different carmakers for nearly half a century. After decades of tinkering, these finally came together in 1984.

It was in 1984 that Renault released the Espace in Europe, and Chrysler introduced the Dodge Caravan and Plymouth Voyager to the US. These were the first modern minivans, or multi-passenger vehicles (MPVs), but their defining elements – front-wheel drive, sliding side doors, adjustable seats, flat floor, and unitized body – had long since existed, scattered among the designs of various compact vans.

Renault and Chrysler were simply the first to put these together in one vehicle. Previous attempts at MPVs, such as the DKW Schnellaster, the Volkswagen Transporter, and the Fiat 600 Multipla (all of which had their origins in the 1940s and '50s) were about the size of modern MPVs, but they lacked today's spacious layouts and user-friendly features. Many were either delivery-van-based, which gave a harsh ride, or had

△ **DKW F89L Schnellaster**
With front-wheel drive and a one-box profile, the DKW F89L Schnellaster is the true ancestor of today's MPVs.

forward control, which obliged the driver and front passenger to climb in and sit over the front axle. The Schnellaster (meaning "rapid transporter") had a transverse-mounted engine and front-wheel drive, and so came close to today's

△ **Espace seating accommodation, 1980s**
The Renault Espace was one of the very first multi-passenger vehicles. It took a total of seven passengers, all of whose seats could be moved; the front two could be swivelled to face the rear.

designs. However, it had an old-fashioned body-on-frame construction (a separate body mounted on a chassis), and hinged, rather than sliding, doors.

Coming of age

In 1981, Nissan introduced the Prairie (sold in the US as the Stanza Wagon). With its sliding doors on both sides, foldable rear seat, and tailgate that opened outward, it almost got the minivan formula right. But there was one problem: the Prairie was based on the Nissan Sunny, which was one size too small for families. It had become clear that minivans should have three rows of seats, be easy to enter and exit, have adaptable cargo space, and be small enough to fit in the average garage, and yet perform like a car. The size issue cut the Prairie's life short, and the Stanza Wagon never really caught on in the US.

The Chrysler minivans, based on a version of the K-Car platform, were an instant hit, not only in the US, but also in Europe. A Dodge Caravan could be ordered nearly any way a buyer wanted it, with either a short or long wheelbase, a 5-speed manual transmission, a turbocharged engine, and, later on, all-wheel drive.

By the late 1980s, the multi-passenger vehicle was the car of choice for millions of families, who could take everyone and nearly everything with them in one comfortable, easy-to-drive vehicle. In the US, through five generations, more than 11 million Dodge Caravan, Plymouth Voyager, and luxurious Chrysler Town & Country minivans were sold.

The market evolves

Since the original Renault and Chrysler MPVs, nearly every major carmaker has created their own versions. In the US, Toyota and Honda presented Chrysler's MPVs with their first real challenge in the form of their Sienna and Odyssey ranges. In Europe, joint ventures between Ford and VW, and Citroën, Peugeot, and Fiat brought a wealth of new rivals to the Espace. Following the success of the MPV, carmakers have sought to offer the same combination of practicality, comfort, and driveability in other classes of car. These include compact MPVs, such as the Renault Scénic and Citroën Xsara Picasso, and urban SUVs such as the Jeep Cherokee and BMW X3. These in turn have spawned a new class of car, the "crossover" (see pp.296–97).

KEY DEVELOPMENT

Stout Scarab

Fifty years before the Renault Espace, Dodge Caravan, and Nissan Prairie arrived, William B. Stout, a Detroit automotive and aviation engineer, created the first minivan: the Stout Scarab. Styled by Dutch designer John Tjaarda, the Scarab had an aluminium spaceframe body, flat floor, a Ford V8 engine mounted over the rear axle, and four-wheel independent suspension.

The Scarab's interior featured seating that could be moved into a number of positions, and even a bed in the rear. The driver and passengers entered and exited from the door on each side. Offered at $5,000 and built to order, the Stout Scarab was one of the most expensive cars available during the Depression, costing the equivalent of $91,000 today. Only nine Scarabs were ever built.

A PROMOTIONAL PHOTOGRAPH **OF THE REAR-ENGINED, ALUMINIUM-HULLED STOUT SCARAB, 1935.**

◁ **Dodge Caravan**
The Dodge Caravan of 1986 had a hatchback and plenty of interior space. Its rear seats were accessed by a single sliding door.

△ **Mitsubishi Minica Dangan ZZ, 1989**
The Dangan (meaning bullet) was one of the most extreme *kei* cars ever. The world's first five-valve-per-cylinder production car, with a maximum of 9,000 rpm, it combined power, sporty styling, and fuel economy.

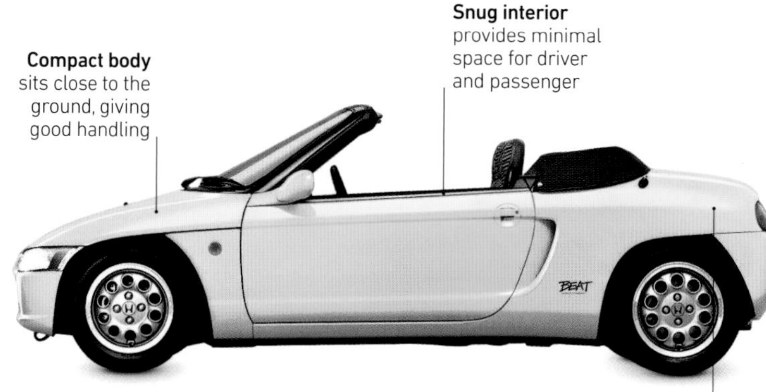

△ **Honda Beat ,1991**
From the same era as Honda's NSX, the Beat was also mid-engined and super specialized. With just two seats and tiny luggage space, the Beat formula was unlike anything else at the time.

Japanese *kei* cars

Japan has long had its own unique sector of mini-cars, known as *kei jidosha* (light cars). These first appeared in the mid-1950s as a new, inexpensive small car series, designed to help get a still-recovering Japan back on the road, following World War II. Since then, the *kei* sector has long moved past basic transport to include fun sports cars, small family cars, MPVs, hybrids, and more. They are strictly limited in terms of size, power, and top speed, but in return, owners get concessions on taxes and parking regulations. The earliest models could have engines no larger than 360 cc, although permitted body size and engine capacity has increased. Today, the limit is 11¼ ft (3.4 m) in length and a 660-cc engine. Some popular models have a boxy "tallboy" body shape, with four doors and a high roof, to provide extra cabin room within the restricted length. With their low running costs and compact size, *kei* cars are hugely popular in urban Japan, especially Tokyo.

▽ **Mitsubishi Minica**
Mitsubishi's Minica series dates back to 1962, when it first appeared with a two-stroke engine. This is the 1975 Minica F4 Super DX model, now equipped with a hatchback but still with two cylinders

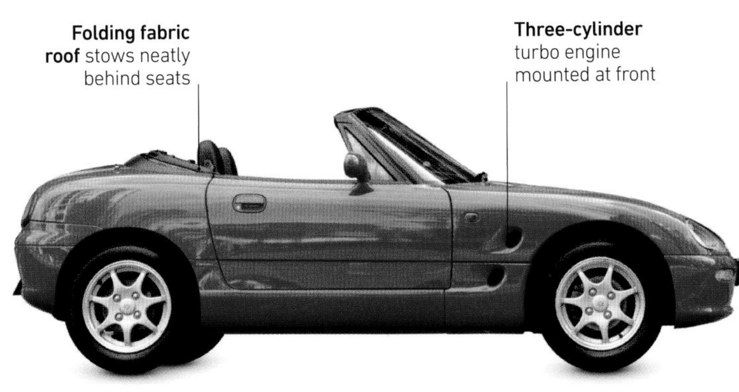

△ **Suzuki Cappuccino, 1991**
Suzuki made a decent stab at reinventing the classic British sports car with the 1991 Cappuccino: front engine, rear drive, and a lot of fun. It was one of the few *kei* cars to be officially exported to Europe.

△ **Suzuki Wagon R, 1993**
Suzuki created a new genre of tallboy *kei* car with the Wagon R, which first appeared in 1993. Through the highly rationalized design of the the Wagon R, Suzuki eked out the maximum cabin space possible.

△ **Suzuki Suzulight**
Launched in 1955, the Suzulight was one of the pioneering *kei* cars. Despite a tempting specification, including a 360-cc, air-cooled, twin-cylinder engine, only 43 were made.

"There is a **need for small**, practical cars that anyone **can afford**."

MICHIO SUZUKI, FOUNDER OF SUZUKI MOTOR CORPORATION

▽ Return of the US convertible
Open-top cars vanished from US manufacturers' ranges in the mid-1970s, as they feared lawmakers would ban them on safety grounds – but this never happened. By 1982, Chrysler was offering wind-in-the-hair driving again, with Cadillac joining the revival with its luxury two-seater Allanté (right) in 1987. The Allanté bodies were handcrafted in Italy by Pininfarina, giving them greater exclusivity, before being air-freighted to the US for assembly.

Safety before speed

As engineers improved the power and performance of their racing and rallying cars, concerns about safety intensified. Their spectacular, four-wheeled beasts had to be tamed.

Renault introduced turbocharged engines into Formula 1 in 1977, but it took several years for turbo cars to become reliable enough to be championship contenders. The technical complexity and sheer expense of turbo engines saw the demise of privately-funded entries to the top level of the sport. Even the small teams had multi-million-dollar budgets, while the bigger ones enjoyed extensive financial support from global carmakers.

The turbo engines brought a huge increase in power. Normally-aspirated cars, mostly Ford Cosworth-powered, raced with a reliable 450 bhp, but the turbo engines were quickly up to 600 bhp and soon producing well over 1,000 bhp in qualifying trim. The cars also had the latest developments in "ground effect" aerodynamics, using sliding skirts to seal an area of low-pressure air under the car to generate grip-inducing downforce. With more power and grip available, cornering speeds increased and lap times tumbled. However, in a few short years the cars were too fast for the circuits they raced on, and rules had to be changed to slow them down. The "flat bottom" regulations banned skirts in 1983, and at the end of the 1988 season turbo engines were consigned to history too.

In rallying, aerodynamics were not as effective in improving grip because the average speeds were lower than in F1. In their Quattro model, Audi had an alternative method of gaining grip using a sophisticated four-wheel drive system – and other manufacturers soon followed. Most works cars had turbocharged engines, and Lancia's Delta S4 had both a turbocharger (for high-rev boost) and a supercharger (for low-rev response), but MG bucked the trend with its Metro 6R4 by using a larger, normally-aspirated motor developed by the Williams F1 team.

The combination of these powerful engines and four-wheel drive in the Group B cars created an awesome spectacle (see box, right), but the cars had to be banned after a series

△ **Standard safety gear**
Mario Andretti wore this helmet and gloves during a testing day at the 1988 IndyCar races.

"If everything seems under control, **you're not going fast enough.**"

MARIO ANDRETTI, 1978 FORMULA 1 WORLD CHAMPION AND FOUR-TIME INDYCAR CHAMPION

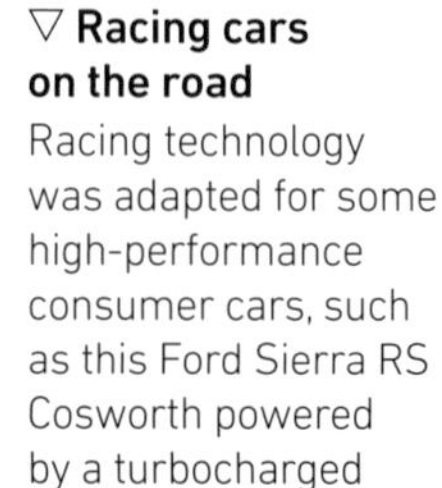

▽ **Racing cars on the road**
Racing technology was adapted for some high-performance consumer cars, such as this Ford Sierra RS Cosworth powered by a turbocharged 2-litre engine.

of high-profile crashes in the 1980s, while plans for the even more extreme Group S were shelved. Instead, rallying adopted Group A cars, which were not so powerful or fast, but captured public interest as they were clearly related to road-going performance cars. That connection was important on the tarmac circuits, too. Touring Car racing grew increasingly popular in Europe, Asia, and Australasia. Cars such as the Ford Sierra Cosworth, Mercedes-Benz 190E 2.3-16, and BMW M3 became as iconic as the Williams-Honda, McLaren-TAG, and Brabham-BMW in Formula 1.

KEY DEVELOPMENT
The "Killer Bs"

Massively powerful turbocharged engines, sophisticated four-wheel-drive systems, and lightweight construction using exotic materials all came together in the Group B rally cars of the 1980s. Their awe-inspiring speed drew huge crowds of spectators, who were sometimes more interested in getting a good view than in staying safe. Inevitably there were accidents, some of them fatal. The last straw came in the Tour de Corse Rally in 1986, when Henri Toivonen's Lancia Delta S4 crashed off the road and burst into flames, killing him and co-driver Sergio Cresto. Group B was swiftly banned.

MIKI BIASION SPEEDS **PAST SPECTATORS IN THE GROUP B LANCIA DELTA S4**

NASCAR racing series

In the US, saloon car racing, in the form of the NASCAR series, also became a huge spectator sport. The highlight was the Daytona 500, which attracted one of the biggest television audiences of any sport worldwide. Meanwhile, American open-wheel racing stumbled through splits, disagreements, and changes of organization, heading into the new millennium with two factions, CART/Champ Car and the Indy Racing League. But the Indianapolis 500 continued as US's biggest motor race, attracting more than 250,000 fans each year.

▽ **Ferrari in flames**
Stefan Johansson's "Ferrari Turbo" Ferrari 156/85 spits flames at the Monaco Grand Prix in Monte Carlo, 1985.

Europe reunited

The world changed in 1989 with the fall of the Berlin Wall, the Velvet Revolution in Czechoslovakia, and other events that signified the end of the Cold War. Drivers from East and West found themselves sharing roads – and comparing cars.

▷ **The fall of the Wall**
On the morning of 10 November, 1989, crowds tore down sections of the Berlin Wall. The revolutions of 1989 led to the collapse of Communism and great changes across Europe and the old East.

As the Iron Curtain fell, the dramatic difference between life in the former Soviet bloc and the West became all too apparent. This extended to the vehicles people drove. In the West, Soviet cars had long been seen as a joke – outdated and poorly assembled although essentially sturdy and cheap.

The East German Trabant was roundly mocked for its two-stroke engine and bodywork made from cotton waste and resin. However, it was not until the "Trabis" – as they were known – came over the border with the fall of the Berlin Wall that their real shortcomings were highlighted. In reunified Germany they shared the roads with larger, more powerful Audis and Mercedes – in a collision, there was no question of a Trabant avoiding damage. Additionally, the Trabant polluted quite heavily. It was not long before sales dwindled to the point that the Trabant factory in Mosel stayed open only through government subsidies, and it was sold to Volkswagen in 1991.

Lada and Škoda

Trabant was not the only company to struggle after the fall of the Wall. Although Lada sales remained strong, behind-the-scenes corruption and alleged involvement with Russian criminality made the company's future more precarious every day. By 1996, Lada's parent company AvtoVAZ was Russia's largest tax debtor and was forced into an agreement with General Motors after a government investigation.

The only Soviet car company to have survived and benefited from the fall of the Iron Curtain is Škoda. A joint venture programme with Volkswagen started in 1991 and led to a full takeover, ensuring the Czech firm had access to vehicle platforms and markets it could exploit. Škoda is still a budget brand – but one that is taken very seriously in the West.

◁ **Two worlds meet**
A West German Mercedes and an East German Trabant stand next to each other in Berlin. The Trabant seems small and old-fashioned in comparison with the more powerful West German car.

KEY EVENTS

- **1932** The Soviet Union and Ford form Gorkovsky Avtomobilny Zavod (GAZ).
- **1957** The first Trabant is made.
- **1959** The Škoda Felicia is imported into the US from Czechoslovakia.
- **1966–70** The Soviet government builds its largest-ever car manufacturing plant.
- **1970** The Lada 2101, a popular car of the Cold War era, is released.
- **1977** Zastava Automobiles in Yugoslavia begins production of the Yugo, its flagship car, under licence from Fiat.
- **1989** The Berlin Wall is torn down, signifying the end of the Cold War.
- **1991** Škoda transfers 30 per cent ownership of the company to the Volkswagen Group.
- **Late 1990s** Ladas are re-imported from the UK to Russia.
- **2001** AvtoVAZ and General Motors form joint venture GM-AvtoVAZ.
- **2012** The Lada Riva ends production.

ŠKODA'S ESTELLE **WAS THE UK'S CHEAPEST CAR IN 1976. THIS MODEL IS FROM 1980.**

△ **East Germans drive their Trabants** across to the West after the fall of the Berin Wall in 1989.

Quality
MEANS BUSINESS
BUSINESS
MEANS YOUR JOB

The decline of Detroit

By ignoring the rise in imports of small cars, carmakers in the US, notably the "Big Three" in Detroit – General Motors, Ford, and Chrysler – almost succeeded in destroying themselves.

By the late 1950s, Volkswagen's Beetle had sold in large numbers in the US, and in 1960 Renault sold 102,000 Dauphines. But Detroit refused to embrace small cars, focusing instead on increasing horsepower and such innovations as high-compression "Rocket" V8 engines, tailfins, turbochargers, and "Ram Air" bonnet scoops. In contrast, carmakers in Europe and Asia were adopting disc brakes, independent rear suspension, rack-and-pinion steering, five-speed manual transmissions, overhead cams, fuel injection, and other advanced technologies that left Detroit's existing cars obsolete.

Token efforts at small cars could not compete with the imports on any level either. Chevrolet's 1959–69 Corvair and 1971–77 Vega, for example, were poorly engineered and lacked the refinement and reliability of a Toyota Corolla, or the value and simplicity of a Beetle. The first credible modern small car to be designed, engineered, and built in the US was GM's Saturn, but it appeared in the autumn of 1990, a full 31 years after the British Motor Corporation's Mini had perfected the template for such cars.

The 1973 oil crisis exposed the depth of Detroit's troubles. Its large cars and trucks sat unsold, and factory workers were furloughed. The United Auto Workers union refused to change with the times, which led to the opening of non-union plants far away from Detroit where labour was cheaper.

Import brands began competing in Detroit's most profitable segments, too. High-quality Honda Accords and Lexus luxury cars were now natural choices for Americans who had grown up with foreign cars. In 2009, GM and Chrysler declared bankruptcy and Ford raised $23.6 billion to survive by mortgaging all of its corporate assets. However, more recently, restructuring, a surge in demand for trucks and SUVs, and cheap petrol have all helped return Detroit's carmakers to financial health.

◁ **On the production line in Detroit, 1982**
A sign above this Ford production line reads "Quality means business / business means your job"– a slogan that would become bitterly ironic as the US auto industry failed to move with the times.

Safer cars, cleaner air

Car design went through quiet-but-radical change during the 1990s with safety and emission control systems coming into widespread use for the first time. Cars were safer and cleaner – even if they were no longer home-mechanic-friendly.

△ **Pollution from unfiltered exhausts**
Children in Milan, Italy, cover their faces in a cloud of car exhaust fumes in 1973. This was the era in which the safety and pollution measures of the 1980s and '90s had their roots.

In the 1960s, cars were often seen as representing fun and freedom, but by the 1970s road deaths and pollution, combined with rising running costs, were worrying legislators, especially in the US. The National Transport Safety Board began pushing laws to address these issues. Some, such as mandatory bumper and headlamp sizes and locations, were resented by carmakers. It was even predicted that open-roofed cars would die out because of roll-over safety requirements. However, the US was such a huge market that European and Japanese manufacturers began ensuring that their cars could comply with the latest US safety and clean air requirements, which eventually inspired legislation in markets across the world. Refinements to laws, as well as creative engineering solutions, meant that cars eventually adapted to legal requirements.

Safety first

Two key safety systems emerged at this time. One was ABS anti-lock braking, which prevented a car's front wheels from locking and skidding during emergency stops by pumping the brakes on and off rapidly, so that the car could still be steered. Some aircraft had used ABS systems since the late 1940s; the 1966 Jensen FF had used a similar mechanism, as had some early '70s American Chryslers and a few luxury General Motors models. However, as computers became smaller and more powerful, full electronic control became a reality. Mercedes-Benz led the way with its S-Class saloon in 1979, and Ford's mid-1980s Scorpio/Granada had standard ABS. A decade later, the technology was widespread.

The other safety advance was the airbag, which had been in development since the 1950s. Both Ford and GM had tried airbags in prototypes by 1973, and GM was installing them in Oldsmobiles two years later. However, serious injuries were caused when the devices self-activated – particularly in cars without hearests. One GM safety engineer suggested that airbags could replace the electric chair because they could break someone's neck if wrongly deployed. However, airbag technology matured. From 1988 every US Chrysler was airbag-equipped, and 11 years later airbags were fitted to all US cars by law. Soon, side, curtain, knee, and seat-located airbags had been developed.

In Europe, progress was slower. Mercedes-Benz first offered airbags in 1981. It was not until the 1990s that European and Japanese carmakers began adopting them *en masse*, with models such as the original Ford Mondeo taking the lead by offering standard airbags. Apart from a few specialist sports cars, almost all new cars had airbags by the mid-2000s, many working with seatbelt pre-tensioners that braced their wearers before an accident. Other safety advances included body crumple zones that absorbed energy in a collision, and side-impact door beams.

Catalyst for change

Catalytic converters, fitted to car exhaust systems, first appeared in the US during the 1970s. They filtered exhaust gases, changing their chemical composition and reducing some harmful emissions. The first catalyzed cars used mechanical carburetors to mix fuel and air. These were not efficient enough to work well with catalysts, reducing power and efficiency, and so using more fuel and producing extra CO_2, or greenhouse gas. From the late 1980s, there was a move to more efficient fuel-injection systems using computers to precisely control combustion. Catalysts worked better, emissions fell, and they soon became a standard feature.

▷ **Crash test simulation, 1997**
A head-on crash test between a lorry and a Renault Mégane. The second generation Mégane was the first car to receive a five-star Euro NCAP rating.

▷ **Engine emissions**
A technician measures a car engine's emissions in 1981. By the end of the 1980s, engine emissions were a major issue for governments, car manufacturers, and drivers alike.

-93
НЯ

△ **Nissan Figaro, 1991**
Inspired by the Peugeot 403 convertible, the Panhard Dyna, and the Nash Metropolitan, among others, the Figaro has endured owing to its distinctive looks. It has the floorpan and drivetrain of the Nissan Micra.

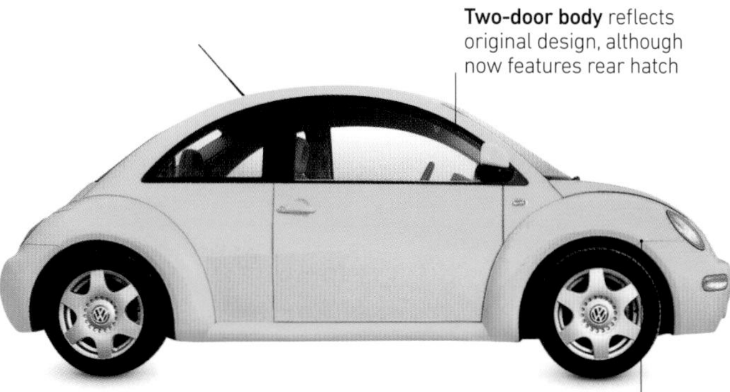

△ **Volkswagen Beetle, 1998**
The new Beetle is front-engine, front-wheel-drive, and based on the VW Golf, but it has become a runaway sales success. Its styling evokes the older model – with modern creature comforts, including air-conditioning.

"The **height** of **postmodernism**."

PHIL PATTON, DESIGN CRITIC, ON THE NISSAN PIKE PROJECT

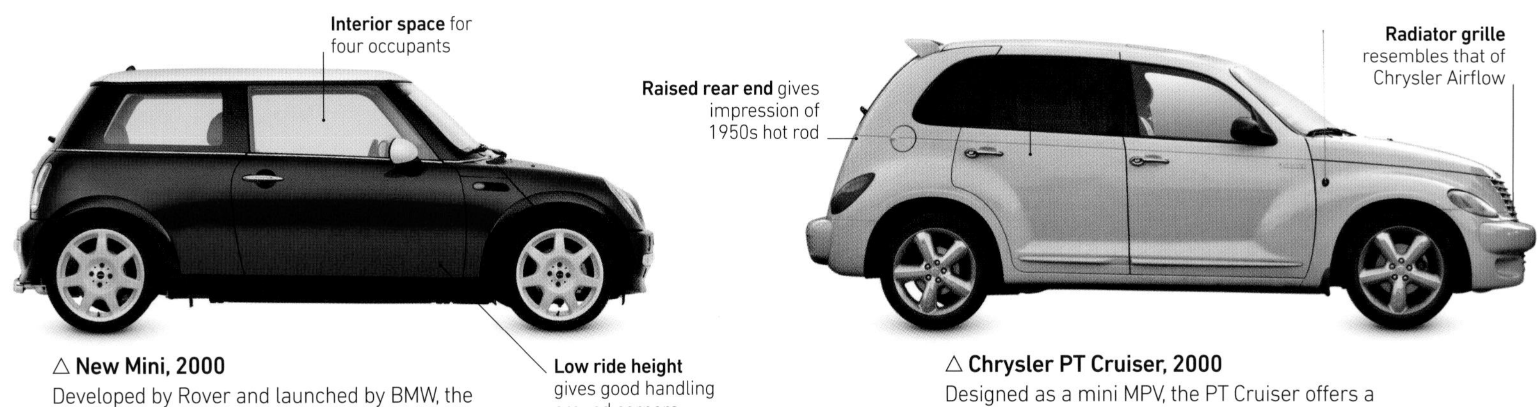

△ **New Mini, 2000**
Developed by Rover and launched by BMW, the new Mini carried over no new parts from the original. Bigger and more comfortable to drive, there are many variations now available.

△ **Chrysler PT Cruiser, 2000**
Designed as a mini MPV, the PT Cruiser offers a spacious, flexible interior, with styling cues that hark back to the 1930s and '50s. It ceased production in 2010, by which time over one million had been built.

Retro designs

Retro styling increased in popularity throughout the 1990s and 2000s. A yearning for the stylistic influences of the 1950s and '60s saw the appearance of a range of cars that bore touches of yesteryear. Many larger cars were styled with nods to the past – such as the Citroën XM's evocation of the SM grand tourer – but the trend was generally for smaller cars that had softer curves and far more comfort than was found in larger models.

Some manufacturers reimagined their iconic marques, such as the Beetle and Mini, while others focused on new, if retro, designs. Nissan's Pike project, for example, spawned not only the popular Figaro, but also the 2CV-inspired S-Cargo van and the 1950s-style Pao hatchback.

▽ **Nissan Figaro, 1991**
The Nissan Figaro was part of the Pike programme, which included the S-Cargo van and the Pao hatchback – replete with elements of Citroën H Van and Autobianchi Primula. The Pike project also produced the Be-1, a small hatchback which evoked the Fiat 600 and Mini.

▷ **Smart tower**
Smart Automobile, a division of Daimler AG, specializes in evoking the days of the microcar. Its signature car has a one-box profile and two-seat layout that ensures it can fit where a normal car cannot.

A world in gridlock

Nose-to-tail traffic jams existed before the invention of the car and usually involved real noses and real tails since most vehicles were pulled by horses. With cars on the roads, congestion only worsened.

Many emerging economies with rapidly rising car populations have their own peculiar traffic problems. In 2010, roadworks and broken-down vehicles on a major highway leading from Beijing created a 62-mile (100-km) tailback, which took a record-breaking ten or more days to clear. Of some consolation were the specialist companies that dispatch motorcycles to rescue hemmed-in drivers on China's roads. The pillion passenger swaps with the driver and endures the traffic jam, eventually delivering the car, while the owner is taken to his destination on the back of the bike.

There was no such help for drivers hitting the road from East to West Berlin over the Easter weekend of April 1990. With the Berlin Wall newly razed, the westward road, which averaged some half a million vehicles per day, suddenly became flooded with 18 million cars. It took days for the chaos to clear, but, as families sought to reunite after generations of political division, it demonstrated the underlying unity of Germany. To date, it still holds the world record for the largest number of cars caught in a traffic jam.

As for the world's longest traffic jam, that happened ten years earlier, in February 1980. It occurred on the Lyon–Paris road, as thousands of French holiday-makers left their ski resorts in the Swiss Alps and returned to Paris in what proved to be an unprecedented number of vehicles. Poor weather exacerbated the situation, creating a traffic jam that stretched 110 miles (177 km) – a third of the distance from Lyon to Paris – and took two days to clear.

Promoters of self-driving cars predict that, once automatic technology is perfected, cars will be able to travel closer together, anticipate hold-ups, and reduce traffic jams. Time alone will tell if these claims work in practice and whether rising vehicle numbers will in effect cancel out any gains.

▷ **The view from above**
The evening rush hour causes a traffic jam at a junction in Beijing. Car ownership has soared in China in recent years, and such congestion is frequent, despite the country's rapidly developing road network and infrastructure.

KEY DEVELOPMENT
Spirit of Ecstasy

Dismayed by some of the mascots that owners fitted to its cars, Rolls-Royce created its own bonnet ornament in 1911. The "Spirit of Ecstasy" was the work of sculptor Charles Robinson Sykes, and depicted a female figure leaning forward into the wind with her robes streaming out behind. At first the Spirit was optional, but it was soon fitted as standard and became synonymous with Rolls-Royce. The Spirit has been subtly redesigned over the years to suit today's lower, wider cars; a kneeling pose was tried for a while, but a scaled-down version of the original was preferred. The mascot on today's Rolls-Royces retracts automatically into the radiator shell when struck, to avoid damage or theft when parked.

THE SPIRIT OF ECSTASY **STANDS PROUDLY ABOVE A ROLLS-ROYCE GRILLE.**

PIG WITH CAMERA

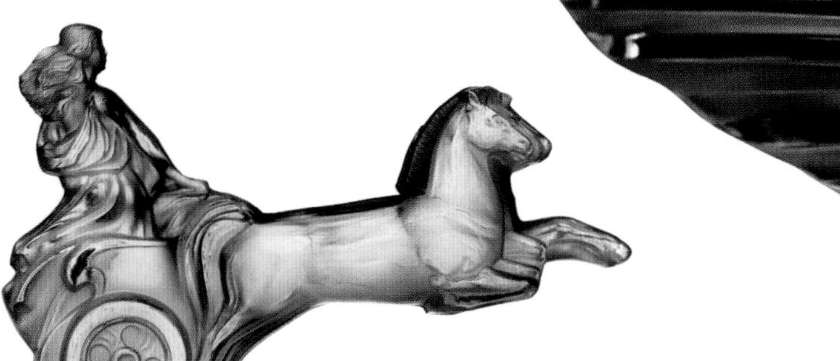

GODDESS RIDING A CHARIOT

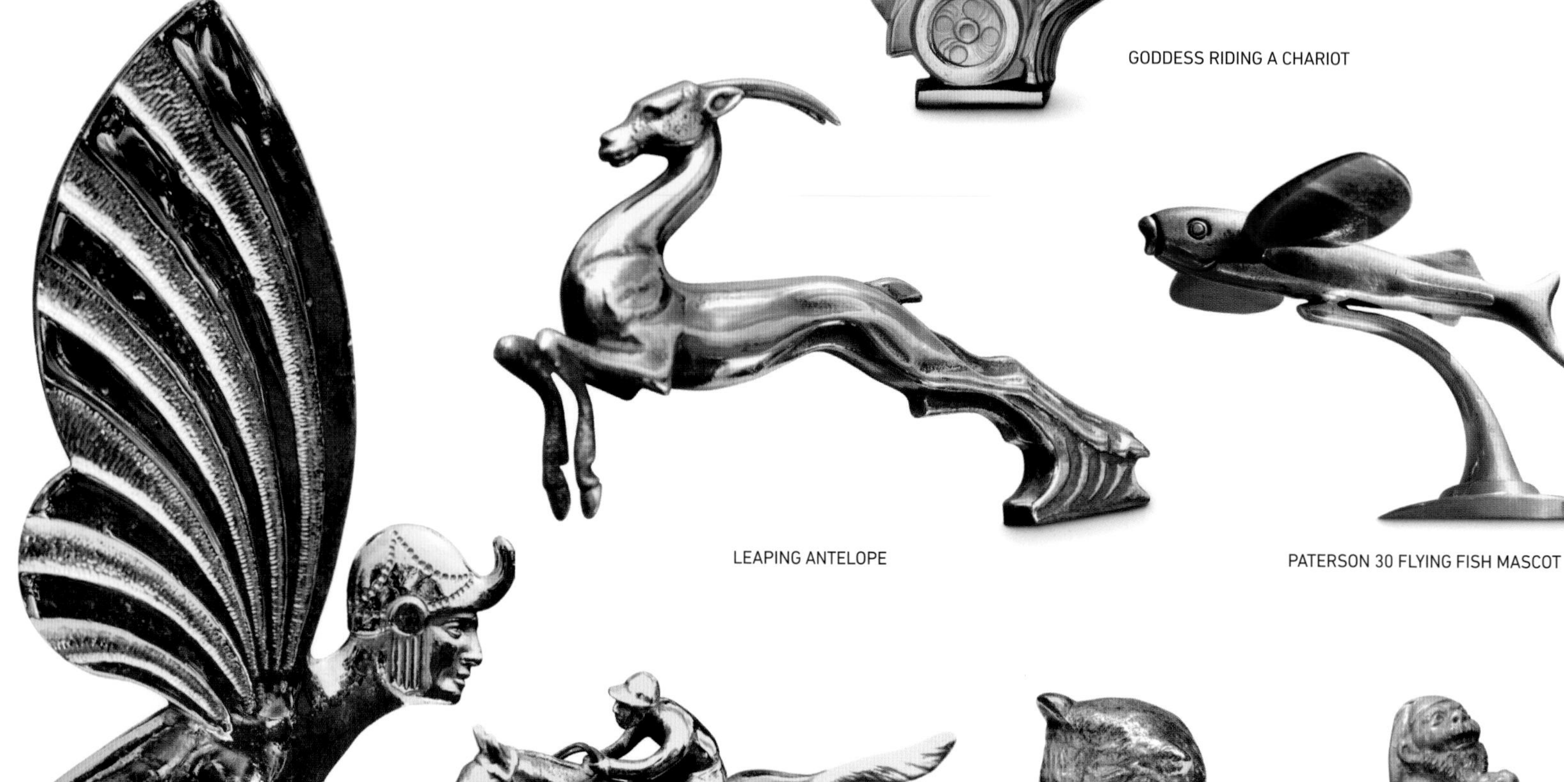

LEAPING ANTELOPE

PATERSON 30 FLYING FISH MASCOT

WINGED FIGURINE

RACEHORSE AND JOCKEY

BEAR EATING HONEY

DRUNKEN FIGURE

Clean lines of mascot typify Art Deco style

AEROPLANE

LOUIS LEJEUNE "OLD BILL"

GLASS LALIQUE "VICTOIRE"

GLASS LALIQUE "LONGCHAMP"

DANCING COUPLE

DANCING LADY

HASSELL POLICEMAN

Motoring mascots

Decorative mascots that fitted onto the radiator cap at the front of the car were popular accessories from the 1910s to the '30s.

Many mascots were inspired by speed. Figures with wings and athletic poses were popular, as were soaring birds and leaping animals. There were also versions of advertising icons such as the Goodyear Blimp and Bibendum, the Michelin Man. As the Art Deco style became popular, mascots adopted angular, geometric shapes – and subjects included modern icons of speed such as aircraft, rockets, and railway locomotives. The best ornaments were usually cast in bronze or brass; cheaper ones were made with nickel and chrome-plated.

René Lalique's glass mascots were expensive and rather damage-prone, but the quality of their design and construction was unrivalled. One of the most famous, Victoire, depicts the head of a woman with her long hair streaming behind her in the wind. Lalique mascots were sometimes lit from below, with a rotating disc of coloured filters driven from the engine to change the colour of the glass as the car moved along. This made the Lalique Libellule (dragonfly), for example, appear to flap its wings.

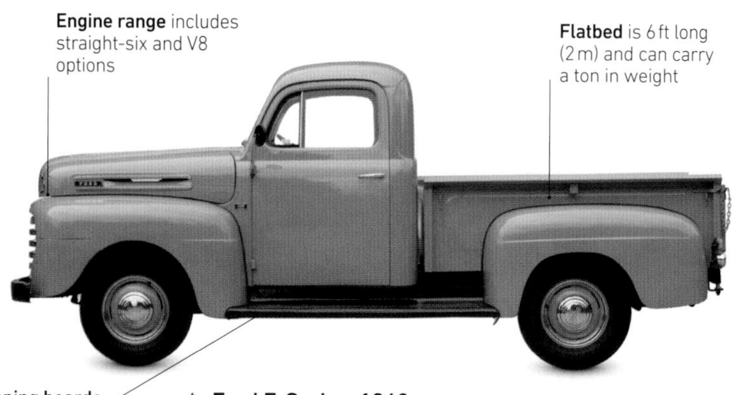

Engine range includes straight-six and V8 options

Flatbed is 6 ft long (2 m) and can carry a ton in weight

Running boards are a prewar design feature

△ **Ford F-Series, 1948**
Ford broke new ground with its purpose-built F-Series range, with their modern cabs and separate load beds. Previously, Ford's pickups had been car-derived. 110,000 were sold in 1948 alone.

Two-tone, red and white paintwork was standard in 1955

Tailgate panel made from fibreglass

△ **Chevrolet Cameo, 1955**
The Cameo offered an early attempt to spread the pickup's appeal from business to pleasure, with its V8 engine, automatic transmission, and plethora of comfort and design flourishes.

▽ **Dodge Ram, 1994**
Today's American pickups come in three sizes – mid-size, half-ton, and heavy-duty. The Dodge Ram seen here falls into the heavy-duty category. Despite its bulk, its top speed is 100 mph (160 km/h).

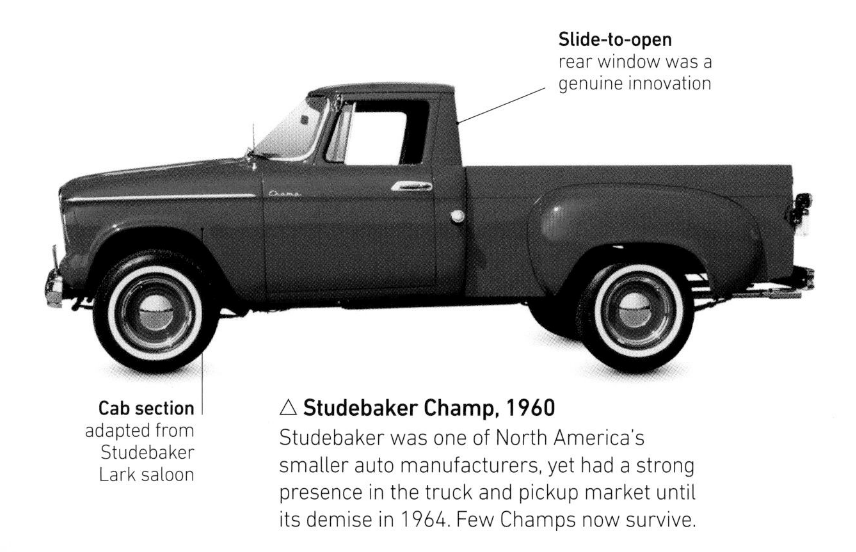

△ **Studebaker Champ, 1960**
Studebaker was one of North America's smaller auto manufacturers, yet had a strong presence in the truck and pickup market until its demise in 1964. Few Champs now survive.

△ **GMC Canyon, 2004**
The Canyon, and its Chevrolet Colorado equivalent are "mid-size" trucks that have gradually grown in scale to have similar dimensions to "full-size" models from the 1980s.

American pickups

The American auto industry is best known globally for its high-powered cars such as its Mustangs, Camaros, and Vipers. However, the vehicles that pay the bills for Detroit's manufacturers have always been trucks, or "pickups" as they are known outside the US.

From the early days of the motor industry through to today, pickups continue to be the US's workhorse vehicles, combining utilitarian functionality with a macho image. From the early 1970s, American consumers began purchasing pickups as "lifestyle" vehicles rather than simply for practical reasons. However, it was in the early 1980s, when General Motors, Ford, and Chrysler started increasing interior room, adding luxury car features and installing performance-orientated engines, that the pickup's appeal exploded – along with sales and profits.

△ **Ford Ranchero, 1957**
Ford's Ranchero started a new trend for capacious pickups derived directly from two-door saloons. It soon saw competition from Chevrolet's El Camino. In Australia and South Africa, this type of versatile, manly, utility vehicle, often called a 'Ute', enjoys massive popularity today.

The rise of the SUV

A new breed of easy-to-drive 4x4s offered an unbeatable combination of on-road refinement, chunky styling, and off-road ability, making them the go-to choice for an increasing number of buyers in search of a family car.

Few people ever aspired to own a people carrier or an estate car – but they sold in their thousands because they suited the lifestyle of their buyers. Then along came the sports utility vehicle (SUV), which was both functional and desirable. As the technology underpinning them improved – more sophisticated suspension and foolproof automatic four-wheel-drive systems – they became viable as spacious and practical family cars with a dash of adventure. The ease of access for passengers, lofty driving positions affording good visibility, and, for the most part, superior comfort and convenience added to their appeal.

In the US, the archetypal SUV was Jeep's XJ-series Cherokee, introduced in 1983 and produced right through to the new millennium. General Motors had the Chevrolet Blazer, and Ford entered the market with the Explorer in 1990, which became the best-selling model in its class. Imported rivalry came principally from Japanese models, such as the Mitsubishi Shogun and Toyota Land Cruiser, and the British Range Rover and Land Rover Discovery.

As the market grew, SUVs broadened in appeal. Full-size machines like the Lincoln Navigator catered for those who wanted even more space, and in the 1990s a new generation of compact SUVs, such as the Toyota RAV4 and Land Rover Freelander, extended their appeal in the opposite direction. "Crossovers" like Nissan's Qashqai blended the best of SUVs and estate cars in a single package and were nothing like utilitarian off-roaders, and other manufacturers offered countrified estate cars like the AMC Eagle and Audi Allroad models. Soon, even Porsche joined the trend with the Cayenne, which became its biggest-selling model.

"I normally drive my **Range Rover** because I feel like **a monster** in it."

EMMA BUNTON, FORMER SPICE GIRL

▷ **Driving through the snow**
Parents collect their children from school in SUVs after a snow storm in Akaska, US, in 1997. Adverse weather conditions were just one of the reasons SUVs became an attractive option for families.

STOP
CCN 602

Mega-mergers

In the 1990s, the automotive industry changed dramatically as car manufacturers sought to maximize profits by buying up smaller competitors. However, things did not always go entirely to plan.

In the mid-1980s, US car manufacturers began to feel threatened by Japanese rivals, who were eroding the market concentration of the established players. The US companies struggled to compete with the lean production style of the newcomers, and responded by using size to try to outmuscle their competitors.

Globalization was viewed as the key to success as a new decade began. By buying up smaller carmakers, and sealing joint ventures for sales, parts supply, and manufacturing, auto companies hoped to achieve more efficient production and increase their profit margins.

▽ **BMW's Munich headquarters**
By the end of the 1990s, BMW had acquired the UK's Rolls-Royce, Rover, Mini, and Land Rover brands.

Strategic alliances played an important part in the consolidation process taking place within the global auto industry. There were some 500 cross-border alliances in the 1990s; 300 of them were joint ventures; the remainder were manufacturing joint ventures, under which many of the major brands bought up production power in overseas markets such as Asia, where labour was cheaper.

The mergers begin

Ford kicked off the 1990s with the acquisition of British manufacturer Jaguar, followed by Volvo in 1999 and Land Rover in 2000, all of which were later sold off to Indian and Chinese firms. In 1996, Ford also increased its share in Mazda from the 25 per cent it had held since 1979 to 33.4 per cent. Meanwhile, General Motors bought a controlling stake in Isuzu in 1999, the same year that Nissan and Renault formed an alliance. Renault took on $5.4 billion of the Japanese carmaker's debt in exchange for a 36.6 per cent stake in the ailing firm. This gave Renault access to markets in which Nissan had a presence, particularly Japan, the US, and Asia. It was the making of what would become the world's largest car manufacturer by the first half of 2017.

One of the most talked-about mergers came towards the end of the decade, when Volkswagen bought British luxury heritage brands Rolls-Royce and Bentley in June 1998, only to sell the rights for Rolls-Royce to BMW a couple of months later. However, the biggest deal of the era came in November of the same year, when Daimler-Benz in Germany and Chrysler in the US orchestrated a merger worth $40 billion.

△ **Cheap, reliable, and popular**
Launched in 1999, the popular Škoda Fabia's parts were developed in conjunction with Volkswagen. Shared production enabled cost-saving, which meant a lower price tag.

Dented dreams

Under the merger agreement, Daimler would control 57 per cent of the newly formed entity and gain a foothold in the US market – where it had only secured less than one per cent to date. Both sides benefitted from the other's strengths. Chrysler boasted low development costs, while Daimler-Benz came with advanced technology and a strong global network. By combining research and development, production processes, and purchasing, the new company could make huge cost savings. However, despite the financial advantages, clashes over management style led to Daimler selling 80 per cent of its stake in Chrysler in 2007.

> "An **odd couple** or a **perfect fit**?"
>
> CNN MONEY ON THE DAIMLER-CHRYSLER MERGER, 1998

△ **Asian powerhouse**
Robots work on an assembly line in a Hyundai factory in South Korea in 1995. The 1990s was the decade in which Western car companies realized that Asian markets were key in improving their profits.

Despite the efforts of the "Big Three" – GM, Ford, and Chrysler – Japanese rivals increased their share of the American market to secure more than a quarter of all US car sales by the end of the 1990s. The decade's frantic mergers and acquisitions activity favoured those companies that were bought, or formed alliances, rather than the big corporations doing the buying. Over the nine-year period, industry sales increased at a rate of 21.8 per cent. Acquiring carmakers increased sales by 15 per cent, while those being acquired increased sales by 38 per cent. Most telling of all, Honda and Toyota, companies that had avoided the mergers, outperformed the Big Three.

KEY DEVELOPMENT

The Big Three, plus Italy

The 1990s merger frenzy was long finished by the time Chrysler found itself filing for bankruptcy in 2009. The company ended up being part-owned by the US and Canadian governments – and by Italian company Fiat. The latter gradually increased its holdings in Chrysler, completing its acquisition of the company in 2014, with the net result that one of the US Big Three car manufacturers is now part-Italian. Fiat Chrysler Automobiles has continued historic Chrysler US marques such as Dodge and Jeep, as well as Italian marques Alfa Romeo and Lancia.

FIAT CHRYSLER AUTOMOBILE **CHAIRMAN SERGIO MARCHIONNE.**

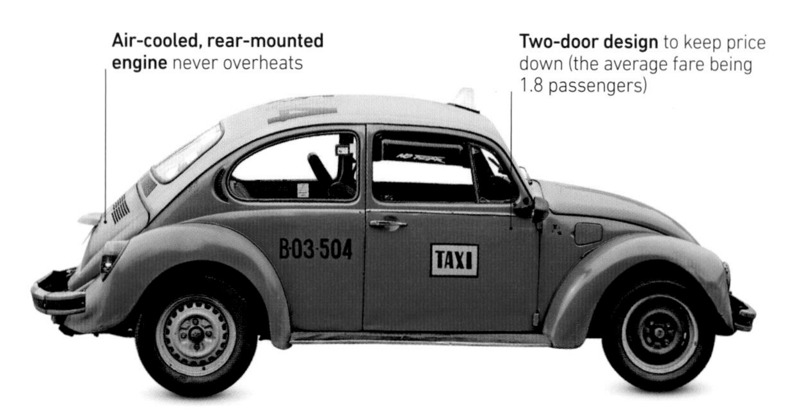

△ **Volkswagen Fusca, 1953**
The Brazilian-built version of the Beetle seemed an unlikely taxi, as it had only two doors, but its drivers thrived on its dependability and low running costs.

△ **Checker Model A8, 1958**
This classic NYC taxi cab first appeared in 1958 as the A8, and continued to be made up until 1982. Since then, standard large saloons have taken its place on Park Avenue and Times Square.

Taxis of the world

Car use is not restricted to people who own a vehicle. In fact, it is possible to travel by car on a frequent basis without ever owning one, thanks to the global taxi trade.

The taxi proudly occupies a place in motoring history, and plays a vital role in the modern world. Taxis have been around almost as long as the car itself, and are so much a part of urban life that it is often possible to identify a city from its taxis alone. New York, London, and Tokyo have had taxis tailored specifically to their ultra-urban city environments. Of these, the Japanese Toyota Comfort, a large and roomy saloon with a diesel engine and durable interior trim, is perhaps the most luxurious. New York's yellow cabs, meanwhile, are almost as iconic as the city itself. Elsewhere, taxis can be basic but no less useful, such as the auto rickshaws of India. Wherever they are, these vehicles are united by the need to be rugged, reliable, and easy to repair.

"They'll all be **riding in your cabs** sooner or later."

BUSINESSMAN WILLIAM RANDOLPH HEARST, TO NEW YORK TAXI PIONEER HARRY N. ALLEN

▷ **Indian taxi, Kolkata**
This Hindustan Ambassador taxi is typical of vehicles in service in larger Indian cities. The Ambassador was made from 1958 until 2014, and was based on the Morris Oxford Series III model.

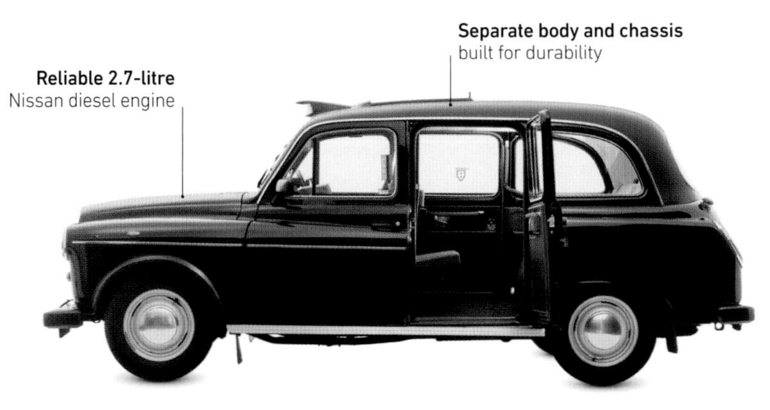

△ **LTI Fairway, 1958**
This black-painted icon of London life began as the Austin FX4 in 1958, and was latterly built by London Taxis International. It was notable for its tight turning circle of 25 ft (7.5 m).

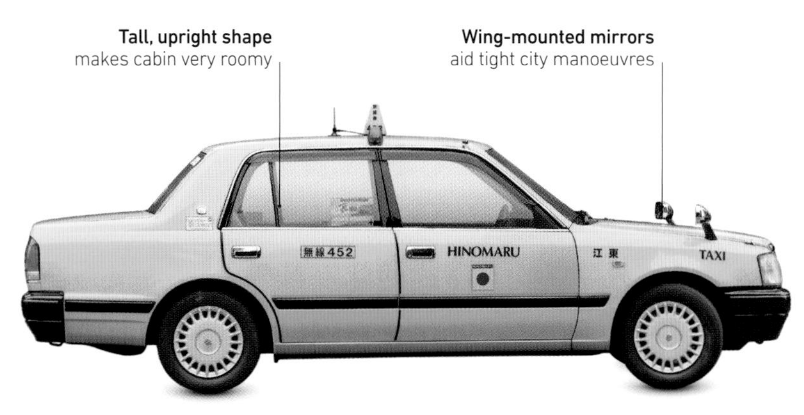

△ **Toyota Comfort, 1995**
Specifically designed for use as a taxi, the Comfort has simple and conventional mechanical parts, offering good longevity. Models powered by diesel or liquid petroleum gas (LPG) are available.

Turning against the car

After years of road planning to accommodate the car, rising traffic levels, together with concerns about the environment, made cities take steps to discourage car use. At the same time, public protests prompted a rethink of road building.

For decades, developed countries had been building roads almost unchecked, seeing them as vital pieces of infrastructure. New cities had been laid out with cars in mind, and older ones had been reworked to provide them with widened access roads, multi-storey car parks, ring roads, and beltways. The connections between cities had also been improved by networks of motorways, autobahns, and interstates. However, road building became an increasingly controversial topic, due to the damage it seemed to be doing to the environment. Cars also represented a danger to public health – not just because of accidents, but because of their various toxic emissions.

Environmental dangers

The environmental movement that emerged in the 1960s and '70s characterized the car as an evil. Certainly, the drawbacks of rising traffic levels – with more noise, more air pollution, more delays, and more road deaths – were plain to see. There was still plenty of support for improving road networks, both from businesses, which could see the economic benefits, and residents, whose neighbourhoods would be improved by relief roads. However, when planning decisions went in favour of new roads, environmental campaigners decided to take direct action against what they

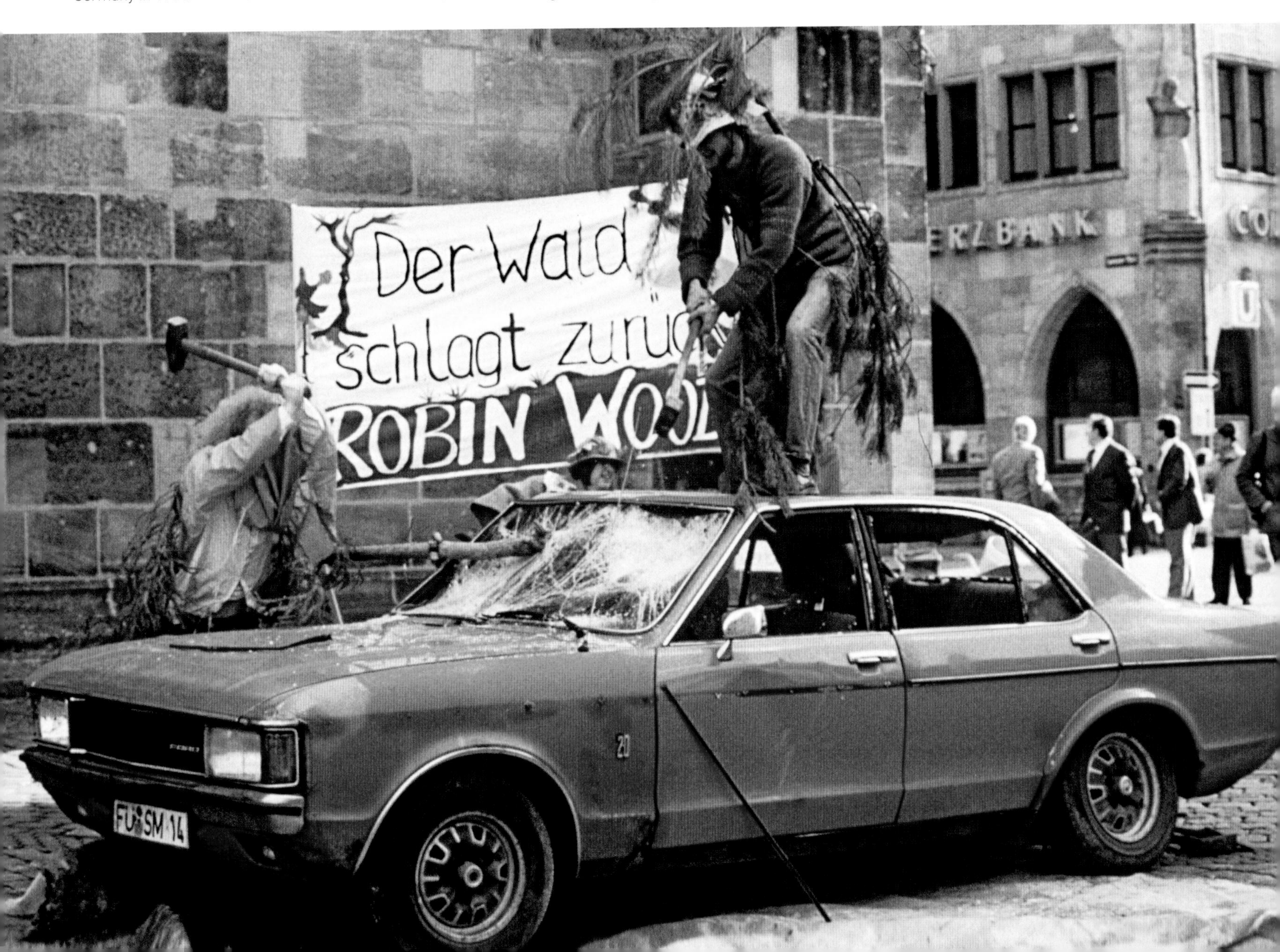

▽ **Pollution protest**
Members of the environmental organisation Robin Wood, dressed as trees, beat an old car in a protest against pollution in Frankfurt, Germany in 1984.

> **"**If I **wrote** a letter **to** my **MP**, would I have **achieved** all this?**"**
>
> SWAMPY, PROTESTOR, 1996

saw as unnecessary and damaging developments. It was the beginning of a war against the car that is still being fought today.

Roads for prosperity

When a new trunk road programme in the UK called "Roads for Prosperity" began in 1989, it was met by a string of protests that made sleepy rural areas such as Twyford Down and Solsbury Hill famous overnight. Protests against the Newbury Bypass resulted in more than 1,000 arrests, and when a protester called Swampy hid in tunnels dug in the path of a major road extension in Devon he briefly became a celebrity. At Newbury, protesters chained themselves to trees and became known as "tree-huggers". As a result, the Newbury Bypass was halted, but it eventually went ahead, at the cost of 10,000 trees, £75 million spent on construction, and £5 million spent on police.

Urban renewal

At the same time there was a growing trend towards improving the living environment in cities. In Philadelphia, a "greening" programme introduced hundreds of public gardens and green spaces to the city, using roofs and vacant lots. In Boston, an enormous project to dig tunnels to bury both the I-93 highway and a new link to the city's Logan Airport became the most expensive single highway project in American history. The "Big Dig" took nine years to plan and fifteen years to complete, and was plagued by delays, cost overruns, and design flaws. But on the ground, the scheme delivered a remarkable new landscape: the Rose Fitzgerald Kennedy Greenway, which follows the path of the I-93 before it was sunk into tunnels. The 1½-mile (2.4-km)-long stretch consists of landscaped gardens, promenades, plazas, fountains, and art.

Also in 1989, years of bitter argument over the ugly, elevated, double-decker Embarcadero Freeway in San Francisco came to a head when the freeway was irreparably damaged by an earthquake. When it was replaced by a boulevard with wide pedestrian walkways and light rail tracks, it prompted a regeneration of the whole area. Similar schemes transformed parts of Portland, Milwaukee, and Seattle, and outside the US there were successful projects in Madrid, Spain, and Seoul. For the first time in decades, cities were no longer in thrall to the car.

◁ **Boston's Big Dig**
The Big Dig under construction in Boston in 1998. The new road rerouted the Central Artery of Interstate 93 into the 3.5-mile (5.6-km) Thomas P. O'Neill Jr. Tunnel.

LIFE BEHIND THE WHEEL

Congestion charging

Charging drivers to enter cities at peak times was proposed in the 1950s, but congestion charging did not become a reality until the 1970s. Singapore introduced a scheme in 1976, and Hong Kong followed in the 1980s, although theirs was not a permanent solution. London's congestion charge scheme was introduced in 2003 and remains in place today. Other ways of managing traffic include restricting the days on which certain drivers can enter the city, increasing the number of passengers per vehicle by using multi-occupancy highway lanes, and introducing park-and-ride schemes, which connect car parks with public transportation.

***A SIGN FOR THE CONGESTION CHARGE ZONE* IN CENTRAL LONDON, WHICH OPERATES ON WEEKDAYS DURING BUSINESS HOURS.**

2001–PRESENT

DRIVING INTO THE FUTURE

2001–PRESENT

Driving into the future

For many years, so-called "concept cars" – those beacons of future transport designed to be leaner, faster, and ever more environmentally-friendly – were revealed to hushed audiences at motor shows across the world, be it in London, Tokyo, Turin, or Los Angeles. However, each of these vehicles seemed to point to a day that never came. Things did not change, and as drivers waited in traffic jams with harmful emissions billowing from their car's exhaust pipes, rueing soaring parking fees and the depreciation of their four-wheeled asset, it was difficult to see the car as anything other than a grim necessity of modern life.

However, fossil fuels are seen as a planet-warming dead end, digital communications have revolutionized the way information is relayed, and brilliant minds have finally applied these facts to the automobile. Real change is currently underway. The practical advantages of new hybrid (petrol-with-electric) powertrains at the turn of the millennium got things going. Everyone seemed to like the reduced pollution, the lower running costs, and the lighter, gentler touch of the hybrid driving experience. Lithium-ion battery technology, so successful in smartphones, was applied to electric cars, enabling them to cover much greater distances. As the scandal of falsified pollution figures for diesel engines was revealed, the credibility of the electric car, after a few false starts, was hugely boosted. Diesel, and fossil fuels more generally, seemed to be nefarious in comparison.

The end of driving?

Today, thanks to the wonders of automation, car owners can even contemplate a future in which the driver is no longer needed – when the car can

NEW CAR GENRES CONTINUE TO EMERGE, SUCH AS CROSSOVERS

THE DIESEL SCANDAL FURTHER FUELS THE CALL FOR GREENER CARS

"We can even **contemplate a future** in which **the driver is no longer needed**."

take care of itself, performing all the mechanical tasks of driving and navigation, while its occupants do other things, such as relax or socialize. The sensation of actually driving, of being in charge of a fuel-burning engine that responds directly and excitingly to the impulses of the driver, will soon be a thing of the past, it is promised. And few can honestly say if that is a good thing or not.

Diverse markets

The shape of conventional cars has also changed considerably since the year 2000. Crossovers have mixed up the car genres. Drivers and passengers generally sit higher in cars with greater versatility, although not always with the four-wheel drive that such designs used to suggest. Satellite-driven global positioning systems, or "sat-nav", have eliminated the need for maps, and neutralized in-car arguments over wrong turnings or excessive tardiness. Cars also keep occupants safer, with in-built technology that stops the vehicle from wandering, even if the driver is distracted, to prevent accidents.

The established order of the carmaking world has changed too. China has joined the top table, making its own cars by the millions and snapping up old, established brands as its roadscape evolves at a rate that beggars the patchwork efforts that have served the West for so long. And yet the hunger for supercars that can bolt for the horizon at 250 mph (402 km/h), and luxury limousines and roadsters customized to their owners' particular tastes and wearing their handmade details out of pride, refuses to go away. Traditional car enthusiasts still have lots to look forward to.

ULTRA SAFE, SELF-DRIVING CONCEPT CARS POINT TO THE FUTURE

DEMANDS FOR TRADITIONAL MOTORING THRILLS ARE LIKELY TO REMAIN

Who killed the EV1?

Between 1996 and 1999, General Motors (GM) spearheaded the mass-produced electric car movement with the EV1. However, in 2002, the cars were recalled and either deactivated or destroyed, for reasons that are still debated today.

An electric car that looked like a vision of the future, the EV1 had its roots in the 1990 GM Impact concept car. Perhaps ironically, the Impact, which was electric, inspired CARB (the California Air Resources Board) to pass legislation that required 2 per cent of the top five car manufacturers' combined output to produce zero exhaust emissions by 1998 – then 5 per cent by 2001, and 10 per cent by 2010. The regulations were designed to improve California's terrible air quality, which was widely attributed to car emissions. Critics suggested that if the Impact was to succeed it would jeopardize GM, which relied on combustion-engine technology for the vast bulk of its products.

Two generations of EV1

Despite claims that it was not in GM's interests for its own electric vehicles to succeed, the company lent 50 Impact cars to drivers for them to review. GM further refined the Impact concept into the EV1 for 1996. The vehicles were then leased to users, whose contracts forbade them from buying the cars outright. These leases ranged from $399 to $549 per month, and lessees were required to live in Arizona or Southern California. By 1999, GM revised the concept into the second-generation EV1, which had quieter operation, lighter batteries, and a lower production cost. First-generation cars were also upgraded to feature the improved batteries, and owners were asked to sign new two-year leases. Then, in 2002, GM announced that the EV1 programme was terminated, and that all of the cars were to be returned to GM for destruction. In explanation, the company claimed that the cost of the mandatory 15-year parts supplies required by the state of California was prohibitive, and that

△ **GM's Impact, 1990**
The Impact was GM's initial foray into electric car technology. It was developed by electric vehicle company AeroVironment and made its debut at the 1990 Los Angeles Auto Show.

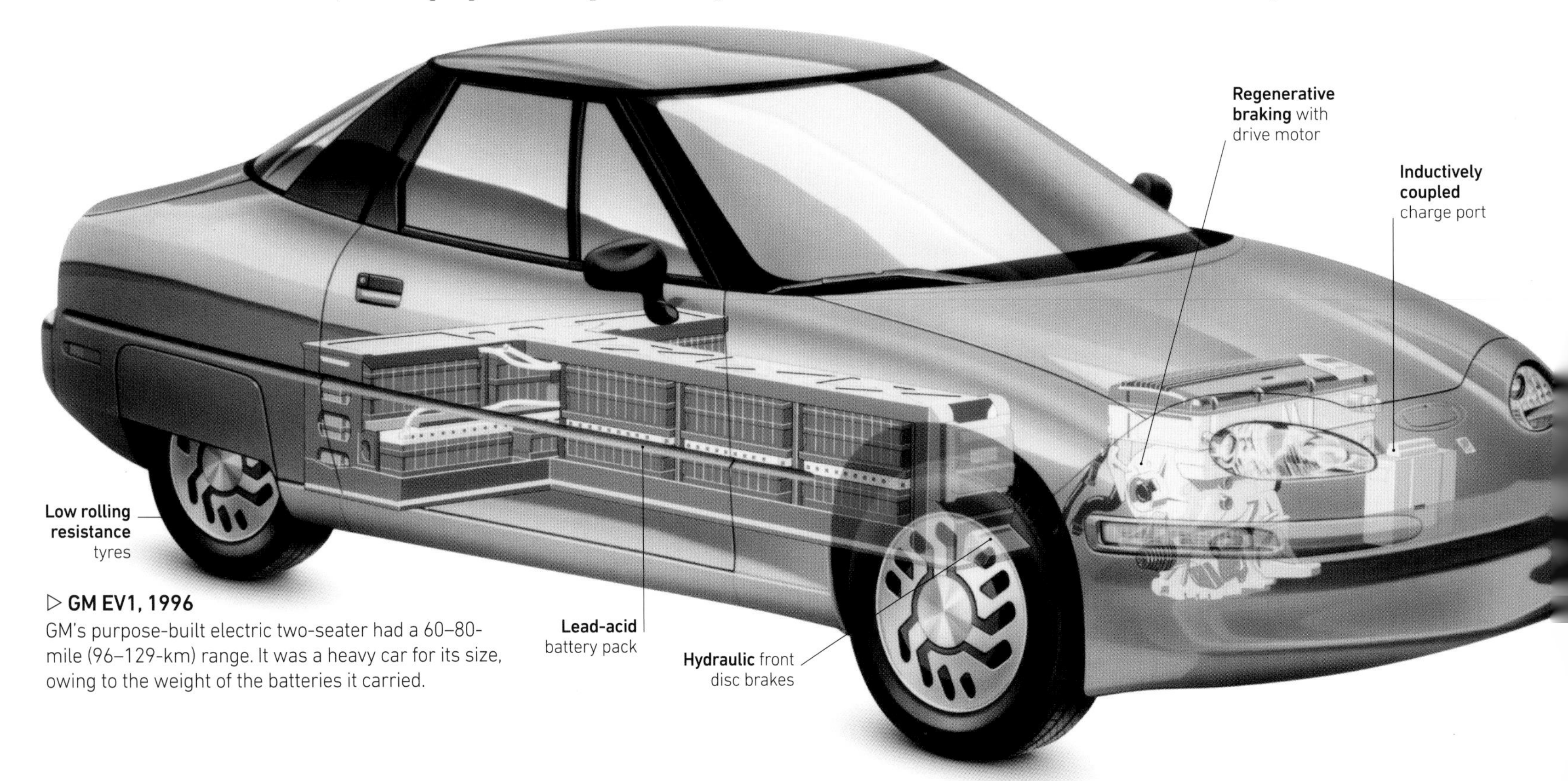

▷ **GM EV1, 1996**
GM's purpose-built electric two-seater had a 60–80-mile (96–129-km) range. It was a heavy car for its size, owing to the weight of the batteries it carried.

the slower-than-anticipated development of battery technology had scuppered its sales projections.

Almost 60 EV1 owners wrote to GM, requesting that they be allowed to continue their leases at no risk to GM, but GM refused, and returned their customers' voluntary deposit cheques. Approximately 40 EV1s were decommissioned and donated to museums – the rest were crushed.

◁ **Crushed EV1s**
A pile of crushed EV1s, repossessed by GM in 2002. Almost all of the 1,117 EV1s ever made were destroyed when their leases ran out – much to the regret of their owners.

The right car at the wrong time

Critics have argued that GM sabotaged its own project for fear that other states might propose similar regulations to California regarding electric vehicles. This was seemingly supported by a study conducted by GM and Toyota, which stated that the market was not ready for electric cars and that hydrogen was the more likely fuel of the future.

It is believed that each EV1 cost GM a total of $80,000–$100,000 to produce, including development costs. One GM official has stated that each car cost the company $250,000. With rental payments averaging $400 per month, depending on model year and rebate, and based on a nominal value of $34,000, the cars never broke even. This was one reason GM cited for abandoning the project. However, the truth is that there was little public interest in electric cars at the time, and hoping that the EV1 would eventually break even – let alone make a profit – was little more than a dream.

The EV1 was the right car at the wrong time. Two decades on, with greater electric car infrastructure, the EV1 would have found success. And yet without the EV1, perhaps that interest in electric cars and their infrastructure might not be advanced as it is today.

△ **Driving the EV1**
On the road the car could accelerate from 0–60 mph (0–96 km/h) in an impressive time of 7.7 seconds, and it had a top speed limited to 80 mph (129 km/h).

> "...the **EV1** is **more** than a **car**, it's a path to **national salvation**."

EV1 OWNER IN A LETTER TO GM CEO RICK WAGONER

LIFE BEHIND THE WHEEL

The movie tie-in

In 2006, Chris Paine directed the documentary film *Who Killed The Electric Car?*, which focused on the EV1 and the story behind it. It posited the theory that GM was encouraged to stymie its project by the oil industry – the traditional supplier of automobile fuel. The film looked at some of the EV1s that survived GM's attempts to reclaim them for destruction, and analyzed GM's attempt to show Californian officials that there was no demand for electric cars at the end of the 20th century. GM responded to the film, outlining various reasons why the EV1 was economically unviable, both for GM and its consumers, and that within three years of manufacture, spare parts had already become difficult to source.

THE FILM'S POSTER **MADE A CLEAR REFERENCE TO THE OIL INDUSTRY.**

△ **Chrysler Pacifica, 2004**
The seven-seater, Canadian-built Pacifica had its flaws but it was novel in offering SUV-like proportions without four-wheel drive. Well-appointed but expensive to buy, it nonetheless started the crossover revolution.

Diesel or petrol engine offers up to 440 bhp

Standard four-wheel drive (air suspension also available)

△ **Porsche Macan, 2014**
After the success of the Cayenne, Porsche introduced the compact Macan in 2014. The V6 engines available at the car's launch were joined by a turbo four-cylinder, entry-level option for some markets.

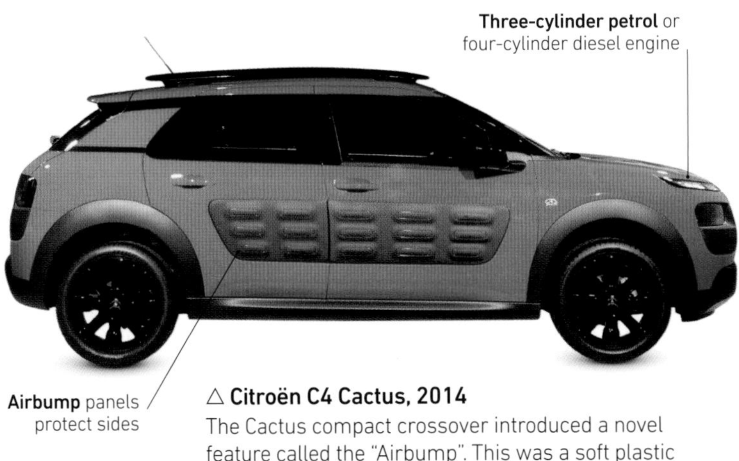

△ **Citroën C4 Cactus, 2014**
The Cactus compact crossover introduced a novel feature called the "Airbump". This was a soft plastic panel along the side of the car that protected the doors from damage in carparks.

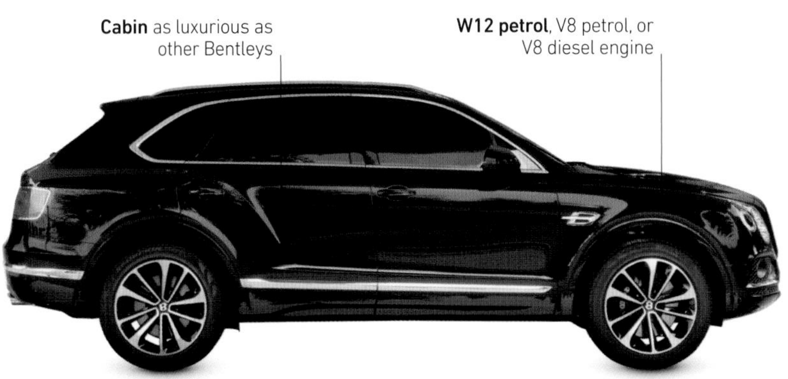

△ **Bentley Bentayga, 2015**
Bentley branched out from its usual saloons and coupés to produce this luxurious, high-performance crossover in 2016. At its launch, it was the fastest and most expensive crossover in series production.

▽ **Nissan Qashqai**
First produced in 2006, the Qashqai was at the forefront of a new breed of car that blended 4x4 and estate car qualities. These crossovers were a hit with families all over the world.

Crossovers for comfort and space

SUVs appealed to families that were looking for safe, adaptable, and above all spacious vehicles. However, the ultra-rugged construction and off-road ability of working 4x4s were unnecessary for this type of buyer, so the "crossover"category evolved. This new type of vehicle combined the best of 4x4 and conventional estate car: unibody construction for low weight, independent suspension for a supple ride and tidy handling, and a tall body offering a commanding driving position. Women drivers were particularly drawn to crossovers, whose success underlined how influential women were (and always had been) in a family's choice of car. Even premium carmakers such as Porsche, Bentley, and Rolls-Royce eventually built crossovers.

△ **Jeep Renegade, 2014**
Jeep is famous for its rugged 4x4s, and the Renegade is one of its entries into the compact crossover market. It shares a platform with the Fiat 500X – Fiat and Jeep both being brands within the Fiat Chrysler group.

Satellite technology improves safety

In recent years, satellite navigation and traffic data systems have revolutionized driving. At the same time, in-car information systems have helped to reduce accidents and have become essential for tracking stolen vehicles.

The Global Positioning System (GPS) was developed for the US military in the 1970s and was fully operational by 1995. At first, the civilian version had a deliberately degraded signal quality, but even this enabled carmakers to include satellite navigation systems in cars for the first time. Alongside the position data, the navigation systems used traffic data to help re-route cars around queues. The result was revolutionary: paper maps became obsolete as drivers received turn-by-turn navigation instructions, which speeded up journeys, lowered fuel consumption, and improved safety. At the same time, it created a new branch of information technology – telematics.

GPS applications

Access to higher-precision GPS data was made available in 2001, by which time GPS applications had become more diverse. General Motors created the OnStar system which combined GPS location data and a dedicated cell phone system to automatically alert emergency services if a car's airbags were deployed. After successful early trials, it was fitted in more and more vehicles.

In the UK, insurance companies began to offer telematic "black boxes" that could determine where, when, and in what manner a car was being driven. This data could then be used to provide insurance cover tailored more precisely to the use a particular driver made of a car. Young drivers faced with rising costs of conventional car insurance could cut

▷ **Connected cars**
This computer-enhanced image depicts telematics in action. Each car has a radar that monitors its distance from other vehicles – and an alarm that sounds if anything gets too close.

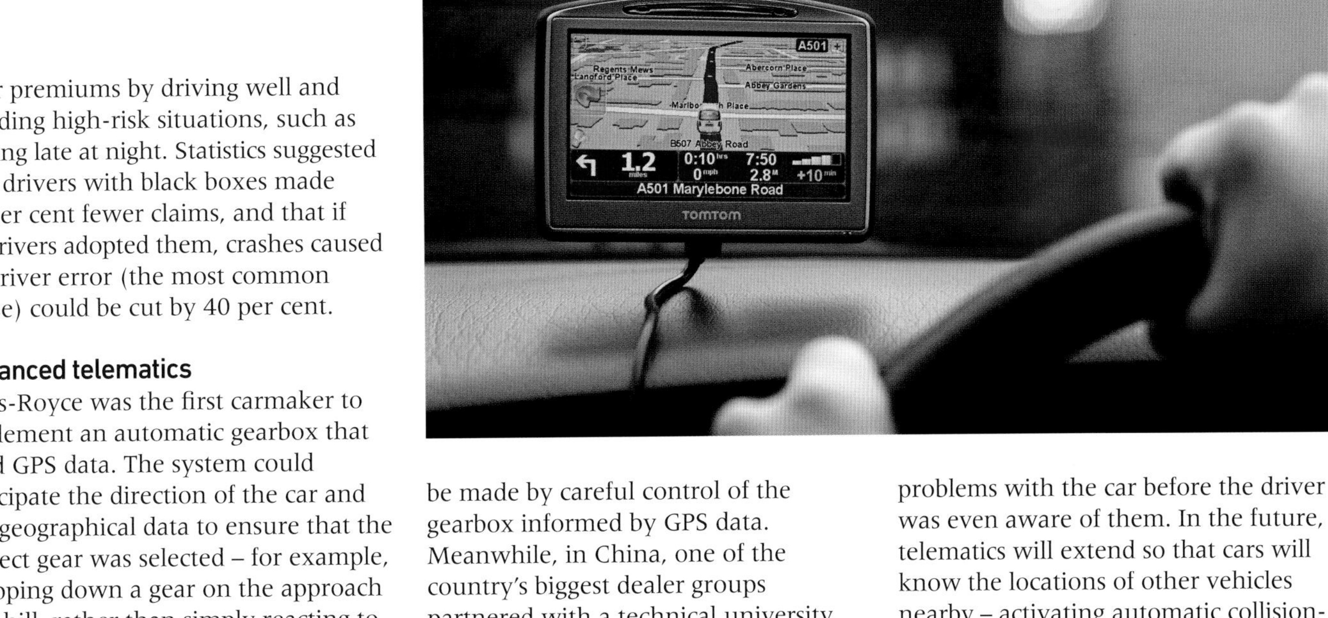

◁ **GPS navigation**
A driver uses a GPS device to navigate the streets of London. The destination is typed into the device, which then gives guidance at every junction.

their premiums by driving well and avoiding high-risk situations, such as driving late at night. Statistics suggested that drivers with black boxes made 20 per cent fewer claims, and that if all drivers adopted them, crashes caused by driver error (the most common cause) could be cut by 40 per cent.

Advanced telematics

Rolls-Royce was the first carmaker to implement an automatic gearbox that used GPS data. The system could anticipate the direction of the car and use geographical data to ensure that the correct gear was selected – for example, dropping down a gear on the approach to a hill, rather than simply reacting to the gradient upon reaching it. The result was a more refined and more responsive drive. A similar principle was investigated for commercial vehicles, where fuel savings could be made by careful control of the gearbox informed by GPS data. Meanwhile, in China, one of the country's biggest dealer groups partnered with a technical university to develop a GPS system to track its customers' cars. The system could provide a rapid response to incidents, and could alert the customer to impending service needs or developing problems with the car before the driver was even aware of them. In the future, telematics will extend so that cars will know the locations of other vehicles nearby – activating automatic collision-avoidance systems if they get too close – and will be able to swap data with roadside features such as street signs and traffic lights to make journeys easier, smoother, and safer.

> "... get **people thinking** about the **car as an information platform**, then ideas... will surface."
>
> VINCE BARABBA, GENERAL MOTORS, 1999

DRIVING TECHNOLOGY

Tracking the traffic

Information is gathered by traffic data systems, such as INRIX, from a variety of sources. Some mobile phone companies provide anonymized tracking of their phones, many major roads have sensors that count vehicles, and large vehicle fleets feed data on their vehicle movements to traffic data centres. Once the data is analyzed, information on traffic speeds and hold-ups is sent to cars using the Traffic Message Channel, a data feed transmitted alongside radio broadcasts that can be decoded by radio receivers and navigation systems.

AN INRIX COMPUTER **ON DISPLAY IN THE COCKPIT OF A BMW I3 AT THE CONSUMER ELECTRONICS SHOW, US, 2014.**

China hits the road

Once known as the land of the bicycle, from the 1990s China rapidly developed its motor industry, initially by forming joint ventures with Western manufacturers. By doing so, it became the world's largest car-building nation.

△ **Dongfeng assembly line** Engineers man the Honda Civic production line at the Dongfeng Honda factory in Wuhan, in China's Hubei Province, 2017.

China's motor industry was first established in the 1920s, but its modern incarnation dates to the 1950s, when the Soviet Union aided its communist ally both in modernizing its factories and providing its designs.

Under the government of Chairman Mao (1949–76), annual production peaked at just 200,000 cars, most of which were sold on China's domestic market. Since private property was banned, most of these cars were for state use only, but occasionally a citizen received one as a reward for exemplary "patriotic" behaviour.

After Mao's death in 1976, China sought to reform many of its political systems and to adopt a market-led economy. Under its new leader, Deng Xiaoping, the country opened up to foreign trade and investment, and encouraged a domestic consumer market, particularly for cars.

As the Chinese people enjoyed new freedoms, the demand for private cars soared, far outstripping China's own ability to produce them. As a result, thousands of cars were imported, mainly from Russia and Japan, leaving China with a huge trade deficit.

In response, China imposed tariffs on imports, but the long-term solution was the fostering of joint ventures between Western companies and Chinese manufacturers. Accordingly, in the early 1980s, American Motors Corporation, Volkswagen, Peugeot-Citroën, and others all built assembly plants in China. This boosted production figures for China, and gave Europe and the US access to relatively cheap labour.

Working in partnership

Today, China's car industry is dominated by five main state-owned groups; SAIC, Dongfeng; FAW; Beijing Automotive (part-owned by the Daimler group); and Chang'an. All have partnerships with major foreign manufacturers, including Volkswagen, Ford, General Motors, Honda, and Peugeot-Citroën, to produce these marques for the Chinese domestic market. In addition to these joint ventures, however, some Chinese manufacturers have purchased Western marques. China's largest car builder, SAIC, now owns the UK's MG brand, while Geely, its tenth largest carmaker, owns the Swedish manufacturer Volvo, the London Taxi Company, and the Malaysian car company Proton, which in turn owns Lotus.

These collaborations have been a huge success, but it has not been without controversy. In the West, copying the creative work of another is unethical, and possibly illegal. But in Chinese culture, replicating the work of a "master" is considered a great tribute. As such, Chinese carmakers have often copied designs from other car companies, leading to the inevitable friction and legal action.

> "In terms of **volume... China** is **building** with **no limit**."
>
> KLAUS ZELLMER, PRESIDENT AND CEO OF PORSCHE

It is estimated that by 2020 there will be 200 million vehicles on the roads of China. The sheer size of this market is already having effects on car design across the world. Western carmakers increasingly include Chinese consumers in their design review process to ensure that their new cars are better suited to the Chinese market. The increase in chrome exterior trim was a response to Chinese buyers' demands for high-status exterior appearance. Increased emphasis on rear seat space caters to the Chinese preference for using a driver and riding in the back, and several Western manufacturers offer long-wheelbase versions of their saloons for the Chinese market. As spectacular motor shows in Beijing and Shanghai prove, China is having a huge influence on the market.

KEY DEVELOPMENT

Take a second look...

English writer Charles Colton once claimed that "Imitation is the sincerest form of flattery" – a phrase that could certainly apply to the Chinese motor industry. Over the decades, many Chinese cars have appeared that bear striking resemblances to existing American and European designs. The Landwind X7 SUV of 2014, for example, was regarded by Jaguar Land Rover as a copy of its Range Rover Evoque. Likewise, Zotye Auto's SR9 of 2016 is a near-perfect clone of the Porsche Macan. Such counterfeits are cheap to make and affordable on the domestic market – and legally, many have proved all but impossible to challenge.

THE DESIGN OF THE LANDWIND X7 **(LEFT) HAS BEEN CRITICIZED FOR BEING TOO SIMILAR TO THAT OF THE RANGE ROVER EVOQUE (RIGHT).**

△ Auto Shanghai, 2017
Auto Shanghai is one of China's premier automotive exhibitions. Here, a Hyundai Celesta saloon stands on display, proof of the increasing presence of foreign brands in the Chinese market.

△ **Mercedes-Benz GLE 850 Brabus, 2016**
Based on a Mercedes-Benz GLE Coupé, German tuning specialist Brabus transformed the standard model, giving it a far more powerful engine and improved suspension. It can still seat four people in comfort.

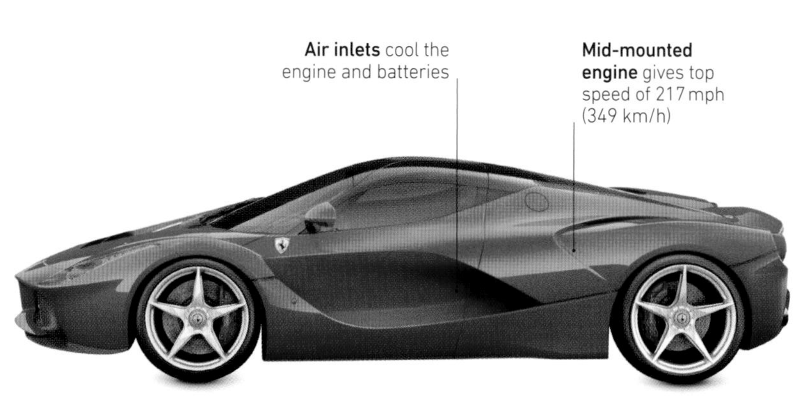

△ **Ferrari LaFerrari, 2013**
Ferrari's first hybrid model, the LaFerrari uses an electric motor to augment its 6.3-litre, V12 engine for brief bursts of acceleration. The combined power of the engine and motor is 950 bhp.

Super-fast supercars

Driving a car at over 200 mph (322 km/h) is something few drivers will ever do, and until 1987 it was not even possible in a standard road car. That changed with the launch of the awe-inspiring Ferrari F40, with its twin-turbo, 478-bhp, V8 engine, which was viewed at the time with the same level of wonder as a 300-mph (482-km/h) car might be seen today. But in the intervening three decades things have moved on, meaning that now not only can supercars exceed the magical "two-ton" figure, but SUVs can too. Technology previously only seen on the racetrack has brought increased power and speed to road-going cars, with many now able to beat the figure that was once seen as a benchmark.

With the Bugatti Chiron limited to 261 mph (420 km/h), yet believed to be capable of reaching 288 mph (463 km/h) with the limiter removed, car manufacturers are edging ever closer to building models that can reach the 300 mph mark.

"**Nothing** is too beautiful, nothing is **too expensive.**"

ETTORE BUGATTI, FOUNDER OF BUGATTI

△ **Nissan GT-R, 2007**
Featuring a 3.8-litre, twin-turbo engine, advanced four-wheel drive, plus aluminium and composite body parts, the GT-R was one of the world's most technologically advanced cars when it was launched.

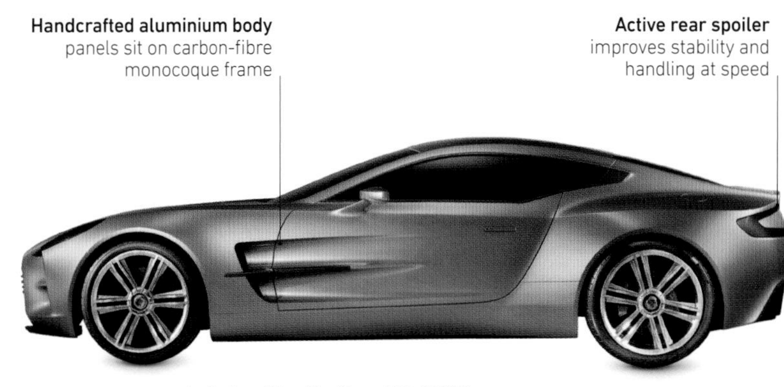

△ **Aston Martin One-77, 2008**
The most powerful naturally aspirated car in the world at its debut, the 750-bhp Aston Martin One-77 uses a 7.3-litre derivative of Aston Martin's venerable V12 engine. Aptly, it can reach 220.007 mph (354.067 km/h).

▽ **Bugatti Chiron, 2016**
Powered by an 8-litre W16 quad-turbocharged engine, the Chiron is the successor to Bugatti's groundbreaking Veyron. Its top speed is limited for safety reasons, but it will still reach 261 mph (420 km/h).

Honda's safety system

After the 1970s, safety assumed an ever more important role. In 2002, Honda launched a new version of the Accord in Japan that offered a sophisticated new safety system to keep the car centred on the road.

As well as being one of the world's most popular cars, the Honda Accord has long been one of the most technically advanced of its class. The Lane Departure System (LDS) that came with the 2002 model in Japan was a milestone in safety and driver assistance. The LDS is an electronic system designed to keep the car "in lane" while cruising on the motorway, and so reduce the burden of work on the driver. It came as part of a package called the Honda Intelligent Driver Support (HiDS) system, which also served to maintain the car's speed and distance from other vehicles on the road. It was the type of extra that might have been found in an expensive Volvo or Mercedes-Benz, but the Accord was an affordable car, and one of the first mainstream models to offer such advanced safety technology.

The core of the system was a piece of technology called the Lane-Keeping Assist System (LKAS). This identified the road ahead, based on an image captured by a digital camera mounted at the top of the car's windscreen. The car's engine control unit (ECU) then calculated the appropriate degree of steering assist to keep the car in lane. The system was set to work at speeds of 40 mph (65 km/h) or higher. If the car drifted out of lane, a series of bleeps encouraged the driver to steer back in. Combining LKAS with a radar system that could automatically regulate speed and the distance to the vehicle ahead gave the HiDS-equipped Accord a new edge in the safety stakes of the day.

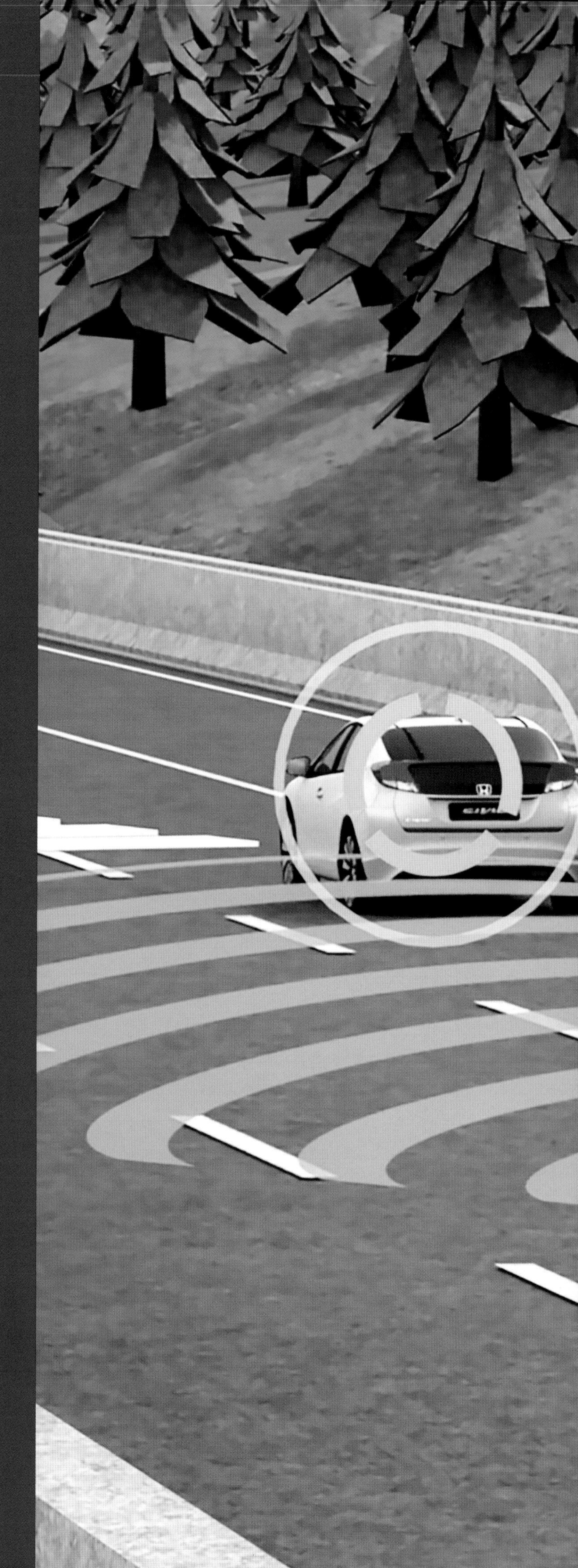

▷ **Evolving features**
The safety systems pioneered by the 2002 Accord in Japan now feature on a range of Honda cars. The 2015 CR-V, simulated here, has a radar built into its grille and a camera at the rear, both of which monitor the vehicle's headway and position on the road.

Dawn of the hybrid

Following the launch of the first Toyota and Honda hybrids in the late 1990s, the "green car" era arrived in earnest in the 21st century. At the same time, electric cars gradually came of age.

△ **Lithium-ion battery cells**
This "li-ion" battery is designed for use in hybrid and electric vehicles. Lithium-ion batteries offer high energy density and are also used in portable electronics.

Hybrids are a modern-day solution to the need for environmentally friendly cars. As their name implies, they have two power sources – a traditional combustion engine and an electric motor attached to a battery – which work together to reduce emissions and boost fuel economy.

Japan leads the way

Across the industry, Japanese manufacturers, particularly Toyota and Honda, have long favoured developing hybrid models, despite the extra costs and complexities involved. Toyota launched the first Prius as its original stand-alone hybrid model in Japan in 1997. The eco-friendly Prius soon became the poster child for the coming green car revolution, and each new generation (in 2003, 2009, and 2015) has been steadily improving on the last. The Prius also paved the way for a growing army of Toyota and upscale Lexus hybrid models, both "standard" hybrid and "plug in" variants, the latter engineered to give a longer driving range on pure (zero-exhaust emission) electric drive. Meanwhile, Honda has established itself as a close competitor to Toyota, launching its first Insight hybrid – a tiny, spectacular, teardrop-shaped coupé – in 1999. Honda's core hybrid technology, called IMA (Integrated Motor Assist), has appeared in a

▽ **BMW i8**
The i8, a futuristic, LED-lit sports car, is a hybrid capable of 75 mph (120 km/h) under purely electric power. Its carbon-fibre passenger compartment reduces weight.

Electric battery used when starting the car and driving at low speeds

Petrol engine used while cruising

△ Inside the Lexus
This cutaway of a 2006 Lexus GS 450h reveals Toyota's Hybrid Synergy Drive (HSD), powered by both combustion and electric motors. The system is a refinement of the one from the 1997 Prius.

number of Fit (Jazz), Civic, and Accord variants, as well as the racy CR-Z sports coupé. Meanwhile, Honda created something bold and new with the highly advanced, three-motor SH-4WD hybrid system in the latest NSX supercar, using state-of-the-art technology.

The spread of the hybrid

While Toyota and Honda have become the most visible hybrid manufacturers, other automotive companies from across the world, including Volvo, General Motors, Peugeot, and Mercedes, have also been developing hydrid technology. Some manufacturers still see hybrids as a part-way solution to having purely electric cars, which for a long time were dogged by cost and "range anxiety" issues – not being able to travel very far due to limited battery capacity – and so were deemed uncompetitive.

> "We have to **decarbonize** the **transport system**."
>
> PROFESSOR DAVID BAILEY, ASTON UNIVERSITY, UK

From hybrid to electric?

GM had made an early move into electric cars with its pioneering EV1 coupé (made between 1996 and 1999) – the 2006 documentary film *Who Killed the Electric Car?* highlighted its early struggles and eventual demise (see pp.294–95).

Today, as the auto industry moves towards wholesale electrification, both hybrid and completely electric cars are set to take the place of the conventional engines of old. The Swedish manufacturer Volvo has pledged that, beginning in 2019, all new models will feature some form of electrification, while other companies, including Renault-Nissan, BMW, and Volkswagen, have announced bold plans for hybrid and electric vehicles. Meanwhile, the first mass-market electric car of the 21st century, the Chevrolet Bolt, was released in 2017, and Tesla has plans to release its Model 3 by 2018. It is a move not only fuelled by manufacturers: in 2017, both the French and UK governments unveiled plans to ban the sale of all new petrol and diesel vehicles. Pioneered by hybrids, the future of motoring looks increasingly electrified.

▷ Peak performance
The dashboard display of the Toyota Prius gives drivers a visual representation of the energy transfer taking place between the battery, the electric motor, and the combustion engine. It also shows fuel consumption and battery life.

△ **Spyker C8 Aileron, 2008**
The Spyker marque had been dormant since 1926 until Dutch entrepreneurs revived it for this 186 mph (300 km/h), handbuilt supercar that was made in small and exclusive numbers.

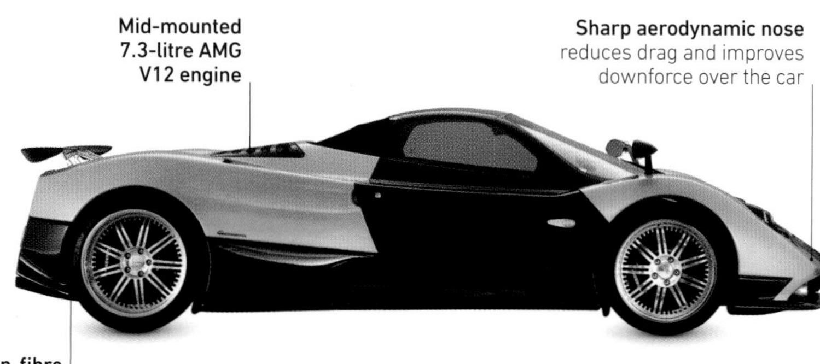

△ **Pagani Zonda Roadster F, 2006**
The Italian-built Zonda was refined over a 12-year period that saw many breathtaking variations built. Lightweight and powerful, it could almost match the Bugatti Veyron for speed.

Handcrafted cars

In the early days of the motor industry, everything was built by hand (see pp.34–35). While mass-production, 1,000-cars-a-day factories are mostly the territory of soulless robots, the upper echelons of the motoring world have always preferred the human touch. Even today, sales catalogues are still well stocked with cars assembled by time-served craftsmen. From the materials used to the ways they are put together, artisan cars offer the discerning motorist an experience similar to that of coach-built cars in days gone by, where customers can deviate from the options list to make the car their own, in colour, trim, or extras.

Such cars also allow the motorist to experience a different type of driving experience from the mainstream motors – whether it's the open top pleasures of a Morgan, the handcrafted luxury of a Bristol, or the "stop and stare" looks of a Pagani. Compared to their mass-produced rivals, artisan cars feel crafted, tailored, special.

▽ **Morgan Aero 8, 2001**
Combining modern engineering and construction with classic styling, the Aero 8 was Morgan's first new design in almost 40 years. Powered by a 4.4-litre, V8 engine, it could reach 150 mph (241 km/h).

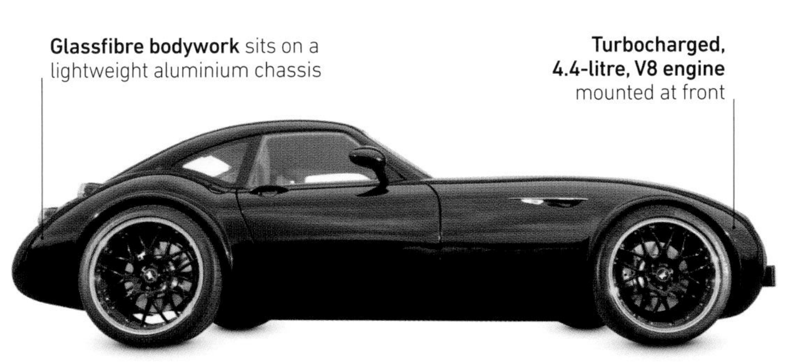

△ **Weismann MF4, 2007**
Weismann's philosophy was similar to Morgan – traditionally styled sports cars with a big engine and plenty of power. Formed in 1993, its last cars were produced in 2013.

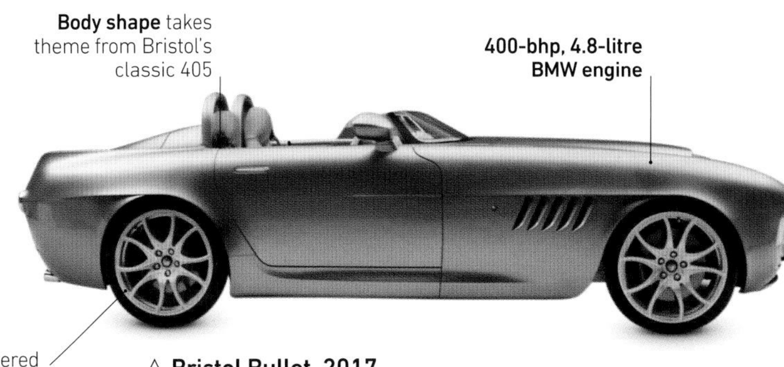

△ **Bristol Bullet, 2017**
The Bullet is Bristol's first new model since the company was reborn in 2011. Built from carbon fibre and aluminium, the body has aerospace-inspired styling, reflecting the company's origins.

"We **make cars** like a tailored **dress or suit**..."

HORACIO PAGANI, FOUNDER OF PAGANI AUTOMOBILI

▷ **Mitsuoka Orochi**
Mitsuoka is Japan's leading bespoke car company, and its mid-engined Orochi, released in 2006, features a sumptuous leather interior.

Under the bonnet

Creating powerplants for cars was a surprisingly diverse science, with a bewilderingly wide range of engines made over 130 years.

Early internal combustion engines tended to be simple affairs – often just a single cylinder – but adding more cylinders made them smoother and more powerful. Arranging the cylinders in a line – the "straight" engine – became the predominant format, easily adaptable to two, three, four, five, six, and even eight cylinders.

Arranging cylinders in a "vee" formation saved space and added refinement. Lancia favoured V4 engines for decades from the 1920s. American manufacturer Marmon launched a V6 in 1905, and would later build a V16. Cadillac built the US's first V8 engine in 1914, and it was destined to become a flexible long-distance performer.

The "flat" cylinder layout was rarer, with pairs of pistons horizontally opposed to each other, such as the Volkswagen Beetle's flat-four, and Porsche's flat-six that continues to this day. Oddball engines included Felix Wankel's rotary engine, first used in the 1964 NSU Spider. Hybrid powerplants with electric motors added to lower emissions and boost efficiency, popularized by Toyota's Prius.

FORD MODEL T STRAIGHT-FOUR CUTAWAY, 1908–27

VOLKSWAGEN FLAT-FOUR, 1936–2003

CHRYSLER FIREPOWER HEMI V8, 1951–58

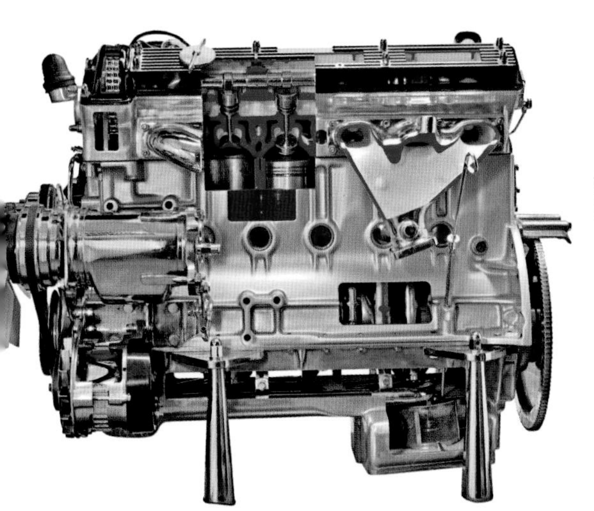

JAGUAR XK STRAIGHT-SIX, 1946–86

NSU WANKEL ROTARY, 1967–77

LOTUS/FORD COSWORTH DFV V8, 1967–86

PORSCHE 911 FLAT-SIX, 1963–98

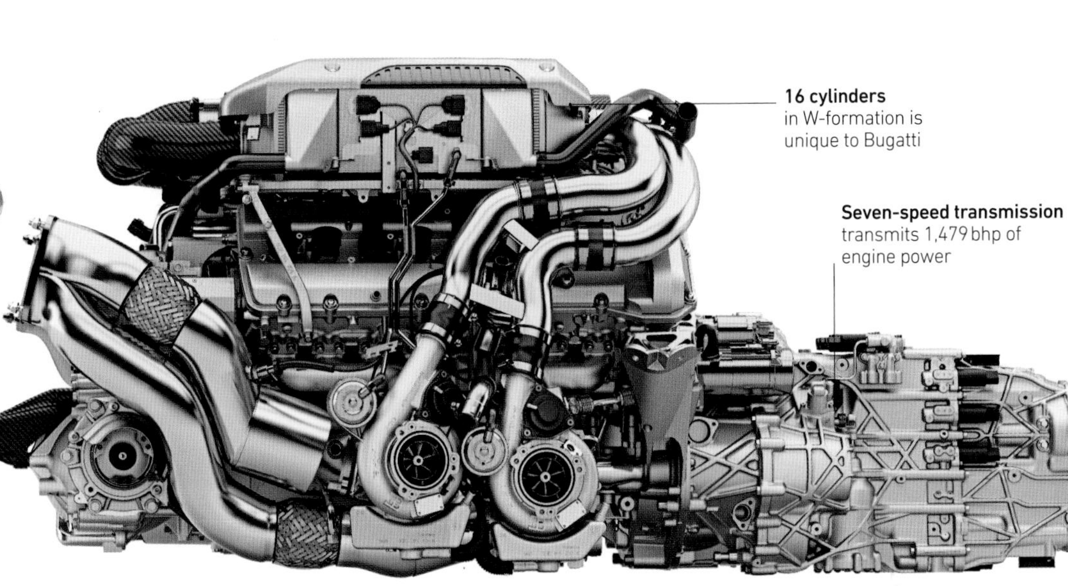

BUGATTI VEYRON W16, 2005–ONWARDS

CHRYSLER/DODGE VIPER V10, 1991 ONWARDS

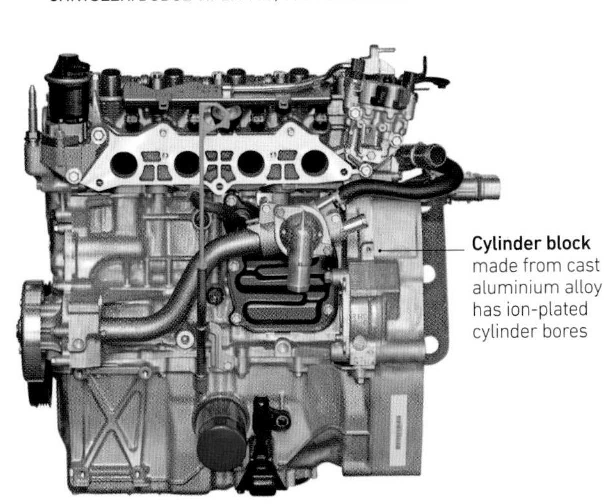

HONDA INSIGHT HYBRID, 2010 ONWARDS

DRIVING TECHNOLOGY
Engine placement

There is no "correct" place for a car's engine. The industry convention of a front-mounted engine sending drive to the rear wheels was established early on, and made for good weight distribution. Volkswagen flouted that norm in 1945, siting the engine for its new Beetle at the rear for a better packaging solution. A front-mounted engine turning the front wheels rapidly became commonplace after the launch of the 1959 Mini; this made roadholding much more predictable and sure-footed. From the mid-1960s onward, sports cars, such as the Lamborghini Miura and Lotus Europa, followed the racing practice of centrally mounted engines for the best handling balance.

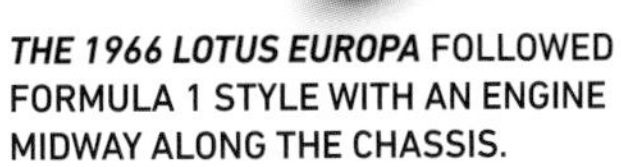

THE 1966 LOTUS EUROPA **FOLLOWED FORMULA 1 STYLE WITH AN ENGINE MIDWAY ALONG THE CHASSIS.**

▽ **Cold-weather testing, 2013**
To ensure that their cars function properly no matter where their customers drive them, manufacturers test prototypes in extreme environments. Here, McLaren puts a prototype P1, covered with markings to disguise its appearance, through its paces on a frozen lake in northern Sweden.

The emissions scandal

The Volkswagen emissions scandal that broke in September 2015 kicked off a heated debate among environmental authorities and in the media about protocols for measuring car pollution and whether diesel engines should be banned.

In 2015, American authorities discovered that VW had tampered with its new diesel cars to ensure that they passed the stringent US emissions tests, setting in motion a chain of events that few could have predicted.

It transpired that VW had programmed some 11 million of its diesel cars with so-called "defeat" software, which activated when the engines were tested under laboratory conditions to produce artificially low emissions. Once on the road, however, those same engines generated nitrogen oxide pollutants up to 40 times higher than was permitted by US regulations. When the fraud was uncovered, VW was forced to pay $17.5 billion, both to compensate affected car owners and to fund clean air projects.

The media dubbed the affair "Dieselgate", and it affected not only the VW marque, but also the Audi, Seat, Porsche, and Škoda brands too, which were all made by Volkswagen. And of course, although the issue was detected in the US, where VW had launched a major pro-diesel promotion for its "clean" machines, the altered vehicles had also been sold worldwide.

The dangers of diesel

The revelations prompted governments to reconsider their stand on diesel car emissions. Complicating the matter was the fact that up until 2012 many governments had championed diesel engines as the engines of the future, as they were thought to emit less CO_2 than petrol cars. In 2001, for example, UK

▽ **Deadly emissions** It is estimated that there have been 38,000 premature deaths as a result of lax diesel emissions testing industry-wide.

△ **Dieselgate demonstrations**
Activists hold up a sign that reads "Stop Lying" while standing on a VW car during a protest in front of VW's headquarters in Wolfsburg, central Germany.

chancellor Gordon Brown had cut duties on diesel fuel in an effort to lure more people to switch over from petrol vehicles. Then, in 2012, authorities were forced to do a U-turn after the European Environment Agency released statistics showing that 71,000 people had died prematurely across the continent that year due to inhaling nitrogen oxide from diesel fumes. Dieselgate reignited the debate about the health risks of diesel.

A few months after the VW fraud was revealed, 153 countries signed up to the UN Paris Climate Accord, pledging to gradually reduce toxic emissions to zero by mid-century. This left car-producing nations, and the governments that had previously encouraged diesel use, in a difficult position, especially in Europe, where diesel cars regularly outsold petrol cars, and demand for diesel fuel had tripled since the mid-1990s. In fact, in 2013, Germany had secretly agreed to block a proposed EU cap on banker bonuses, of great concern in the City of London, in return for British help in resisting Europe's proposed stricter emissions policy.

Lowering emissions

Nevertheless, the Paris Accord spurred diesel-guzzling Europe into action. France, for one, pledged to ban not only diesel cars but also petrol cars by 2040. Global carmakers agreed to comply, although the Japanese already had a head start with a high market penetration of hybrid cars, resulting in 16 per cent lower CO_2 emissions than in Europe. On a more immediate time scale, anti-diesel action unfolded at a local level. Berlin banned older diesels from the city centre, while Paris, Oslo, Hong Kong, Seoul, and Mexico City, among other cities, all took steps to restrict or ban diesel cars. While the death of the diesel car may now seem inevitable, there could still be a stay of execution. Renewable diesel fuel, produced from waste fats since 2008, and now an evolving, expanding energy source, could prove the last hope for keeping diesel engines on the road.

DRIVING TECHNOLOGY
Biofuel

Traditionally, motor vehicles have been powered by fuels derived from crude oil, which, like coal, is a fossil fuel that took millions of years to form. Being non-renewable, oil reserves are depleting, but the fuels derived from them can also be harmful when burned, and so alternatives are currently being sought. One such alternative is "biofuel", which is made from plant material, and so is endlessly renewable.

One of the best-tested biofuels is biodiesel. This is commonly made from rapeseed oil and can be used in any diesel-powered engine without modification. Another is bioethanol, which is made by fermenting the sugars in sugar cane or wheat. On its own, bioethanol requires engine modification, but most modern petrol engines can burn petrol mixed with a 10 per cent dose of ethanol.

THE CHASSIS **OF A CHEVROLET VOLT, AN ELECTRIC CONCEPT CAR THAT CAN RUN ON ELECTRICITY, PETROL, OR BIODIESEL.**

> "The **irregularities**... go against **everything** VW **stands for**."
>
> DR MARTIN WINTERKORN, CEO, VW

◁ **Recalled VW diesel vehicles**
Following the scandal, VW made buy-back offers on nearly 330,000 of its cars, and stored them in regional facilities such as this parking lot in Pontiac, Michigan.

Autonomous cars

Touted as the transportation of the future, autonomous cars are edging closer to taking on public roads as carmakers sink huge sums into their development. But there are still issues to resolve before the driverless dream becomes reality.

△ **Firebird II**
The Firebird II (above left) was a concept car produced by General Motors in 1956. Designed by Harley Earl (above), it supposedly had a guidance system that was intended to read sensors in the road.

The idea of a car that can drive itself has existed in motoring culture for over a century. Almost as soon as cars had established themselves as the primary mode of transport, artists and engineers began imagining vehicles that did not need a human to drive them.

The first attempt

As early as 1925, the first self-driving car made its debut – a radio-controlled vehicle made by Houdina Radio Control. The brainchild of inventor and ex-army engineer Francis Houdina, the car made its way through the traffic of New York City without incident, taking radio signals from a second car travelling behind it. Houdina's idea was to take radio control one step further so that the autonomous car would receive radio signals from telephone wires at the side of the road to guide it, rather than from an escort vehicle.

Handing over control

Despite the early promise of Houdina's concept, the prospect of an automated vehicle (AV) made little progress until a few decades later, when GM dabbled with the idea. Its 1956 Firebird II concept car was fitted with sensors intended for use on a "highway of the future" – one that would be seamed with cables that sent signals to guide the driver.

The idea was developed further in the 1960s, with projects funded by the US Bureau of Public Roads and Transport and the Road Research Laboratory in the UK, but it failed to bear any fruit. The first real step towards automated driving was taken in 1958, when the Chrysler Imperial became the first car to feature cruise control. This enabled drivers to cruise at the same speed for the first time without having to hold the accelerator.

Although there were further developments in the 1970s, the production of the first true AV was set in motion with the appearance of the VaMP in Germany in 1995. A re-engineered Mercedes S Class, the VaMP (or Versuchsfahrzeug für autonome Mobilität und Rechnersehen, meaning "experimental autonomous mobility vehicle with computer vision") drove from Munich to Copenhagen at speeds of up to 108 mph (175 km/h). Funded by the EUREKA Prometheus Project – a pan-European, inter-governmental organization dedicated to research and innovation – VaMP laid the foundation for today's generation of autonomous cars.

The DARPA Grand Challenge

Spurring on the innovation was the DARPA Grand Challenge launched in 2004, offering a prize of $1 million to the best self-driving car. An initiative of the US Department of Defense, it garnered interest from engineers around the world, and was followed up by further events in 2005 and 2007, each of which pushed self-driving technology further. A team drawn from the DARPA challenges was recruited by Google, which launched the world's most advanced self-driving car in 2009.

Meanwhile, various car manufacturers, including BMW, Volkswagen, and GM, had invested heavily in autonomous car technology. Honda announced it would release a near-fully autonomous vehicle in 2020,

◁ **DARPA Grand Challenge, 2005**
The autonomous robotic vehicle from the Stanford Racing Team reaches the finish line at the DARPA Grand Challenge in 2005. The event trials emerging technologies that could be of use to the military.

△ **Tesla Model 3**
The Tesla Model 3, produced in 2017, offers semi-autonomous driving in certain situations. It also features an all-electric range of 220 miles (350 km) and a dashboard with a central touchscreen instead of dials.

and BMW promises to launch its own AV in 2021 – one that only needs the driver in emergencies.

Uncertain future

After taking almost a century to evolve, truly autonomous cars are now on the verge of becoming a mass-market reality. However, while the pace of self-driving technology is accelerating, the infrastructure required to support them is not. One reason for this is the issue of driver liability. In the case of an accident, who is to blame if a human driver is not in control? AVs may also have a negative effect on the insurance industry. Insurers are likely to start treating AVs as a new class of vehicle – one that is safer, involved in fewer accidents, and so subject to lower premiums.

> "It will be **unusual** for cars to be built that are **not** fully **autonomous**."
>
> ELON MUSK, CEO, TESLA

DRIVING TECHNOLOGY
Motoring future

In a world in which driving is computer-controlled, and in which car systems have wireless connections to the outside world, security is a risk that needs to be addressed to prevent hacking and exploitation. In response, carmakers are focusing on developing secure networks and over-the-air software updates that can be installed as soon as they are required in real time. Industry experts also foresee that car ownership will fall, as drivers use AVs as a flexible, mobile service that can be delivered on demand from a third party to suit specific needs, such as a business trip or a shopping run. This is just one of the ways in which the driverless car may bring with it a complete change in motoring culture as we know it.

IN THIS GRAPHIC, A DRIVER READS A BOOK **IN THE COCKPIT OF AN AUTONOMOUS CAR, WITH THE OPTION OF DRIVING MANUALLY.**

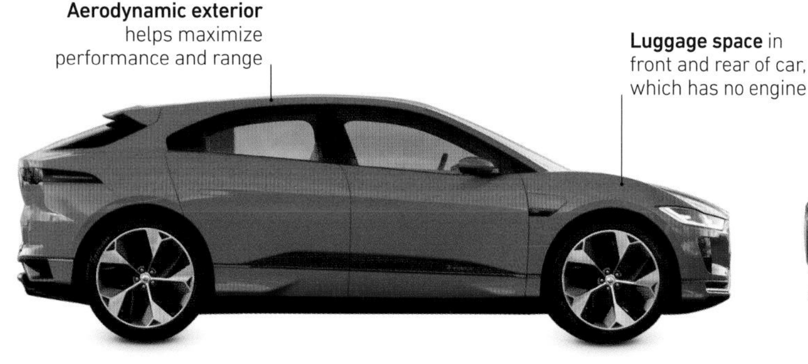

△ **Jaguar i-Pace, 2016**
This radical concept will become Jaguar's first production electric vehicle in 2018. It has a range of over 310 miles (482 km) on a single charge, and will accelerate from 0-60 mph (0–96 km/h) in four seconds.

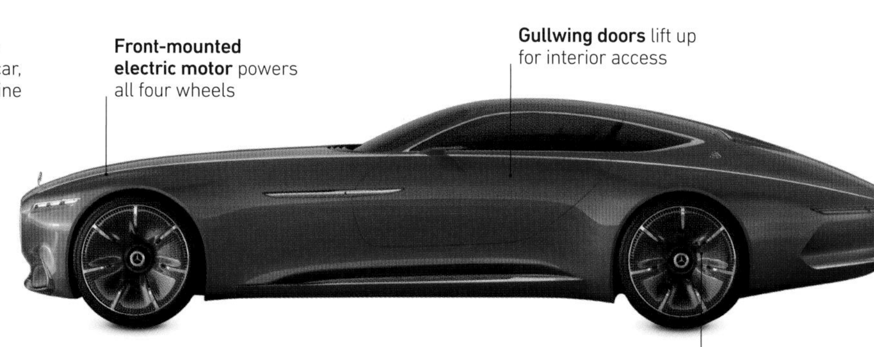

△ **Mercedes-Maybach 6, 2016**
Unveiled at the prestigious Pebble Beach Concours d'Elegance in 2016, this extravagant, electric coupé is nearly 20 ft (6 m) long, yet is only a 2+2. It can be recharged wirelessly via an electro-magnetic field.

Future concepts

Not since the pioneering days of motoring has there been such an intense period of change in the motorcar as there is today. New forms of propulsion and the development of autonomous driving are creating new challenges, opening up design possibilities that have never been tested before. Carmakers are using concept cars to explore innovative methods of operation, construction, and propulsion – and even question fundamentals of car design, such as the need for glass windows and the positioning of occupants within the vehicle. Some of these concepts are close to forthcoming production cars, others are more like fantasies intended to get people talking – but each is fascinating in its own way. Only time will tell which will hit the roads.

▷ **Volkswagen Sedric, 2017**
This is the Volkswagen Group's idea of a self-driving vehicle. It is operated by voice commands, and so has no steering wheel, pedals, or conventional controls.

◁ **Airbus Pop.Up**
Separate ground and air modules convey Pop.Up's passenger capsule through and over city traffic. Artificial intelligence systems plan the most efficient route and which module to use.

"The **time** is **right** for **electric cars** – in fact, the time is **critical**."

CARLOS GHOSN, CHAIRMAN AND CEO OF RENAULT

△ **Toyota i-TRIL, 2017**
Unveiled at the Geneva motor show in 2017, the i-TRIL electric vehicle is Toyota's alternative to city cars, small hatchbacks, and motorcycles. It aims to bring passion and driving pleasure to urban mobility.

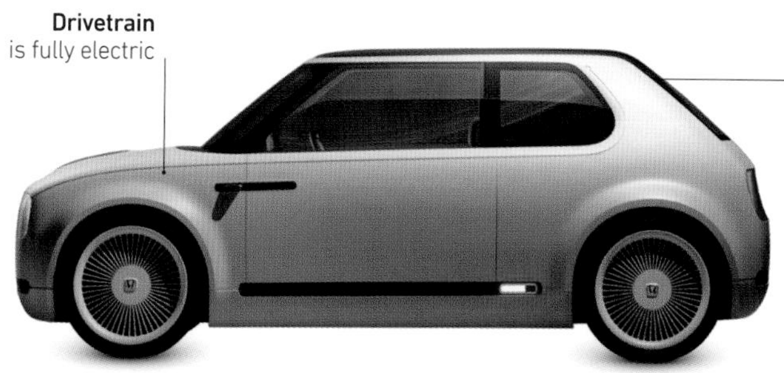

△ **Honda Urban EV, 2017**
This electric city car, design with room for four, has a chunky retro appearance. The wood-trimmed dashboard features a full-width information screen.

GREAT
DRIVES
INSPIRING ROUTES FROM
AROUND THE WORLD

North America

▽ RIVERHURST CROSSING

CANADA

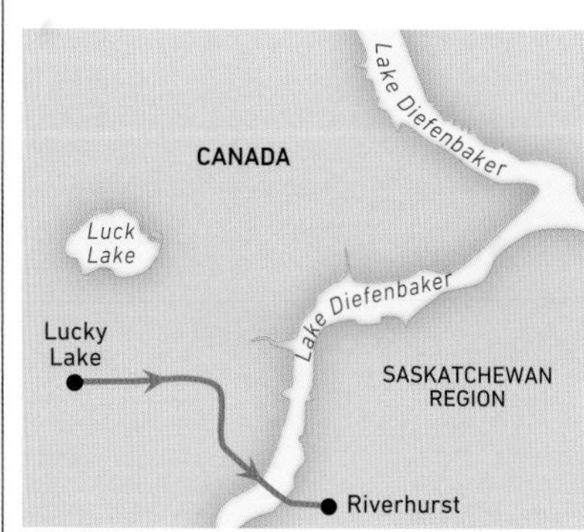

Highway 42 runs across the Canadian province of Saskatchewan – a flat prairie region dotted with 100,000 lakes – linking Highway 2 with Highway 15, 124 miles (199 km) away. About halfway along Highway 42, the road must cross Lake Diefenbaker – a 200-ft (60-m)-deep, man-made reservoir, which is 1 mile (1.6 km) wide at the point where the road meets it.

In summer, drivers making the 16-mile (26-km) journey from Lucky Lake on the west bank to Riverhurst on the east must wait for the traditional cable ferry boat that crosses the lake. It is a toll-free, 24-hour service, which carries up to 15 vehicles on its hourly sailings.

During the winter months, however, the lake freezes over and is often covered with snow – and that is when this route is at its best. Since there is no waiting for the next ferry, Highway 42 is simply diverted across the frozen lake, creating the Riverhurst Crossing, one of Canada's famous ice roads. Driving it is an exhilarating experience, and the flat, snow-covered landscape has a beauty all of its own.

Crossing a frozen lake may sound dangerous, but the route is perfectly safe. It is carefully marked by the Ministry of Highways each year as winter approaches. The snow is cleared away, and the frozen carriageway is lined with cones – plus signs that remind drivers of the 30-mph (50-kph) speed limit. A maximum weight limit of 5 tons (10,000 lbs) is also enforced, which is well over the weight of the heaviest family car or SUV.

It is forbidden to drive on the ice road before the Ministry of Highways officially declares the crossing open, and those who break the law in search of adventure can expect to pay hefty fines if they are caught. Throughout the winter months, experts regularly measure the thickness of the ice, and the route is closed if it is deemed unsafe. At such times, local drivers face an enormous road trip to get between the lakeside communities.

During the river's freeze and thaw periods, the road is closed and the ferry is unable to operate. At these times, Highway 42 is impassable.

A CAR CROSSES THE FROZEN LAKE DIEFENBAKER, SASKATCHEWAN

▷ YOHO VALLEY ROAD

CANADA

THE YOHO VALLEY ROAD WINDS THROUGH BRITISH COLUMBIA

As the 7-mile (11-km) Yoho Valley Road branches off to the north from the Trans-Canada Highway, next to Shuswap Lake, a sign warns that the road is not suitable for motorhomes or trailers. The way ahead is narrow and winding, with many steep corners, and no-one should tackle it in a ponderous RV. Even in the kindest weather, it is best suited for smaller vehicles and motorbikes.

Canada is full of epic road trips, Arctic driving adventures, and serious motoring challenges. Some follow famous long-distance highways, and plenty offer great driving experiences. This is a lesser-known, smaller route, but it is one that gives you a chance to take a difficult road into a fabulous part of the Rocky Mountains without undertaking a major expedition.

The Yoho National Park lies in British Columbia, on the western slopes of the Continental Divide of the Americas. "Yoho" is the Cree word for "awe" and "wonder", and the Park forms part of the Canadian Rockies World Heritage Site. A visitor centre is situated in the small village of Field near the Yoho Valley Road. Field is the only settlement in the whole National Park, and has a population of around 200. The Yoho Valley Road passes into a magnificent valley along the course of a wild mountain river. It is the only driving route that takes you to one of the major sights of the Park – the Takakkaw Falls. Fed by the melting ice of the Daly Glacier, this spectacular cascade plunges from a mountainous cliff over 1,000 ft (300 m) high, making it the second-highest waterfall in Canada. During peak season, the fall is torrential, and the forested valley below becomes damp with the spray, making rainbows a common sight. A carpark marks the end of the route. From there, tracks lead off to waterfall viewing points, or turn into hiking trails that take you up into the mountains. If you follow one of the trails, many more waterfalls await. Back in the car, there is nothing for it but to turn and retrace your route to the Trans-Canada Highway.

Due to heavy snow in winter, the road is only open in the summer months – usually from June to October. However, whenever you go, it is advisable to check that the road is open.

▽ FLORIDA KEYS

US

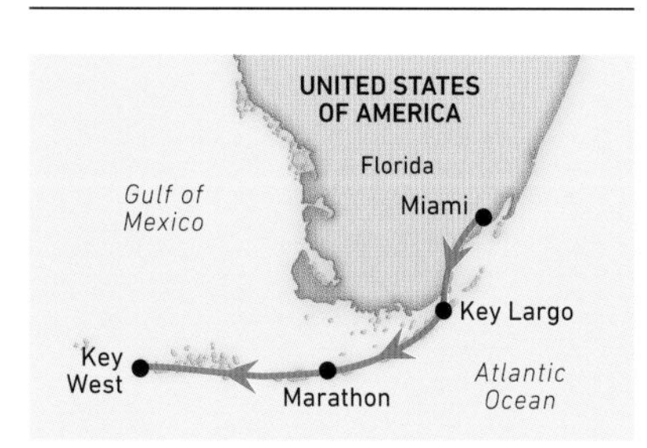

Begin this fantastic drive a little before the official starting point, in the epicentre of Florida's beach life. Driving along Miami's Ocean Drive in South Beach is a chance to compare the trendy Art Deco bars and hotels of this ultra-cool city beach with what is to come on the Keys. Pass the glossy skyscrapers of Miami itself and then head south on Highway 1. The Scenic Highway officially starts at the tip of the Florida peninsula. This is where a gentle arc of low-lying islands stretches into the Gulf of Mexico, like a string of pearls. It is a world-class drive, and the closest any road-going car can get to island hopping in the Caribbean. The smooth, flat highway leads onto the first island, Key Largo, and then swings 166 miles (267 km) to Key West, the final island of the chain.

There are 43 islands, or keys, in all – the word "key" deriving from the Spanish word "cayo", meaning "small island". The route is a sequence of long, low bridges and causeways – 42 bridges in all, the longest of which stretches for 7 miles (11km) across the Gulf of Mexico. For most of the drive the road seems to be floating just above the sparkling blue tropical sea, the views alternating between island geography and seascapes.

The islands themselves are a mixture of coral and limestone and are covered with mangroves, which flourish in the brackish water. There are also plenty of palm trees, which shade the sandy beaches. All kinds of wildlife can be seen, ranging from pelicans and turtles to dolphins and manatees – or "sea cows", as they are known locally.

At Key West, the road reaches the southernmost point of the US. Here, the luxurious beachside retreats of the rich and famous stand among the swaying palms. Period homes with exotic gardens can also be found, plus a scattering of quirky arts-and-crafts shops.

The island is a popular spot for all kinds of water sports and fishing expeditions. You can go diving or snorkelling in the coral reefs, take boat trips, or go swimming. Otherwise you can just find a hammock to lie in, or relax at atmospheric venues such as Sloppy Joe's Bar – a former haunt of two historic characters: Ernest Hemingway, the writer, and the rum runner Habana Joe.

More than anything, Key West is a great place to park up, sit by the ocean, and reflect on the fact that you are now closer to Cuba than to mainland America.

THE FLORIDA KEYS SCENIC HIGHWAY, HEADING OUT TO SEA

VISITOR INFORMATION ON THE LEWIS AND CLARK TRAIL

△ LEWIS AND CLARK TRAIL

US

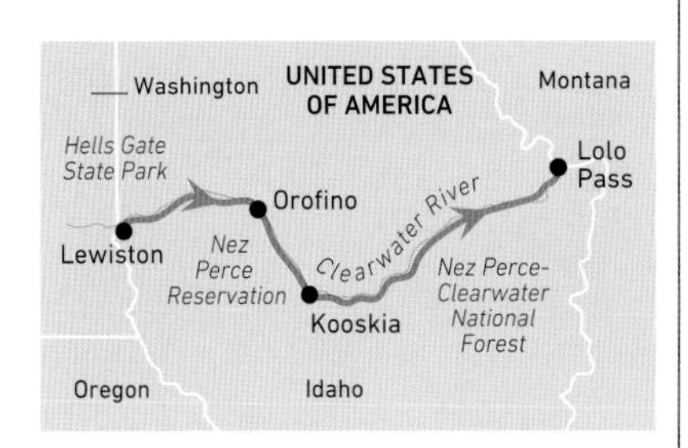

Driving routes designated as official US Scenic Byways have long been considered the most attractive roads on the continent. This 174-mile (280-km) backcountry route through the mountainous landscapes of Idaho, following the Clearwater River Canyon, is one such road.

The well-marked journey is easy to follow and gives you the chance to retrace at least some of the footsteps of the early-19th-century explorers Lewis and Clark. The intrepid duo were commissioned by US President Thomas Jefferson to explore the enormous tract of land west of the Mississippi River that the US had recently bought from the French – an area almost a third of the size of today's United States – for $15 million.

The road gives you access to landscapes that have changed little since the explorers battled through them in 1804. The journey is a lot smoother today, thanks to US Highway 12, which was built in 1925.

Highlights of the drive include Hells Gate State Park and sites celebrating Native American history. It is believed that a Native American guide helped Lewis and Clark struggle through snow to cross the Lolo Pass, which stands at 5,233 ft (1,595 m). The current driving route crosses the same point to reach Montana, and the scenery is still impressively rugged and remote.

▽ TAIL OF THE DRAGON

US

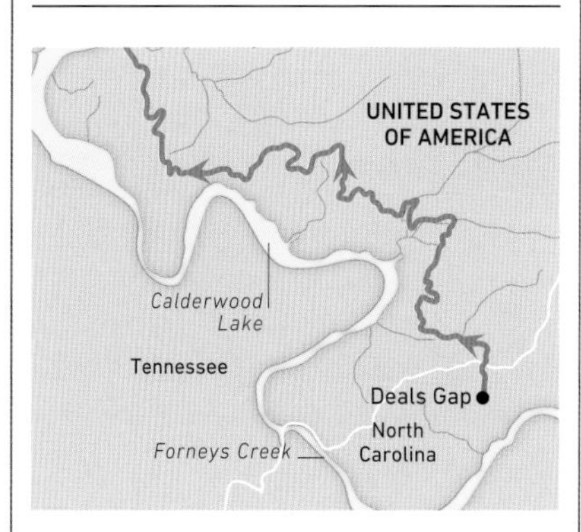

The Tail of the Dragon (or simply The Dragon) is a stretch of US 129 that winds through a mountain pass called Deals Gap on the border of Tennessee and North Carolina. It is popular with enthusiastic drivers because it features 318 bends in 11 miles (18 km) as it winds through a beautiful hilly forest. It is widely considered one of the US's finest driving roads, and in a country famous for its straight, flat tarmac, it is certainly unique. Some of its more drastic curves have memorable names, such as Gravity Cavity, Sunset Corner, Beginner's End, and Mud Corner.

The road passes right through the Great Smoky Mountains National Park, and it is open all year round. There are no side roads or junctions to worry about, but the route is still an accident blackspot. This is due partly to the terrain, and partly to the area's unpredictable weather: fog and rain can appear without warning. In 2005, the speed limit was reduced to 30 mph (48 km/h), and local police mount regular traps to catch reckless drivers. If you go off the road, you may even face humiliation: parts of your wrecked vehicle may end up hanging beside other bits of auto debris on the roadside "Tree of Shame", which acts as a reminder of the dangers on this road.

The Dragon has featured in several Hollywood movies, including Robert Mitchum's *Thunder Road* (1958), Monte Hellman's *Two-Lane Blacktop* (1971), and *The Fugitive* (1993), starring Harrison Ford.

THE TAIL OF THE DRAGON, DEALS GAP

THE IMPOSING BIXBY CREEK BRIDGE, BIG SUR

△ BIG SUR, HIGHWAY 1

US

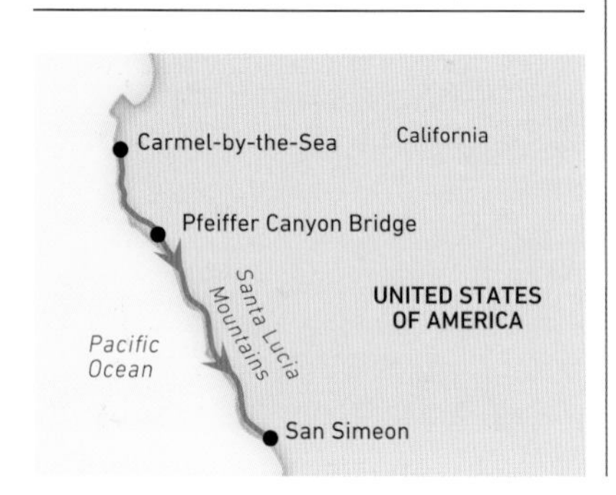

Following the curves and contours of the Central Californian coastline, which runs between San Francisco and Los Angeles, Highway 1 is widely considered one of the world's great drives. However, the real heart of this 400-mile (645-km) road journey is the ruggedly spectacular section known as Big Sur. Its name comes from the Spanish "el país grande del sur", meaning "the big country of the south".

The road here hugs the cliffs for 85 miles (137 km) in a region where the Santa Lucia Mountains reach the Pacific, between the seaside towns of Carmel-by-the-Sea and San Simeon. Peaks of 5,000 ft (1,524 m) stand just 1 mile (1.6 km) from the shore, creating the US mainland's longest stretch of undeveloped coastline – and maybe its most scenic.

The road lies at the heart of high-tech California, but there are few signs of civilization here. For most of the route there is no phone signal, and petrol stations are few and far between, so it is best to set out with plenty of fuel. The region is even geologically unstable in places. In the spring of 2017, a landslide destroyed the 320-ft (98-m)-long Pfeiffer Canyon Bridge. The bridge has since been restored and the road reopened for business.

Despite its challenges, the drive is worth the detour from the faster inland highways. Much of the road is almost at sea level, but some of it rises to 1,000 ft (305 m) above the Pacific. It sweeps through forested gorges that tumble down to the sea and up again past waterfalls and rocky outcrops that overlook the pristine beaches below. Seals are a common sight on the rocks, while out to sea whales can sometimes be spotted. In the heights, giant redwoods and cacti punctuate the landscape.

Many drivers simply tick off this route without stopping. However, if you want to explore the area, there are various hiking trails that lead off into the mountains. Take one of these, and soon you are in the Californian wilderness.

Several stylish, ultra-trendy eco-hotels cling to the slopes, and restaurants offer terraces overlooking the sea. Otherwise, there are plenty of stopping points where you can just pull over and have a rest. At these you can stop worrying about the bends and the dizzying drops to the ocean, and simply relax and take in the natural beauty of the area.

▽ MOUNT WASHINGTON

US

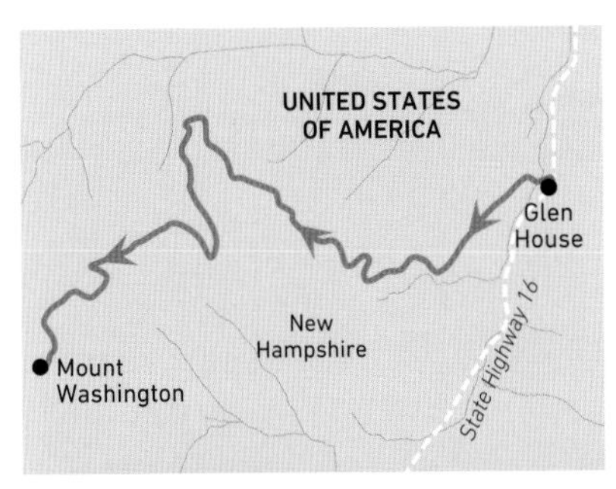

Located in New Hampshire, Mount Washington is the highest peak in the northeastern United States. The summit stands at a lofty 6,288 ft (1,917 m), making it the most prominent mountain east of the Mississippi River.

Drivers expecting a gentle cruise around a pretty part of New England are in for a big surprise. This is an uphill drive over one of the continent's toughest mountains, the upper reaches of which are fearfully exposed to the elements. Mount Washington holds the record for the strongest wind speeds in the Northern Hemisphere – 231 mph (372 km/h) – and it endures hurricane-force winds for some 110 days of the year. The temperatures are also extreme – ranging from 72°F (22°C) to a freezing -60°F (-51°C). However, if you choose the right day, preferably in summer, the drive is easy enough, and the views from the top are stunning.

An 8-mile (13-km) toll road climbs from State Highway 16 almost to the summit, ascending 4,618 ft (1,408 m). For the most part it is steep and narrow, but it is eminently drivable, and has turned this otherwise inhospitable region into a major tourist attraction. If you prefer not to drive, coach tours are available in summer. In winter, caterpillar buses creep all the way to the top.

THE MOUNT WASHINGTON ROAD

CARS CREEP DOWN THE HAIRPIN BENDS OF LOMBARD STREET

△ LOMBARD STREET

US

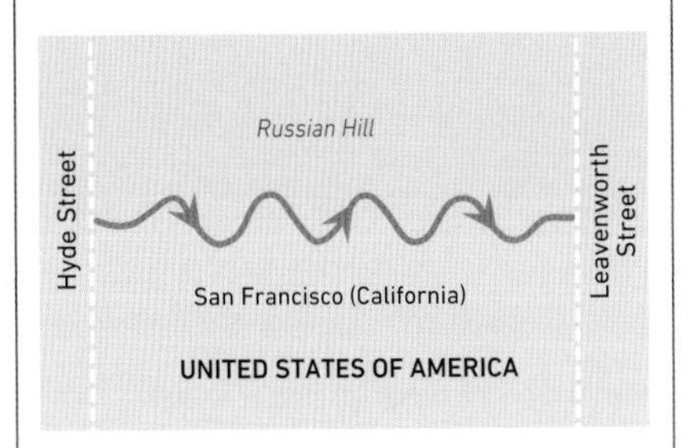

Lombard Street, in San Francisco, is a busy thoroughfare that stretches from The Presidio east to The Embarcadero, and it is unexceptional but for a single 600-ft (180-m) section between Hyde Street and Leavenworth Street that attracts thousands of tourists each year. Its view across San Francisco Bay, Alcatraz, and the Bay Bridge is fantastic, but that is not what draws the crowds. Most people come to experience the eight steep hairpin bends that make Lombard Street "the most crooked street in the world".

The road was built in 1922, and the hairpin bends were designed to lessen the gradient for the cars of the time, which struggled on hills. Today, paved with red bricks and lined with well-tended gardens, the road cuts through a rather upmarket neighbourhood called Russian Hill. Local residents are resigned to the constant stream of drivers who come here to test their brakes. On busy holidays, cars follow bumper-to-bumper down the hill, many having come from miles around to do so. Not surprisingly, there is a 5 mph (8 km/h) speed limit, and cars can only go one way.

Due to its quirkiness, the street has featured in numerous Hollywood films, including Alfred Hitchcock's *Vertigo* (1958) and Bill Cosby's *Driving in San Francisco* (1969).

EVENING TRAFFIC ON SUNSET BOULEVARD

△ SUNSET BOULEVARD

US

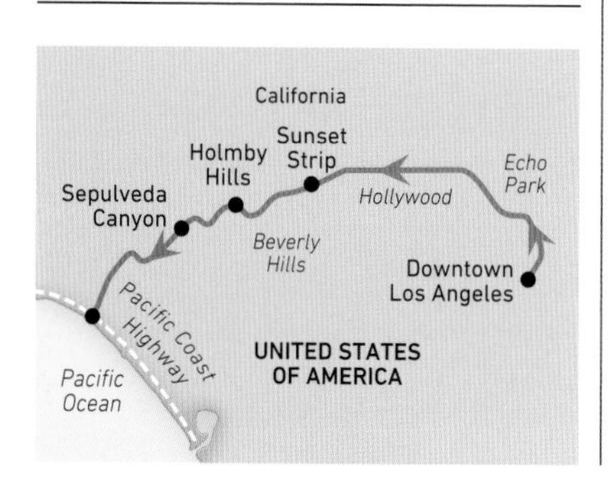

It starts near the intersection of Hollywood Freeway and Harbor Freeway, deep among the office blocks and skyscrapers of downtown Los Angeles. At first it seems like another busy, multi-lane commuting strip as it heads up through the bars and clubs of Echo Park, skirting the arc of high ground to the north of the city to reach better-known areas such as Hollywood and Beverly Hills.

No wonder drivers tackle this urban road with some degree of expectation. Is there another street in the world with such an evocative name? Sunset Boulevard is such a part of popular culture that there is both a film and a musical of that name, plus a TV series called *77 Sunset Strip*.

However, unlike other great drives, the Boulevard is not full of major sights or extraordinary views, and it certainly is not a challenge to any car or its driver. Instead, it is a route to relax on. The driver's arm should be resting on the open window, and the sound system should be playing something appropriately West-Coast, be it punk, rock, or hip-hop. Sunglasses are an optional extra.

The entire Boulevard stretches for 22 miles (32 km), but its best-known section is Sunset Strip, which runs through the heart of the edgy West Hollywood area. Traffic here may be slow, but it gives you the chance to soak up the atmosphere of a place that has been the location of hundreds of films and TV shows.

You may not see celebrities strolling down the pavements, but some of the best-known bars, clubs, and studios in the world stand between the neon signs and huge colourful billboards. Every aspect of

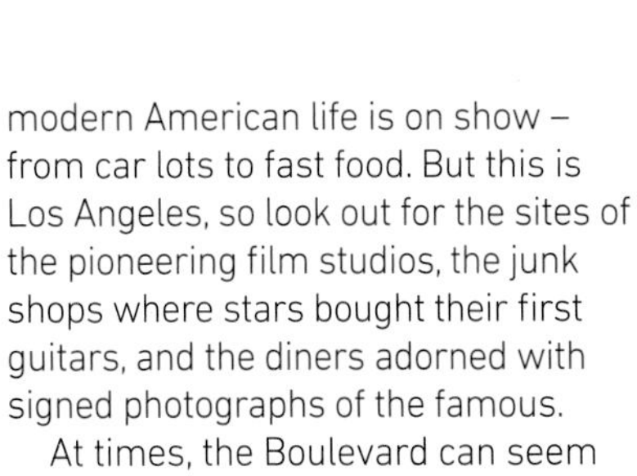

modern American life is on show – from car lots to fast food. But this is Los Angeles, so look out for the sites of the pioneering film studios, the junk shops where stars bought their first guitars, and the diners adorned with signed photographs of the famous.

At times, the Boulevard can seem like a tour through the dark side of modern American culture. On the Strip, you pass Johnny Depp's old club, the Viper Room, where actor River Phoenix died of a drug overdose on Halloween morning in 1993. The club was built on the site of a former jazz bar called The Melody Room, which was a popular hangout of the infamous mobster Bugsy Seigel. Then there's the Whisky a Go Go, the club that helped launch numerous bands, including The Doors, The Byrds, Van Halen, Mötley Crüe, and Guns N' Roses.

Finally, do not miss the Rainbow Bar and Grill – a timeless hangout for rock musicians and groupies. Keith Moon and John Lennon were regulars here, as was Lemmy from Motörhead, who sat at the bar playing poker games daily. Before becoming the Rainbow, the venue was the Villa Nova restaurant, where baseball player Joe DiMaggio and Marilyn Monroe had a blind date in 1952.

▽ CADILLAC RANCH ROAD

US

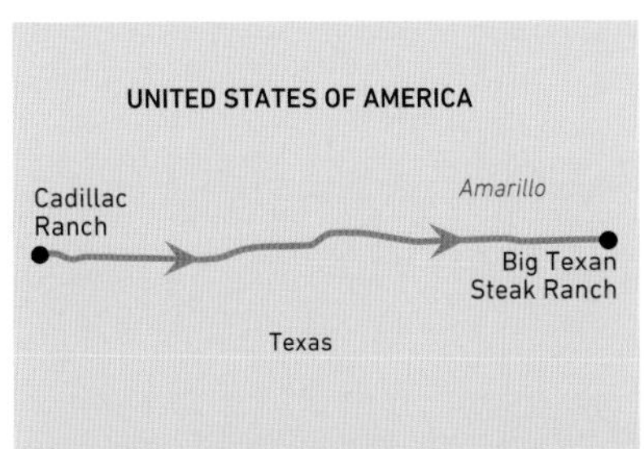

Route 66, the historic road from Chicago to California, offers one of the best-known road trips in the world. Since its origins as a migration route for those hit by the Midwest's dust-bowl deprivations of the 1930s, it has become a classic long-distance American car journey, and is famously celebrated in popular songs, films, and literature. Perhaps the best-known works inspired by the road are John Steinbeck's novel *The Grapes of Wrath* (1939) and the song *(Get Your Kicks on) Route 66*, which was first sung by jazz legend Nat King Cole in 1946.

However, the entire 2,448-mile (3,939-km) journey is somewhat daunting, and those who attempt it find much of the original route is lost, bypassed, or, even worse, not very interesting. A good idea is to take a short, iconic section of the route, such as the Cadillac Ranch road, which gives an authentic taste of the historic highway, plus a couple of classic roadside sights.

The Cadillac Ranch is 10 miles (16 km) west of Amarillo, Texas, and stands alongside the south side of the Interstate 40 highway, where it overlaps the old line of Route 66. One of the most memorable automotive sights in the world, the "ranch" consists of 10 Cadillac cars buried nose-first in a row in the desert. The Caddies are all vintage models from 1948–63, but they have all been covered in multicoloured graffiti. Originally set up as a temporary art installation in 1974, the site quickly became an attraction for travellers looking for Route 66, and so it has remained ever since. It is free-of-charge and open all day, each day.

From the Cadillac Ranch, drive east along Route 66 into the old Texas cattle-trading city of Amarillo. Follow the old road to find another famous Texas attraction on the east side of town – the Big Texan Steak Ranch. This shrine to over-indulgence is another part of Route 66 cult legend: an enormous yellow roadside motel and restaurant with a large model bull and white Cadillac stretch-limos wearing bullhorns outside. Its motel features a Texas-shaped swimming pool, and its restaurant has been voted the best steakhouse in the state – and steaks are taken seriously in Texas.

The Steak Ranch famously offers a 72 oz (4.5 lb/2 kg) steak and chips for free to diners who can finish it in an hour – or a bill for $72 if they cannot. The human record for eating the giant steak is currently held by a woman who then proceeded to eat a second one within 15 minutes. The overall record, however, is held by a lioness. She devoured the enormous slab of meat in 80 seconds.

THE BIG TEXAN STEAK RANCH, HOME OF THE BIGGEST STEAK IN THE STATE

South America

▷ YUCATÁN PENINSULA

MEXICO

This sunny and safe 80-mile (129-km) drive is a chance to explore the coast of the fabulous Yucatán Peninsula along the blue waters of the Caribbean Sea. From the mega-resort of Cancún, with its dazzling array of clubs and bars, the road passes along the Punta Nizuc beach road – a narrow strip of sand with crystal-clear water on both sides. The route then joins Highway 307, heading south along the coastal plain. Unkempt roadside bars line the road to the right; a continuous strip of pristine, palm-lined beaches lies to the left.

At Playa del Carmen, the road passes more gleaming white beachfront hotels overlooking the white sand and sea. The gorgeous beach and watersport island of Cozumel lies in clear view, just offshore, and can be reached by a short boat trip. The 307 continues south, arrowing straight across the flat landscape between the lush roadside vegetation. Finally, the road reaches Tulum, which has a bustling tourist craft market and some extraordinary archaeological treasures.

The remains of an entire walled Mayan city stand high in the jungle on a rocky headland above a glistening beach. Highlights include the clifftop Castillo, built as a watchtower, and the Templo de las Pinturas, which has a partially restored mural.

▷ TRANS-ANDEAN HIGHWAY

VENEZUELA

The Carretera Transandina, or Trans-Andean Highway, runs for almost 1,000 miles (1,610 km) across Venezuela. It was built by prisoners before World War II, and used by the army to quell regional revolts. Today, Troncal 7, to give it its official name, is one of South America's great road trips. It provides an automotive adventure that runs from the centre of the country up to San Cristobal near the Columbian border, and crosses the heights of the northern Andes along the way.

However, if you are unable to do the entire marathon, you can always concentrate on a central section of the highway, such as the 30-mile (48-km) route from Apartaderos to Timotes. This takes you along the highest road in the country, peaking at the Collado del Cóndor mountain pass (sometimes called Pico El Aguila), which lies at 13,510 ft (4,118 m). Given the terrain, the road is extremely bendy, but its tarmac surface is

A BIRD'S EYE VIEW OF THE CANCÚN HOTEL ZONE

THE ROADS AND RUGGED LANDSCAPE OF THE TRANS-ANDEAN HIGHWAY

smooth, making for an easy drive. If the weather is clear, you can expect stunning, dramatic views. A small chapel, a café, and a gift shop stand at the top of the pass, plus a landmark statue of a condor.

The latter commemorates the crossing of the Andes by Simón Bolivar and his army in 1813. From the pass, you can drive down to the beautiful Lake Maracaibo.

▽ PUNTA OLIMPICA

PERU

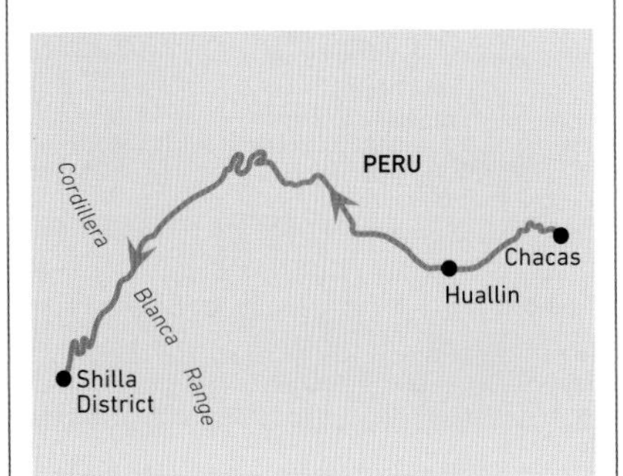

Ruta departamental AN-107, or Punta Olimpica, in the Andes, crosses one of the highest driveable mountain passes in the world – and a new branch of the road also runs through the world's highest road tunnel. Located at a staggering 15,535 ft (4,735 m), and running almost three-quarters of a mile (1.2 km), the Punta Olimpica Tunnel provides a thrilling drive.

Before the tunnel was built, the route through the snow-topped Cordillera Blanca, with its multitude of harsh bends and unpaved surfaces, was considered one of the world's most dangerous drives. The tunnel was built to avoid the climb to the top of the pass, and the Peruvian authorities have since paved and vastly improved the road. Today, Punta Olimpica is a great mountain drive with breathtaking views, although the locals still call it the "Road of a Thousand Curves" – even if only 46 curves remain since the road was improved.

For a more adventurous trip, you can still bypass the tunnel and take the old gravel road that climbs 1,000 ft (305 m) above it and contains another 21 hairpin turns. However, this section has not been maintained since the tunnel was built, so it really is dangerous. The neglected old road surface is unsuitable for all but the sturdiest of off-road vehicles, which should be driven with the utmost care. Better still, swap the car for an off-road motorbike, or even go on foot – preferably in strong hiking boots with cleats.

THE PUNTA OLIMPICA SNAKES INTO THE PUNTA OLIMPICA TUNNEL

THE CARRETERA DE YUNGAY TRACES THE SHORES OF THE LLANGANUCO LAKES

△ CARRETERA DE YUNGAY

PERU

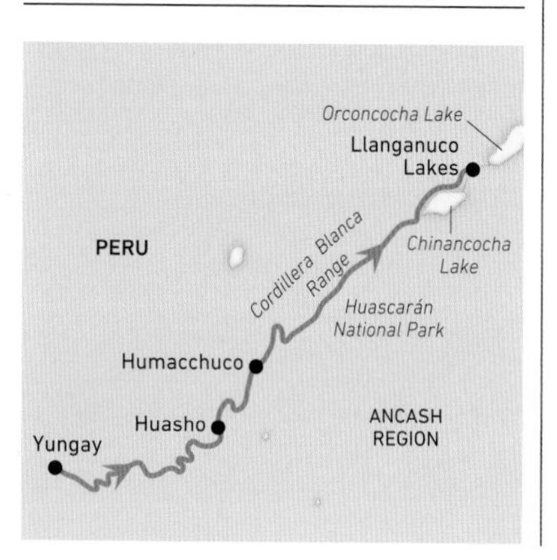

The Llanganuco Lakes – Chinancocha and Orconcocha – are a well-known tourist sight in the snow-capped Cordillera Blanca mountain range of the Peruvian Andes. The two beautiful, turquoise lakes are located at a high altitude – 12,630 ft (3,850 m) – very close to Huascarán, the highest mountain in Peru. They lie side-by-side beneath sheer rock faces and are separated by a slim neck of land.

You can reach the lakes via the Carretera de Yungay, or Carretera 106, as it is also known. This is an unpaved, stony track that winds up into the mountains from the provincial capital, Yungay, the site of a tragic earthquake in 1970 that killed 20,000 people. The track threads up into the peaks that loom above the town.

Not surprisingly, the road is a challenging one, and it is a great example of what drivers in developing countries have to cope with when travelling long distances. If you are only used to smooth tarmac, you may be in for a shock. Expect steep, tight, narrow corners, sand-and-gravel surfaces, and frightening, unguarded drops, and be ready to perform some awkward reversing manoeuvres when oncoming traffic appears.

▷ SERRA DO RIO DO RASTRO ROAD

BRAZIL

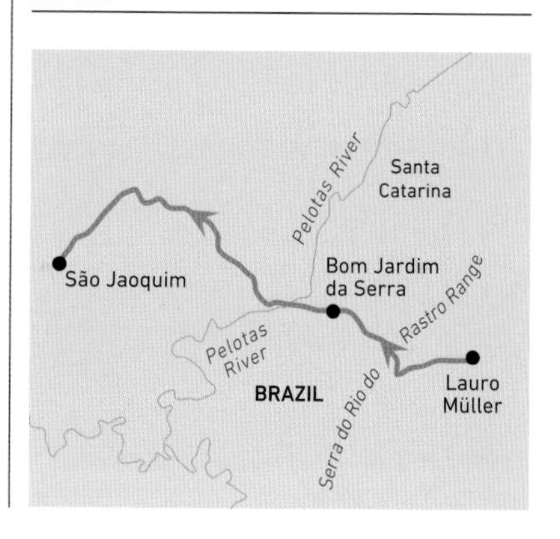

Serra do Rio do Rastro is a dramatic mountain range in Brazil's southeastern state of Santa Catarina. The Serra's jagged green peaks rise sharply from the coastal plain around 50 miles (80 km) in from the sea.

The landscape quickly climbs to thousands of feet above sea level – on a good day the Atlantic is visible in the hazy distance. The steep cliffs, rocky pinnacles, and deep valleys are thick with lush rainforest, but eventually lead to a more barren, lofty plateau that offers great views in all directions.

Several eco-hotels provide accommodation on the summit plateau. However, to reach the top, motorists have to navigate the only road available – the SC390, which is known locally as the Serra do Rio do Rastro Road. This is one of Brazil's most famous, or infamous, highways, and it wriggles up through the trees for 16 miles (25 km) from Lauro Müller to São Joaquim.

Tackling the road has become something of a national institution. Car and bicycle races, festivals, and demonstrations are common, and people drive for hundreds of miles here to enjoy the route. The drive is both challenging and beautiful. The steep, cliff-hugging climb through the rainforest involves some 250 hairpin bends. The toughest part is a section that climbs from sea level to 4,790 ft (1,460 m) in just 8 miles (11 km).

The forest, with its colourful birds, exotic plants, racoons, and screeching monkeys, is part of the attraction, but so is the road itself. Its course is frequently revealed as it twists and turns in the distance.

Being made of concrete, the road is generally smooth, but it pays to watch out for cracks caused by the extreme variations in climate. The weather ranges from tropical sunshine to snow storms, which together cause considerable erosion. Heavy snowfalls and avalanches occasionally block the road, and ice can be a problem in winter.

The bends in the road are easy to see, but often they are only guarded by a shallow parapet wall, and the drops into the valley are unforgiving. Perhaps the greatest surprise of all, particularly for a remote jungle road, is the fact that the SC390 is completely lit at night by overhead lights powered by a series of windmills. After dark, from a distance, the road appears to be a mysterious, illuminated helter-skelter coursing across the mountains.

THE UNEARTHLY GEYSER FIELD OF EL TATIO

THE SERRA DO RIO DO RASTRO WINDS THROUGH THE BRAZILIAN RAINFOREST

△ ATACAMA DESERT ROAD

CHILE

The Atacama Desert – some 50,000 sq miles (129,500 sq km) of treeless, barren rock, dotted with salt lakes, volcanic sites, and a few mining communities – is the driest place on Earth. This 55-mile (88-km) route starts from San Pedro de Atacama, where the altitude is already exacting: at 7,000 ft (2,130 m), you may need a few days to acclimatize before going any further. The town of El Tatio, at the end of the route, stands at 14,000 ft (4,267 m).

When you are ready, take the unmade B245 north. It is a smooth enough highway – without rain, its damage is limited – but it does climb deep into the Andes. It takes you through some extraordinary desert landscapes, and terminates at the greatest concentration of active geysers in the world: there are 80 at El Tatio alone. This steaming geyser field looks best at sunrise, when you can bathe in the naturally heated water, surrounded by Andean peaks.

Africa

▷ AVENUE OF THE BAOBABS

MADAGASCAR

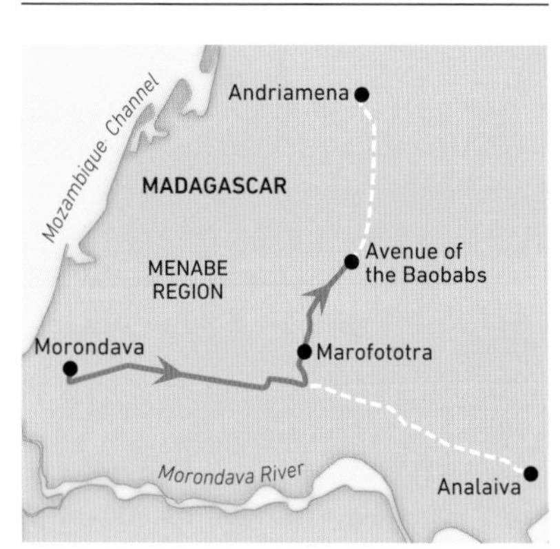

A botanist would call it *Adansonia grandidieri* – the biggest and most spectacular plant on the island of Madagascar off Africa's east coast. These giant baobab trees are a unique and wonderful variety that grows only on this tropical island.

The fat, smooth, and shiny trunks grow up to 10 ft (3 m) wide and almost 100 ft (30 m) high. They are branchless until the top, where dainty bonsai-like crowns sprout large flowers and fruits. These strange, dinosaur-like trees are up to 800 years old, and locals call them "upside-down trees", "bottle trees", and even the "mother of the forest" – lone relics of a dense forest that once covered the island. The best place to see them is an 853-ft (260-m) stretch of rough, dusty road known as the Avenue of the Baobabs. Here, some 25 baobabs flank the road, creating one of the world's most memorable and photogenic sites.

Today, the trees have protected status, and the Avenue has become an unofficial national heritage area. It is not an easy spot to get to, but it has become a major tourist attraction, so there are plenty of local guides to take you if you do not have your own vehicle. The most popular mode of transport is the taxi brousse – the Madagascan version of the bush taxi – which can take 15 people or more, plus supplies. If you are driving, take an SUV, particularly in bad weather.

THE AVENUE OF THE BAOBABS AT SUNSET

The town closest to the site is Morondava, on the west coast of the island. From there, take the road east towards Analaiva, but turn north at the Avenue sign after 6 miles (10 km) or so. From here, the road becomes an unpaved, uneven, and bumpy dirt track. It leads between wet paddy fields, sugarcane plantations, and untamed scrubland. Depending on the season, it can either be hard, dry mud or a sticky, wet marsh. After 4½ miles (7 km), the road reaches the Avenue, where you can leave the car and explore the site on foot.

The perfect time to visit is at sunrise or sunset, when professional photographers and amateur selfie-takers converge to capture one of the great African images of the trees against a colourful sky. April is the best month, after the rainy season, when the trees burst into fresh greenery at the top.

It costs nothing to visit the site, but more and more villagers gather at a small carpark at the start of the trip, hoping to sell tree-shaped wood carvings, which make excellent mementoes. The trees have a special significance for the locals, who have helped to conserve them. They are believed to be the sacred homes of their ancestors – and they produce a highly nutritious fruit.

Europe

HARDKNOTT PASS

ENGLAND

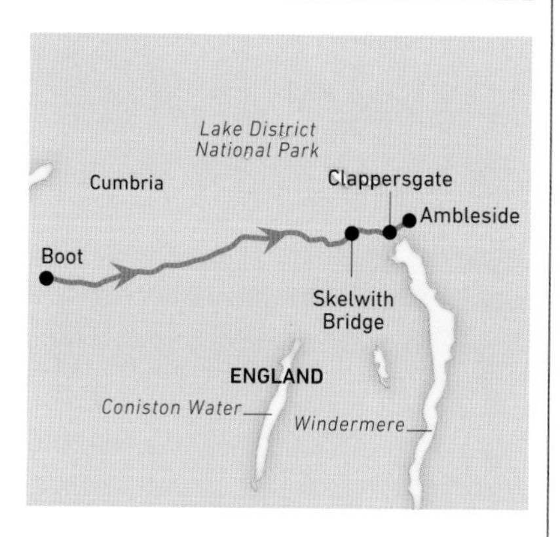

Drivers expecting a gentle, picturesque cruise through England's Lake District National Park are in for a surprise on this arduous 15-mile (24-km) route. The way through Hardknott Pass in Cumbria has been dubbed "the UK's most outrageous road", and it will test your car as well as your driving skills. In good weather, most cars can handle the terrain, otherwise a powerful four-wheel-drive vehicle is advised.

As the road climbs from the pretty lakeside, signs warn drivers of the challenges ahead – then soon it is too late to turn back from the oncoming sequence of extreme hairpin bends, unguarded drops, and gradients of up to 33 per cent. It is challenging enough on a sunny day – but that is rare in the Western Fells. An average day features horizontal rain, buffeting sidewinds, and slippery tarmac. On a bad day, the road is impassable.

The reward for all the strenuous steering and gear-changing is access to an untouched mountain landscape of rare savage beauty – one of roaring waterfalls, sheer rock faces, and stunning views suddenly opening across the fells. Dramatic terrain soars into the clouds on either side as hardy sheep wander confidently across the road. They know that cars are the outsiders here.

▽ WESTERN ISLES

SCOTLAND

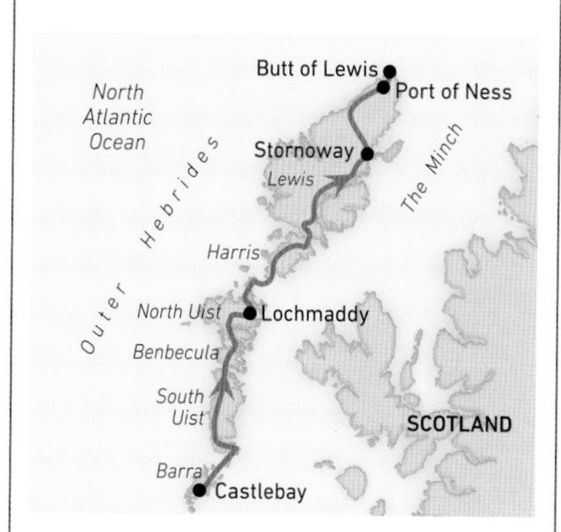

Exploring this 174-mile (280-km) chain of remote Scottish islands linked by dramatic causeways, bridges, and ferries is one of the most exciting driving adventures in the UK.

The road through these small islands of the Outer Hebrides includes a sequence of inspiring seascapes, beautiful heather-clad moors, and fabulous sandy beaches. This is wild, untamed land that is often inspiring and always memorable. Look out for otters and seals, standing stones, bagpipe players, blackhouses (traditional, one-storey, dry-stone buildings with thatched roofs), and some of Scotland's best fresh seafood. Roads are quiet, smooth, and well signed. The biggest hazards are wayward sheep and unpronounceable Gaelic place names.

By the time you have sailed from the mainland and edged up the ferry slipway at Castlebay in Barra you will already have had a hint of what is so special about these islands. Standing in the sea in the middle of the bay is the medieval Kisimul Castle. Kisimul has been the impregnable fortress of the MacNeil clan for a thousand years – although the clan recently leased it to Historic Environment Scotland for the nominal fee of a single bottle of whisky per year.

THE MEDIEVAL KISIMUL CASTLE, BARRA

MILLAU VIADUCT

FRANCE

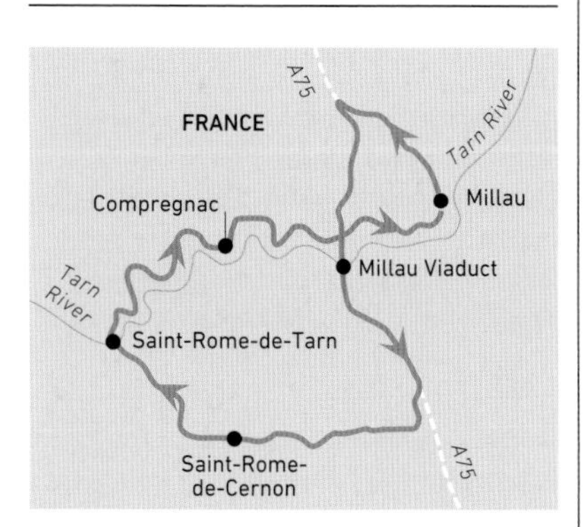

The A75 motorway heads south from Paris towards Montpellier like a river of concrete sweeping through the landscape. Traffic on the road speeds without stopping, passing the mountainous Cevennes region of southern France, until it reaches the wide valley of the River Tarn. This is where even the most hardened drivers have the heart-stopping experience of crossing the world's tallest road bridge – the Millau Viaduct.

While crossing the bridge it is hard not to be distracted from the important job of steering. The highest part of this extraordinary cable-stayed bridge is a dizzying 1,125 ft (343 m) above ground, and most of the support towers are taller than the Eiffel Tower. Built at a cost of almost 400 million euros, this 1½-mile (2.5-km)-long engineering masterpiece has also been judged one of the world's most beautiful bridges.

However, the experience of crossing the bridge is all too brief, even if you slow down to enjoy the view. To get the complete experience of the Millau Viaduct and see its spectacular span from underneath, take a circuitous route around the local countryside. One such route is a magnificent figure-of-eight starting and finishing at the picturesque old town of Millau. The town was bypassed when the new bridge opened in 2004, and since then all long-distance traffic whizzes past without stopping. Although the locals welcome the peace, traders mourn the loss of business. Millau is worth exploring before setting out; it is a traditional market town with barely a tourist in sight.

From Millau, the route twists to join the A75 as it heads south across the bridge. After the crossing, it leaves the motorway for a scenic loop through charming French villages such as Saint-Rome-de-Cernon and Saint-Rome-de-Tarn. The road curves along the river valley, passing right under the Millau Viaduct, from where the views of the bridge are even more impressive.

The bridge is so high that occasionally it soars above the clouds that gather in the valley. On such days, driving the bridge is an unforgettable experience.

▽ ROAD OF A THOUSAND BENDS

SPAIN

Although the Costa Brava in Spain seems like a busy section of coastline, full of predictable, well-organized, family-friendly tourist resorts, it hides a short stretch of truly sensational road. This is the section of the GI-682 between Tossa de Mar and Sant Feliu de Guíxols – two of the main resorts – and it is known by locals as "The

THE TOWN OF TOSSA DE MAR ON THE COSTA BRAVA

Road of a Thousand Bends". There is another longer, more conventional route on the modern road that loops around far inland, but if you are looking for a real driving experience, take this more direct way, which clings spectacularly to the clifftops. The Road of a Thousand Bends provides a glimpse of what this part of northeastern Spain was like before the Costa Brava became a popular tourist destination.

Watch out for local motorcyclists testing their machines on this notoriously wild driving route. It is a sequence of dizzying drops, hair-raising corners, and amazing views of the Mediterranean, which lies hundreds of feet below. Thankfully, plenty of lay-bys allow you to pull over and take pictures – and take a deep breath before driving on.

THE STELVIO PASS TWISTS AND TURNS THROUGH THE ORTLER ALPS

▷ STELVIO PASS

ITALY

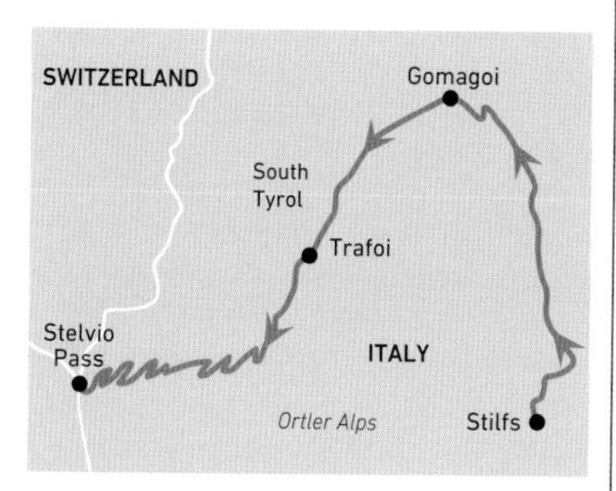

The Stelvio Pass is a 12-mile (19-km) route through the Ortler Alps from Italy into Switzerland, and its road is one of Europe's highest. It may not be the prettiest, but it is often chosen as the world's best driving road. BBC TV's famous former *Top Gear* trio tackled it in three supercars, and declared it their "ultimate drive". The pass has since become an iconic route for drivers from all over the world.

Its reputation is all due to the "wall" of 48 hairpin bends that ascends to the giddying 9,045-ft (2,757-m) summit of the pass, and the 34 bends on the other side. Negotiating these involves a lot of arm-twirling and tyre-squealing, and a good deal of concentration.

In places, the road is narrow, steep, and intimidating – a real test for driver and vehicle alike. But any driver in any car can appreciate the views and the experience of tackling such a dramatic route. There is certainly no requirement to drive as fast as possible.

▷ STRADA CRISTO REDENTORE DI MARATEA

ITALY

The 72-ft (22-m)-tall statue of Christ the Redeemer near the town of Maratea, Italy, is a more casually dressed, 1960s-style version of the one in Rio de Janeiro, Brazil. Created by Florentine sculptor Bruno Innocenti, and completed in 1965, it stands dramatically on a mountainous pinnacle called Monte San Biagio, 1,942 ft (592 m) above sea level.

Amazingly, a road winds all the way up to the summit of the narrow finger of rock on which the statue stands. It is one of Europe's most incredible stretches of roadway. The scribbled sequence of 18 switchback bends are elegantly supported by concrete pillars that jut from the side of the cliff.

A prodigious amount of hard cornering and gearchanging gets you to the small carpark at the top. Arriving there gives you a chance to recover and then admire the views – a panorama that includes the town below and the Basilicata coastline leading off to the north and south. Looking back down the sheer face of the mountain, the most impressive sight is the extraordinary road itself – which, of course, has to be tackled again to get down.

CHRIST THE REDEEMER WELCOMES VISITORS WITH OPEN ARMS

▽ STRADA DELLA FORRA

ITALY

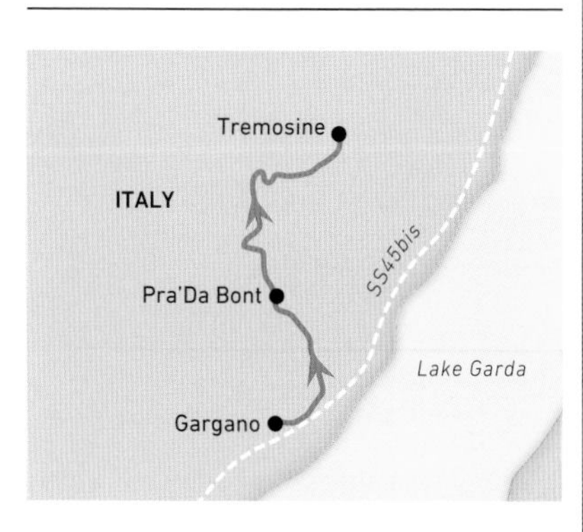

James Bond's Aston Martin tackled this road at tyre-squealing velocity in the opening car chase of the movie *Quantum of Solace*. Of course, it looked dramatic on screen, but some idea of the nature of the route itself can be gleaned from the fact that three experienced stuntmen crashed during the filming. One ended up plunging into Lake Garda and another had to be airlifted to hospital. So it is not a road to be taken lightly, at any speed. Nevertheless, the Strada is often called one of the world's most spectacular drives.

The SP38, to give its official name, is a 10-mile (16-km) twisted tagliatelle of a route, spiralling up from the shore of this beautiful Italian lake in the Alpine foothills, following a cleft in the mountain worn by the River Brasa. It is not just a James Bond location – many car manufacturers have used it for commercials and photo-shoots.

The route winds between brutal, sheer cliffs, through tunnels blasted through rock, and jagged gaps barely wider than a car. Then, every so often, you emerge into the glaring sun, with a stunning view of the lake below.

The Strada della Forra (meaning "Road of the Gorge") attracts drivers from all over the world, but it is not so loved by the locals. Popping to the shops or driving to work along this narrow single-carriageway road involves innumerable stops to let cars pass or to reverse around blind rocky corners when faced with on-coming trucks.

Wide vehicles can end up with scraped wings, or in extreme cases, like 007's Aston Martin, no door. Narrow cars fare better in this hazardous helter-skelter, as do motorcycles, which arrive here from all over Europe.

With only a small rail protecting you from drops into the ravine, and the possibility of having to reverse through a narrow rocky tunnel, the Strada demands total concentration. Unsurprisingly, visitors are often left breathless by the time they reach the pretty village of Tremosine, which clings to the mountain at the top of the climb.

The unlikely motoring adventurer Sir Winston Churchill once tackled the Strada della Forra. His verdict? He said that it was the "eighth wonder of the world."

THE STRADA DELLA FORRA SNAKES AWAY FROM LAKE GARDA, ITALY

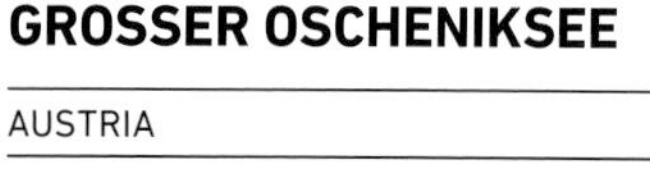

GROSSER OSCHENIKSEE

AUSTRIA

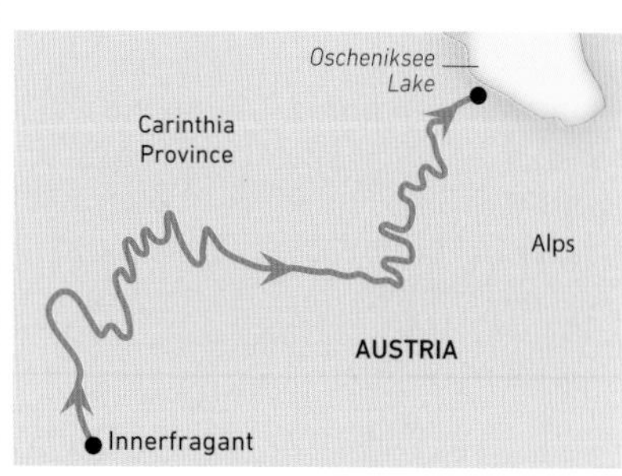

With gradients of up to 20 per cent, this route is effectively a case of mountain climbing in a car. The Grosser Oscheniksee is a quick way of ascending almost 5,000 ft (15,000 m) into the Alps.

In winter, of course, the route is impassable due to snow. In summer, however, enthusiastic drivers love the challenge of more than 40 hairpin bends in only a few miles. As you whizz past, spare a thought for the monumental efforts of the cyclists who seem to love trying to conquer the hill too.

The route starts on an innocent-looking small country road in a pretty wooded valley in the mid-Austrian region of Corinthia. After a small turning on to a concrete lane, things start to change. The roadway rises like a cable car through the trees and up into the bare mountains beyond. In just 7 miles (11 km) the road reaches the dam of the Oscheniksee reservoir at a lofty 7,854 ft (2,394 m). The reward for all that driving effort is a fantastic 360-degree view of the Alps.

▷ ØRESUND BRIDGE

SWEDEN

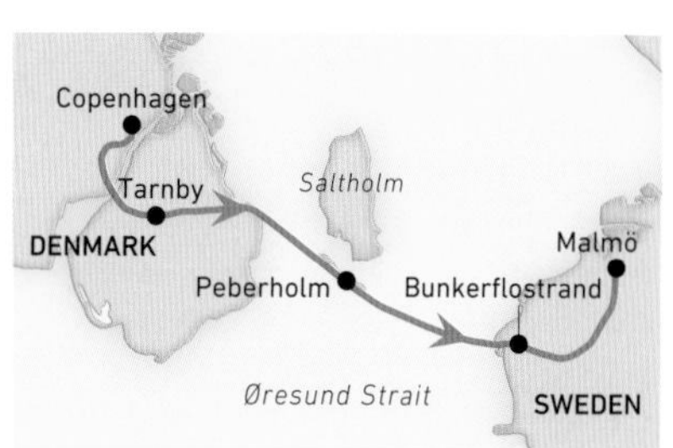

Øresund is the strait that lies between Denmark and Sweden and connects the North and Baltic seas. It is one of the world's busiest waterways for

THE LYSEVEGEN ZIG-ZAGS DOWN TO THE VILLAGE OF LYSEBOTN

shipping, and one of its most-crossed waterways due to the fact that two major cities – Copenhagen and Malmö – stand on either shore.

The most exciting way to cross the strait is to take the Øresund Bridge, which, at 28 miles (45 km) long, is Europe's longest road-and-rail bridge, and one of the continent's most breathtaking structures. The road bends for 5 miles (8 km) across the sea on concrete pillars to reach a small man-made island between Sweden and Denmark called Peberholm. After that, the highway plunges underground through the middle of the island into a tunnel to reach the Danish shore 2½ miles (4 km) away. This creates a clear stretch of water for shipping to use. The tunnel is built from 60,600-ton (55,000-tonne) concrete tubes, which are the largest such structures in the world. Øresund Bridge has become a television celebrity as well. It is the titular location of the popular Scandinavian crime drama *The Bridge*.

THE ØRESUND BRIDGE PLUNGES INTO THE DROGDEN TUNNEL

△ THE LYSEVEGEN

NORWAY

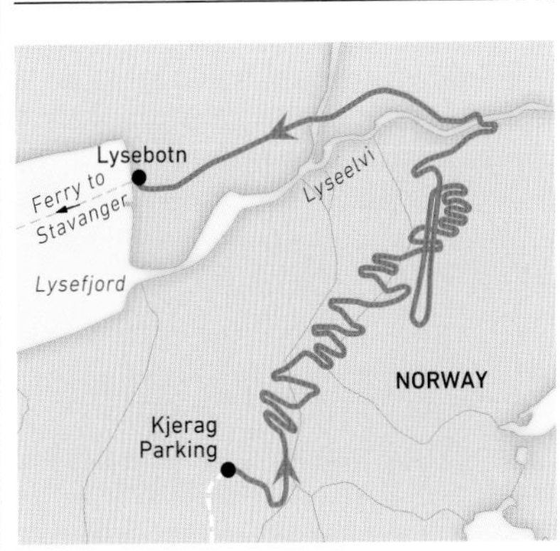

At first glance, the map of the Lysevegen seems to have been scribbled by a child. However, a bird's eye view of the route shows that this is the actual path of this mysterious squiggle of tarmac. The 18-mile (29-km) road begins as a sensational series of 32 hairpins and severe gradients as it heads down the sheer side of a Norwegian fjord. Then it enters a corkscrew tunnel that turns through 360 degrees before emerging at the village of Lysebotn, where the road ends. From here, drivers can either wait for a ferry to Stavanger or drive back up the Lysevegen.

The road was built in 1984 to move stone during the construction of a hydroelectric plant in the mountains. Before that, Lysebotn was cut off from the Norwegian road network, and the only way to reach it was by boat.

Today, the Lysevegen has become an automotive attraction. Videos on the Internet show rally drivers racing down it in less than ten minutes. Conventional drivers take a little longer.

THE STORSEISUNDBRUA AT SUNSET

△ ATLANTIC ROAD

NORWAY

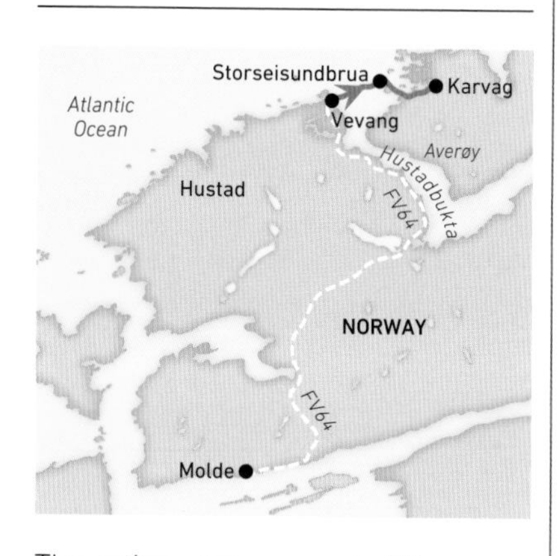

The entire western coast of Norway is a spectacular sequence of fjords, islands, and snow-capped mountains linked by one of the world's most expensively engineered networks of roads, bridges, tunnels, and ferries. However, a single 22-mile (36-km) stretch of highway has been dubbed the "Atlanterhavsvegan", or "Atlantic Road", by tourist chiefs anxious to boost visitor numbers to some of Norway's most remote areas. But they are right: the Atlantic Road is one of the most sensational driving routes in the world, and genuinely takes you into the sea.

Just north of the small coastal town of Molde, so sheltered and full of flowers that it is known as the "Town of Roses", the tarmac twists away from the mainland. It winds from rocky islet to wave-splashed causeway, across sweeping bridges and remote island communities, leading from a wild mainland shoreline out into the bleak islands of the Atlantic Ocean. The road has been voted Norway's "engineering feat of the century" – which is some accolade for a country that can boast hydro-electric dams, mountain tunnels, and ingenious examples of Arctic-proof architecture.

The complete route opened in 1989 to connect villages previously reliant on boats to reach the mainland. It is now designated a National Tourist Route, but it remains toll-free. Best of all, the highway snakes across eight bridges in an environment where a storm from the northwest sends waves right across the road. In calmer times, watch for soaring sea eagles or seals lounging on rocks next to the road.

The highlight of the road is the cantilevered Storseisundbrua, the road's longest bridge, and its logo. Its 850-ft (260-m) span curves high to allow boats underneath and twists in the middle. In a storm, it is a photographer's dream.

Various rest stops and hiking trails can be found along the route, and some of the viewpoints are sheltered against the heavy sea spray. If you want to explore, you can follow raised wooden walkways through boggy moors, or climb to the high points of each island. Other areas have protected fishing spots for anglers.

For a more surreal experience, you can stop and take the coastal path on the island of Haga to find parts of a memorable marble sculpture by modern artist Jan Freuchen. As intended, his creation appears to be the remains of twisted white, ancient pillars discarded among the rocks of the Atlantic seashore.

BAKHCHYSARAI HIGHWAY

CRIMEAN PENINSULA

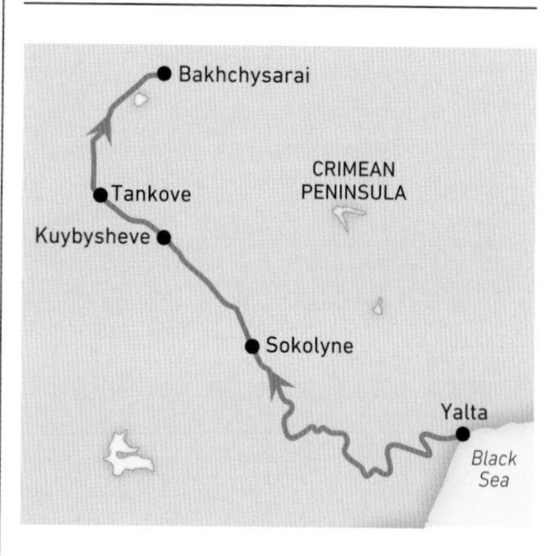

The journey from the popular Black Sea beach resort of Yalta inland to the glittering historic town of Bakhchysarai, once the capital of Crimea's Tatar kingdom, is popular among tourists. However, drivers face the classic route-makers' decision: to take the modern, safe, smooth highway or the much more direct, older road. In this case, however, the older road is the notorious Bakhchysarai Highway. The T0117, to give it its official title, is another of the world's hardcore, multi-hairpin routes. Drivers have to negotiate more than 50 switchback corners as the road spirals for 48 miles (77 km) through steep rocky valleys and gorges through the mountains. Add in vertiginous unguarded drops and blind narrow bends, and this route contains all the ingredients of an extremely challenging road journey.

With care and in normal conditions, the road is, of course, merely a slow route that requires concentration. It is certainly a very scenic journey through the Crimean landscape. However, when conditions are bad and drivers are less than careful, it becomes a considerable challenge.

THE VILLAGE OF LAHIC, AZERBAIJAN

▽ KEMALIYE TAŞ YOLU

TURKEY

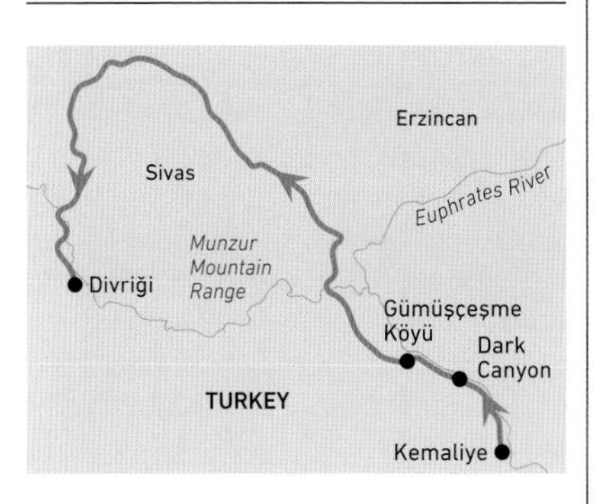

Motorists planning to drive from the town of Kemaliye through the Munzur Mountains to see the UNESCO-recognized World Heritage Site of Divriği Great Mosque have a choice of taking the long route via the modern, paved roads skirting the mountains, or a shorter route through the highlands. The latter is Turkey's famous Kemaliye Taş Yolu, or Stone Road of Kemaliye, which is 50 miles (80 km) long and took over a century to build. Labourers toiled for a lifetime cutting the road through the otherwise impassable cliffs by hand, even digging a 3-mile (5-km)-long tunnel through a mountain.

Today, the Stone Road is a well-known adventure trail, and is not for the faint-hearted. It is narrow, crumbling, and un-paved, and winds through a canyon following the course of the Euphrates River. In the deepest part of the gorge it enters the Dark Canyon, whose walls are so high they almost blot out the sun. The drive is truly daunting, and you can only feel humbled as you pass the monument to the men who died building it.

THE STONE ROAD HUGS THE EUPHRATES RIVER VALLEY

△ LAHIJ MOUNTAIN ROAD

AZERBAIJAN

The wonderful ancient village of Lahij is one of the tourist treasures of the Republic of Azerbaijan. This remote labyrinth of stone cottages and cobbled alleyways stands high in the foothills of the Greater Caucasus Mountains.

Situated amid a daunting, barren, rocky landscape, Lahij is an extremely remote and isolated village. Over the centuries, its distance and isolation has led the inhabitants to develop their own infrastructure and amenities, including what is now a 1,500-year-old sewage system – thought to be the oldest in the world. They also developed their own distinctive crafts style, and are still renowned for their intricate copperware and rugs.

Although the distance from Tazakend to Lahij is only about 14 miles (22.5 km), it is a difficult place to reach. The main road into the village is a challenging route for drivers who are used to smooth tarmac highways. The route follows a dry river through a deep and dramatic ravine. It is very photogenic, but some stretches of the road are simply a narrow, gravel track halfway up a sheer rock face. The views are impressive, but care should be taken when driving – to avoid potholes, and the various places where the road edge crumbles to a daunting, unprotected drop. To be on the safe side, only take an off-road vehicle through this remote and rugged landscape.

Asia

▽ CHUYSKY TRAKT

RUSSIA

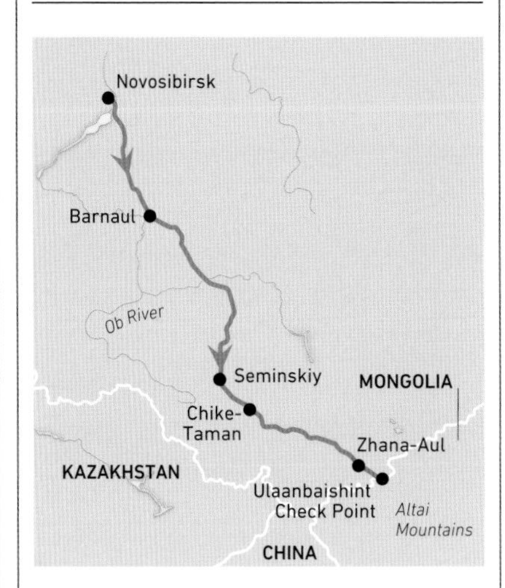

From Novosibirsk, Russia's little-known third city and the capital of Siberia, this fine highway runs 596 miles (959 km) south across the Altai Territory and the Altai Republic down to Russia's border with Mongolia. It will never be jammed with tourists, but for those in the know, Chuysky Trakt is one of the world's great long-distance drives.

The route passes through some little-known regions and offers a fabulous range of spectacular scenery, and the road itself – officially the M52 or P256 – is of unusually good quality for this part of the world. It is wide, smoothly paved, and it curves and rises gently. Sadly, much of that is due to the efforts of up to 12,000 prisoners who were brought here from the notorious Siberian gulags. For 20 years, before World War II, the prisoners were forced to toil in horrific conditions to build the road. According to local legend, the verges are lined with their skeletons.

The route follows an ancient mule caravan trail that linked the Far East with Russia, and which was used by Genghis Khan when he invaded the region. It is effectively a northern branch of the Great Silk Road.

Today's driver will find that at first the road arrows south along the Ob River, which flows north. The Ob is the world's seventh-longest river and leads all the way up from the Altai Mountains to the Arctic Ocean. The road follows the water as it froths around forested islands and hurries past picturesque log-cabin villages with golden-domed churches.

The glorious Sekinskiy and Chike-Taman passes lead into a higher landscape of conifer forest, sparkling lakes, dramatic rocky cliffs, and towering snow-capped peaks. These are the Altai Mountains, where it is possible to spot herds of reindeer and distant glaciers, and even visit caves adorned with prehistoric art.

SIGHTS ALONG THE CHUYSKY TRAKT

A VIEW OF THE ROCKY PANORAMA EN ROUTE TO DIANA'S POINT

It is worth stopping at the village of Zhana-Aul to visit the Museum of Kazakh Culture, which is housed in a real yurt tent. Finally, the route reaches high, arid steppes where horsemen herd their animals in wide-open landscapes under huge skies.

The highest points of the route are 6,560 ft (2,000 m), but thanks to the quality of the roads this does not pose a challenge. Still, this is a long journey through remote regions, and winter conditions can be harsh. During the summer, hotels, petrol stations, and cafes open along the route.

△ NIZWA TO DIANA'S POINT

OMAN

Sayq
Diana's Point
Wadi Ghul
OMAN
Al Hajar Mountains
AD DAKHLIYA REGION
Nizwa
Birkat Al Mouz

The Sultanate of Oman lies on the southeastern corner of the Arabian Peninsula. Most of its cities are near the coast, but recently tourists have started exploring the unspoilt landscapes inland. This route leads from the intriguing oasis town of Nizwa, with its shady souk market sprawling around an intact medieval fort, high into the Al Hajar Mountains. It does so via a new highway, which is toll-free but has a checkpoint at which you have to confirm that you have a four-wheel-drive vehicle. The parched mountain views are mesmerizing as the road climbs to the remote town of Sayq on the edge of Oman's version of the Grand Canyon – Wadi Ghul, a vast dry gorge. A raw pinnacle of rock some 6,500 ft (2,000 m) up on the rim forms a natural viewing point. It is called Diana's Point because the UK's Princess of Wales visited the spot by helicopter in 1986. Locals say she quietly sat here for several hours, reading a book.

Today, Diana's Point is a popular attraction and forms part of a glamorous terrace bar. It is an enjoyable place to watch the setting sun cast its light across the canyon.

SHEEP CAUSE A TRAFFIC JAM ON THE ROHTANG PASS

△ LEH-MANALI HIGHWAY

INDIA

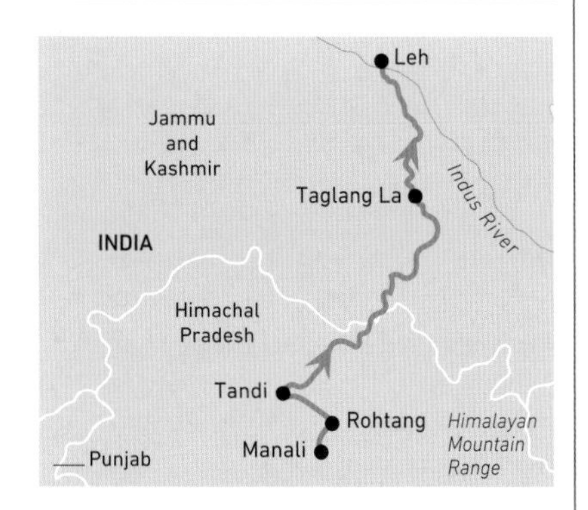

Most of the routes selected for this section are short, exciting, "destination drives", but there is nothing short about the Leh Manali Highway, and its destination is neither here nor there. This is a serious, life-changing road trip, and anyone attempting it should be fully prepared for a hazardous 300-mile (488-km) journey.

The Leh-Manali Highway is a notoriously challenging route connecting two northern Indian states through very high and difficult Himalayan terrain, and driving it is more like an expedition than a casual drive. The road is only open for four-and-a-half months of the year (in summer), and any journey will have to be made in this brief window of good weather. The rest of the time, heavy snow renders the road impassable.

The route of the highway crosses some extremely high-altitude mountain passes. The highest of these is Taglang La, which, at 17,480 ft (5,328 m), is one of the highest roads in the world. The most dangerous pass, however, is Rohtang, which literaly means the "pile of corpses". Here, the gravelly road skirts some frighteningly unguarded drops.

The road is maintained by the Indian Army, and includes at least 12 temporary bailey bridges and various fords that cross streams of meltwater. Some of these fords have to be crossed by midday, before the sun melts too much snow. Maintaining the road is quite a feat, since much of it is at heights at which it is difficult to breathe. Travellers can be affected by altitude sickness and are advised not to linger too long in the highest passes.

The average height of the route is over 13,000 ft (4,000 m). This is a hazardous height, but one that affords some of the best mountain views from any road in the world. Sights include year-round snowy peaks stretching into the distance, remote camps of tents, interesting rock formations, small roadside canteens, ancient mountain monasteries, and landscapes dotted with fluttering Buddhist prayer flags.

Fellow travellers are usually an interesting bunch, ranging from dauntless explorers to local bus drivers. Motorcycle expeditions are also common, as are convoys of trucks, which often cause tailbacks as they struggle on the steep, unguarded sections. Allow at least two days to complete the highway, but also be prepared for delays caused by weather and poor road conditions. Fuel stops should be planned; there is a 225-mile (362-km) gap between the Tandi and Leh fuel stations.

▽ THREE LEVEL ZIG-ZAG

INDIA

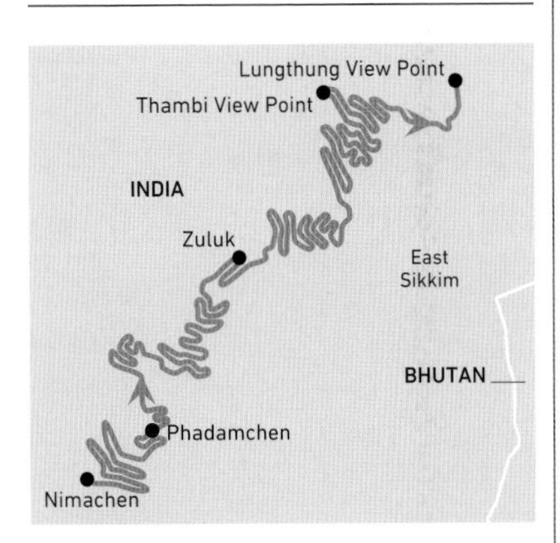

The Three Level Zig-Zag, in East Sikkim, India, is such an unusual piece of engineering that it looks like an optical illusion as it crosses to and fro up the steep slopes of the Himalayan foothills. The landscape falls away into a hazy row of mountain ranges, but the neighbouring hills appear to be densely striped with roads, which are all part of The Three-Level Zig-Zag. From afar, these stripes look like geological strata, as if the hills have had their interiors exposed.

Consult a map and it looks like a printing error. The road turns into an apparently random series of squiggles. It only extends for 20 miles (32 km), but this section has more than 100 hairpin corners, making it perhaps the world's most convoluted driving route. From above, it looks as if the mountain is almost entirely made of tarmac.

For drivers, this means a lot of hard steering, braking, and concentrating on the road. For passengers, if they can relax and trust the driver, it means a series of incredible views, including glimpses of Kanchenjunga, the world's third-highest mountain.

There is an historical significance to this route as well. Lying close to the Chinese border, it was once part of the Silk Road, an ancient network of trading tracks that connected Japan to the Mediterranean.

Today, the road creates its own unique views. Cars can pull over into viewing points to look back at the road they have just tackled and see it snaking up through the sloping landscape. The best place to stop is right at the top, at the Thambi View Point, preferably at dawn. The route seems to twist back on itself over and over again, up and down the steep sides of this row of mountains. It is such a bizarre panorama of road construction that the Zig-Zag itself has become a visitor attraction. The road is surprisingly well paved, but it has a huge number of loops, most of which are unguarded. It stands at such a severe altitude – 11,200 ft (3,414 m) – that it is prone to sudden snow and torrential rain at any time of the year. There is also the danger of ice on the surface, which is a serious danger. May to September is the best time to go; from October to March the weather can be atrocious and the road is frequently closed. Whenever you go, check conditions before travelling, and take plenty of spare fuel.

THE PRECARIOUS KEYLONG-KISHTWAR ROAD

THE THREE LEVEL ZIG-ZAG FROM THE THAMBI VIEW POINT

△ KEYLONG-KISHTWAR ROAD

INDIA

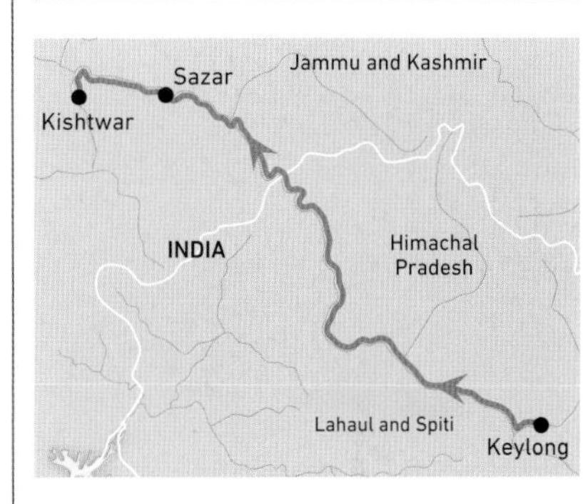

Any list of the world's best driving routes has to include one of the extreme roads that are invariably found in the Third World. The 152-mile (244-km) Keylong to Kishtwar road is certainly one of the most terrifying. Many people would not wish to walk much of this high-altitude route, let alone drive it.

Think of winding around an unmade, narrow single-carriageway track carved from the side of a mountainous gorge. Waterfalls plunge onto the rutted gravel roadway, the edges of which crumble away into a 1,000-ft (305-m) drop. Round a blind bend a truck appears coming the other way. There is no room to pass, and there is a deadly, unguarded drop on one side. Vehicles either have to reverse – or try to squeeze past by dangling wheels off the edge of the gorge. The mountain weather can be brutal too at 10,000 ft (3,050 m). That's the reality of a journey on the innocent-sounding National Highway 26.

▽ LATERAL HIGHWAY

BHUTAN

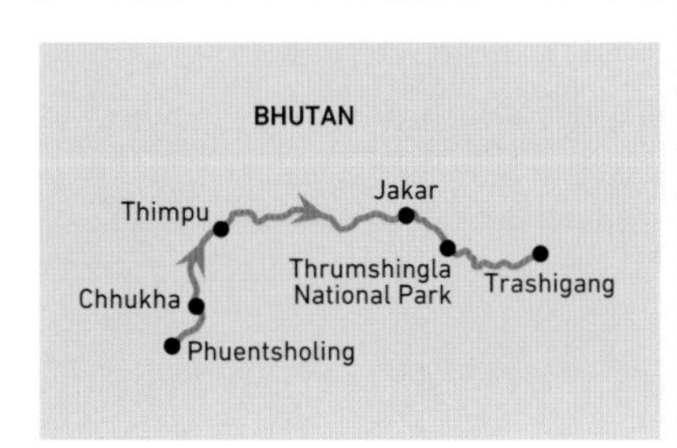

There is only one national highway in the poor Himalayan nation of Bhutan. Locals call it the "Lateral Highway", and it runs from west to east across this remote country. Bhutan covers around 15,000 square miles (38,850 sq km), but until 1962 it only had a few miles of tarmacked road. The Lateral Highway was begun in the 1960s with the help of neighbouring India, and it connects the town of Phuentsholing with Trashigang, 346 miles (557 km) away. Due to financial constraints, the road is often just an 8-ft (2.5-km)-wide single carriageway that links the main settlements between the vast mountain valleys. There are sections that are just gravel and stone, the signage is irregular, and severe weather often causes landslides, flooding, and erosion. The payoffs are the frequent views of forested mountains rising into the clouds from a landscape dotted with temples.

It is a long, tough journey for any vehicle, and driving it demands constant attention – not least to dodge motorbikes, buses, and overloaded freight trucks. However, it will give you a snapshot of a nation like few other roads can. Bhutan is one of the world's most peaceful and corruption-free countries, and the Lateral Highway is a great way to see it.

PARO TAKTSANG MONASTERY, OFF THE LATERAL HIGHWAY

THE PICTURESQUE BOATYARD OF THE ANCIENT TOWN OF HOI AN

DAHA ATA WANGUWA

SRI LANKA

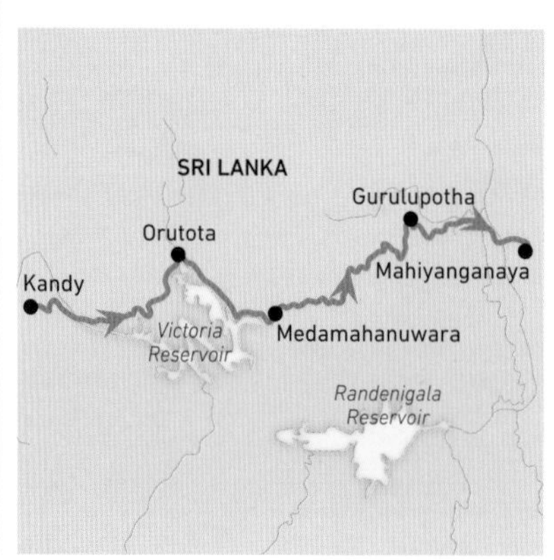

This 48-mile (77-km) road on the island of Sri Lanka has a fearsome reputation. The Daha ata Wanguwa, or "18-bend road", as it is known, linked the former capital of Kandy to Mahiyanganaya, a holy city said to be the site of the Buddha's first visit to Sri Lanka.

The road was originally just a rough, narrow track across the thickly forested mountains that rise up out of the central plain between the cities, and it was renowned for its sequence of 18 sharp corners.

Today, the name and the reputation remain, but the road itself has been transformed, thanks to the Asian Development Bank. It still links these

two interesting cities, and climbs over the mountains, but it is no longer the liability it once was. It is still a challenging drive, and may not be suitable for novice drivers, but it is now quite popular with tourists and has plenty of stopping points and cafes at which you can get out and admire the views across the plain.

Not only has the road surface been improved, the bends have been given sturdy stone safety barriers, and convex mirrors have been installed to help visibility. Most importantly, one of the bends has been removed – so technically it is now the 17-bend road.

△ DA NANG COASTLINE

VIETNAM

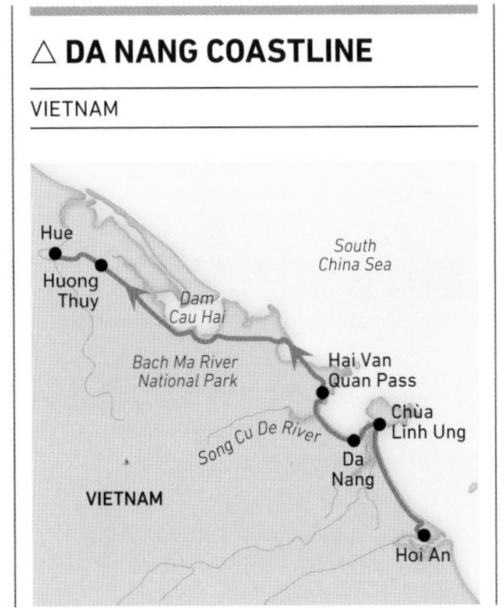

From one World Heritage Site to another, this drive along Vietnam's central coast is a short, 80-mile (129-km) road trip full of colour, character, and scenic coastline.

The journey starts in the medieval town of Hoi An, where elaborately carved wooden houses line the harbour. The route runs along the coast road, between paddy fields backing the sandy, palm-fringed beach. At Chua Linh Ung, where the Lady Buddha statue dominates the shoreline, the route turns on to the Da Nang seafront promenade and follows the shore around the bay to cross the Song Cu De River on the Highway 1 bridge.

Drivers can stick to the old Deo Hai Van road as it winds up into the forested mountains and stop at various points to take in the views of the bay and the city. Busy shops and cafés can be found among the ruins of US fortifications at the Hai Van Quan Pass, where the views along the coast to the north are stunning.

Rejoin Highway 1 for a spectacular drive around huge seawater lagoons before entering the ancient Imperial capital of Hue, with its UNESCO-recognized grid pattern of historic streets and monuments. Its 19th-century citadel is still surrounded by a moat and thick stone walls.

▽ HALSEMA HIGHWAY

PHILIPPINES

The Halsema Highway, which passes over the Cordillera Central mountain range in northern Luzon, is the highest-altitude road in the Philippines. At some points it rises to 7,400 ft (2,255 m) above sea level.

The road was originally built as a footpath, and is named after the American engineer Eusebius Halsema who constructed it, with local help, between 1922 and 1930. Since then, the 93-mile (150-km) path has been widened into a road, and has gained a reputation for being hazardous.

On a sunny day, motorists weaving through the terraced farms of the mountain slopes may wonder how the highway earned its notoriety. The surface is smooth and its potentially lethal corners are protected by sturdy stone safety walls. Useful services such as garages and cafés also dot the route. The landscape is an inspiring mix of rainforest and tiny fields. To all appearances, it is one of the great scenic drives of the area. However, in wet weather it earns its reputation. Rain falls frequently, and heavy downpours can block the road with landslides or floods. The smooth asphalt can turn slippery, and mist from the forest can reduce visibility severely. It is for these reasons it is considered one of the world's most perilous roads.

THE SUNGAI SELANGOR DAM, ON THE ROAD TO FRASER'S HILL

THE HALSEMA HIGHWAY CUTS INTO THE CORDILLERA CENTRAL MOUNTAINS

△ FRASER'S HILL

MALAYSIA

This is a delightful short drive into the forested hills north of Malaysia's capital city of Kuala Lumpur. It starts at the charming colonial town of Kuala Kubu Bharu, or KKB as it is known locally, which is about an hour's drive from the capital.

Route 55 leads through rolling landscapes to the huge Sungai Selangor Dam and reservoir, which was completed in 2002. A visitor centre and water activities can be enjoyed at the reservoir, and various carparks nearby serve as starting points for hiking routes. One popular route is to the small mountain of Bukit Kutu. Another is to Sungai Chilling, a pretty waterfall that plunges from the rainforest into a circular pool.

Beyond the reservoir, Route 55 gets bendier and steeper as it climbs into the Pahang Mountains. The views are spectacular, although mists are common. Eventually the road reaches the pretty highland resort of Fraser's Hill, part of Malaysia's "Little England" region of former colonial hill stations. Sights for visitors include the town's quaint, ivy-covered clock tower and neatly tended beds of geraniums.

▷ KELOK SEMBILAN

INDONESIA

The Indonesians have created one of the world's most complicated road systems – Kelok Sembilan, which sprawls in the middle of a tropical rainforest. The previous old Dutch colonial road was built more than 100 years ago, and used to wind laboriously through the steep-sided gorge in a challenging series of hairpins. Traffic jams were common as vehicles got stuck on the tight, steep curves, and journeys from the town of Payakumbuh up to Riau province could take half a day or more. Since then, the Sumatran government has intervened, and Kelok Sembilan is the result.

This 1¾-mile (2.8-km) section of road is a sensational system of bridges, flyovers, and underpasses that has transformed the gorge and the journey between the two destinations. According to locals, the journey time is up to four hours shorter than it used to be.

To outsiders, however, the new road system can seem like a strange fairground attraction. People gather at dedicated viewing points to watch the traffic down below – many amused and puzzled by the extraordinary tangle of roads. Various food stalls stand at the top of the canyon, catering to the multitudes that gather to view the spectacle. At first it can look like a junction – perhaps the worst intersection on Earth – but it is in fact a single road, albeit one that has been subject to the grandest attempt at getting a highway over a hill.

From a motoring perspective, it is certainly a unique experience. You enter at one end and simply hope, fingers crossed, that you emerge at the other end. It is not easy grasping the logic of the six bridges on stilts and the triple S-bends that dominate the gorge, and at times it can feel like you are driving through an optical illusion – the way ahead apparently taking you in the wrong direction. However, if you just stay on the road and trust its engineers you will reach your destination. In West Sumatra, the road is nicknamed "Kelok 9", or "9-curve climbs", because of its nine major bends – features that drivers come from far and wide to experience. It is a smooth, enjoyable place to practise your steering, particularly in a mid-engined car. Motorcyclists in particular seem to love tackling the most convoluted section, with its swirling bends winding in the air above this vast area of rainforest.

KELOK SEMBILAN, A TREAT FOR DRIVERS AND ONLOOKERS ALIKE

Australasia

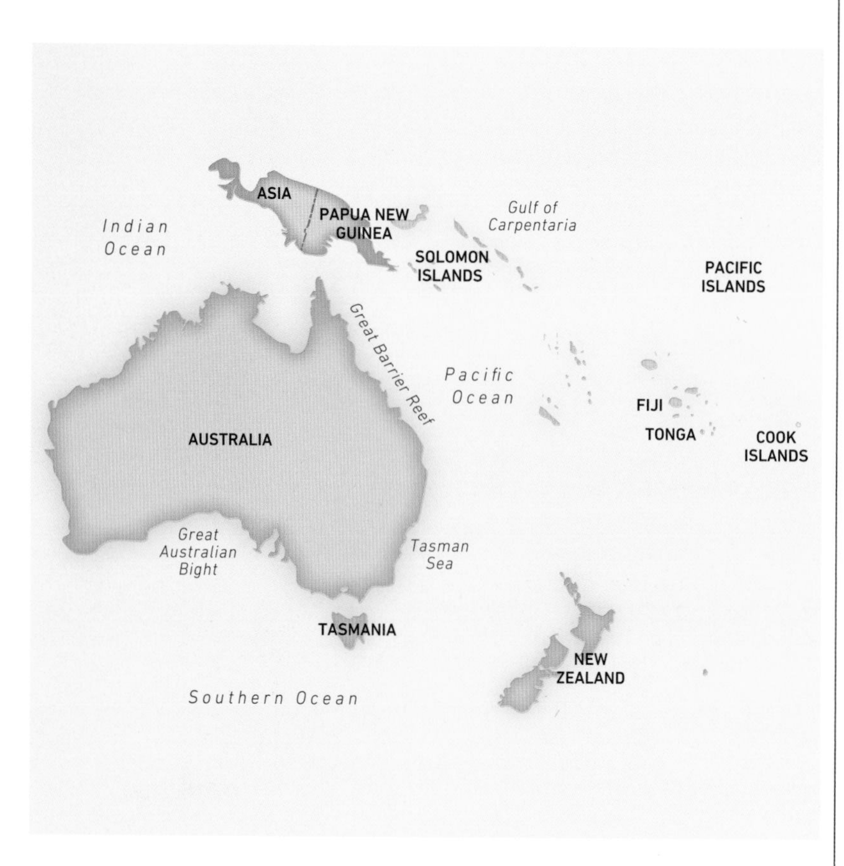

▷ GLOW WORM TUNNEL ROAD

AUSTRALIA

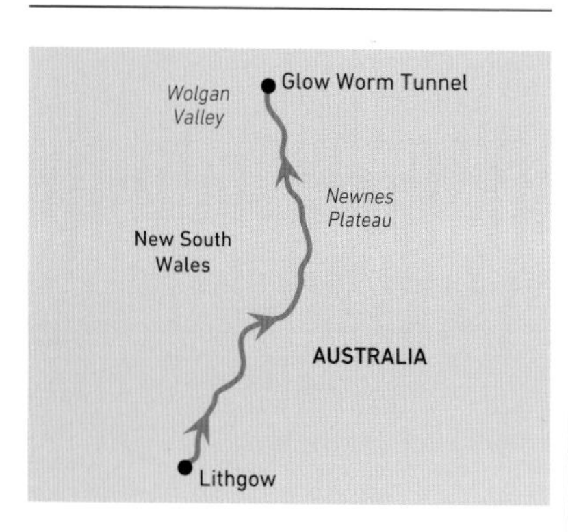

Located in New South Wales, Australia, this interesting 22-mile (35-km) route from Lithgow to the Glow Worm Tunnel follows a railway line that was built more than 100 years ago to serve the local oil shale industry. The rails have long since been removed, but you will still pass through the railway cuttings and even drive down an old abandoned railway tunnel. This is just one vehicle wide – so be prepared to reverse if necessary.

The route gets progressively more challenging as it passes through thick rainforest and deep gorges. Travellers should watch out for kangaroos and wombats among the dense vegetation. The rutted, pot-holed track becomes slippery and muddy in the monsoon season and can be almost impassable.

The drive ends at a carpark from which a short walk takes you to the renowned Glow Worm Tunnel. This is one of the best places in the world to see the eponymous creatures. The site is a former railway tunnel, and it has a very dark and damp central section, which is perfect for spotting the worms.

The worms are the larvae of the fungus gnat, and their glow, which is blue, is caused by a chemical reaction in their bodies. Its purpose is to lure mosquitoes and other insects, which the larvae use as food. For the full "Milky Way" effect, try to keep quiet and turn off all torches.

AN OLD RAILWAY TUNNEL ON THE LITHGOW ROAD

▷ GREAT BEACH DRIVE

AUSTRALIA

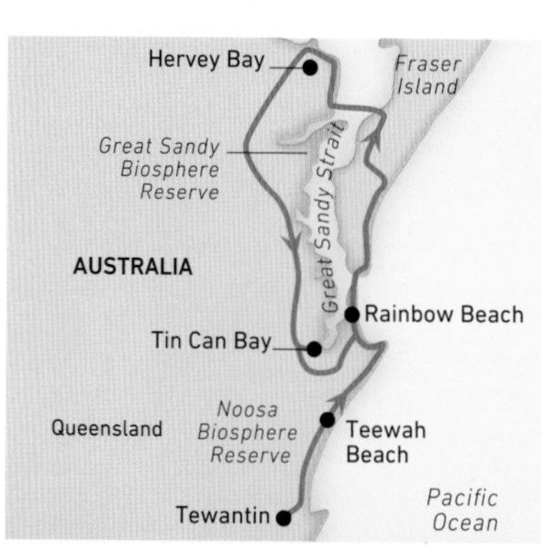

DRIVERS SKIM THE EDGE OF THE PACIFIC OCEAN ON THE GREAT BEACH DRIVE

The 235-mile (378-km) Great Beach Drive, along Queensland's Nature Coast, provides a unique road trip – except it does not use a conventional road at all. It offers the experience of driving on the pristine white sand of some of the world's most stunning beaches. It is also one of the longest beach drives anywhere in the world.

The route connects Australia's Sunshine Coast with the World-Heritage-listed Fraser Island and the Fraser Coast. It is renowned for its abundant wildlife, which includes kangaroos, whales, turtles, dingoes, koalas, platypuses, and a huge variety of bird species.

The highlights of the "highway" are crossing two UNESCO Biosphere Reserves, a World Heritage Marine Park, and the largest sand island in the world. Travellers need to

use a four-wheel-drive vehicle, and permits are required for some sections. These can be quickly arranged at Tewantin at the start of the journey, and several companies in town also offer tours of the entire route.

If you are driving yourself, the advice is to stay on the harder sand between the waterline and the high-tide mark and to keep off the fragile sand dunes. Reducing tyre pressure to maintain traction in soft sand is also recommended, as is travelling within two hours either side of low tide.

The route runs along the Pacific shore, past the colourful sands of Teewah and the high dunes and massive surf of Rainbow Beach. A couple of ferry crossings are required along the way. Then the circuit returns by looping around Tin Can Bay and Great Sandy Strait through the inland bush and protected parkland, including the acclaimed Kondalilla Falls National Park. This is home to more than 100 species of birds – as well as the 56-ft (90-m) waterfall that gives the park its name.

Most travellers spend several days enjoying the route, which offers various accommodation options, from luxury beach hotels to camping sites. Apart from the protected wildlife, sights on the drive include pretty creeks, freshwater lakes, beached shipwrecks, lighthouses, an old deserted logging camp, and patches of tropical rainforest. The area is also steeped in indigenous and pioneering settler history.

If you love the sea, there are opportunities to kayak with dolphin pods and migrating whales just off the shore. Otherwise, you can just stop the car and paddle or sunbathe on the miles of empty beaches.

MCKILLOPS ROAD

AUSTRALIA

The C6ll, also called the McKillops Road, is the main road in the eastern mountains in the state of Victoria, Australia. It runs along the edge of the Snowy River National Park for about 50 miles (80 km), through deep river gorges and steep, thickly wooded hills dotted with spectacular viewpoints. This is officially a trunk road linking Wulgulmerang to Bonang, but much of the road is rough, with unpaved, pot-holed surfaces, and sheer drops on one side. The route is therefore not suitable for inexperienced drivers, towing caravans, or low-slung sports cars.

The highlight of this "backcountry" route is the McKillops Bridge, which crosses high above the point where the Snowy River and the Deddick River merge. This basic wood-and-steel structure is more than 800 ft (244 m) long and is the only crossing over the river for many miles. Built in 1935, the bridge was a significant engineering feat in its day, and locals are still proud of it, even though driving across the creaking wooden planks of this remote landmark can be an unnerving experience. It is a popular spot for canoeists, who use the bridge as an embarkation point for exploring the local gorges.

THE TWISTS AND TURNS OF JACOB'S LADDER

△ JACOB'S LADDER

TASMANIA

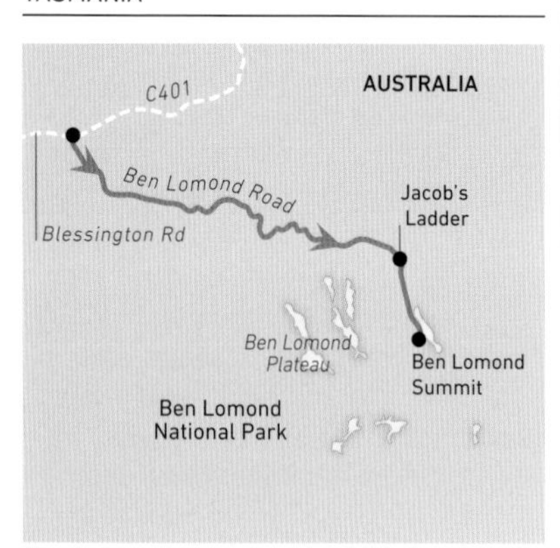

A sharp, zig-zag road situated in the Ben Lomond National Park in northeast Tasmania, Jacob's Ladder is one of the most thrilling hairpinned roads in the world. As you approach this steep, winding road, either from the top or the bottom, yellow-and-red warning road signs prepare you for the potential problems ahead – "Risk of rock fall", "Severe hazard area", "Avoid brake fade", "Use low gears", and "Keep two car lengths between vehicles", to name a few. It is therefore no surprise that this narrow, meandering route has a speed limit of just 19 mph (30 km/h) over its 7-mile (11-km) length.

Despite its notoriety, Jacob's Ladder is a route that motorists travel long distances to enjoy, and it is certainly a memorable drive. The gravelly road snakes up a fearsome, rocky amphitheatre. Many of the worst drops are unguarded, and the area is prone to hazardous rock falls. However, the views are magnificent. It is the only route to a stunning viewpoint at a lofty elevation of 5,000 ft (1,524 m) above the sea level on the Ben Lomond Plateau. The only uncertainty is the weather, which can change from bright sunshine to moderate snowfall in very little time.

▽ LINDIS PASS

NEW ZEALAND

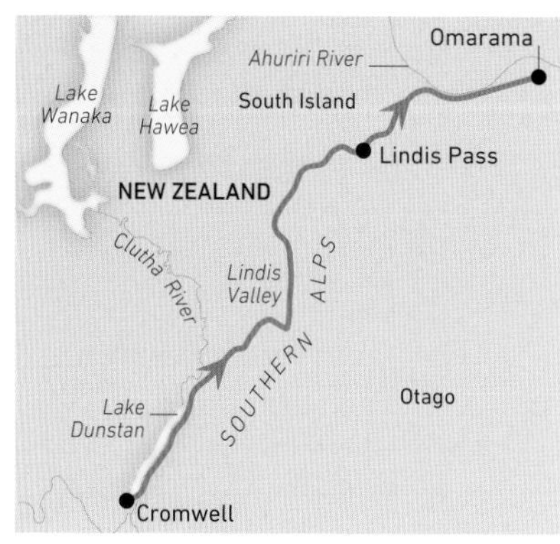

Deep in the Alps of New Zealand's South Island is a smooth, two-lane, 39-mile (63-km) highway that crosses a mountain pass at more than 3,000 ft (914 m) above sea level. This route, commonly known as the Lindis Pass, is the main link road between the Mackenzie Basin and the plateau of Central Otago, and it is the highest road on the South Island.

Officially called State Highway 8, the road is wide, undulates gently, and has a smooth, tarmacked surface. It offers a comfortable drive, and the views are a constant, if welcome, distraction. The road weaves between sharp-ridged hills that regularly open to reveal a backdrop of alpine peaks. The hills are clad in brown tussock grass, which from a distance seems like a covering of fur. Depending on the season, there can be snow right down to the roadsides or just on the tops of the distant peaks. In summer, flowering lupins line the roadside.

The Lindis Pass is a protected wilderness area with no developments or towns. However, there are picnic spots and hiking trails along the route.

CROWN RANGE ROAD

NEW ZEALAND

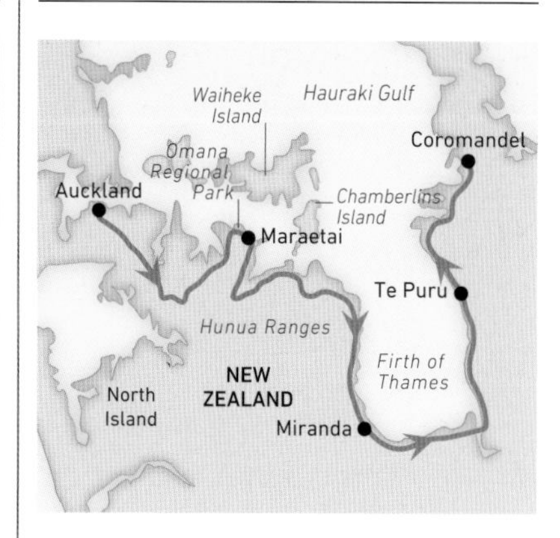

This is a classic road trip through some of the best places in New Zealand's North Island. The 120-mile (193-km) journey begins in Auckland, the largest city on the island, at the glamorous, high-rise waterfront, next to the famous Harbour Bridge.

Leaving the city on State Highway 1, New Zealand's most important road, the route soon heads towards the coast, passing through the attractive woods of Omana Regional Park along the way. The route continues through to Maraetai, with its sandy beaches, then goes along the East Coast Road. Here, it comes dramatically close to the gulf shoreline, with views of the mountains on the far side of the Hauraki Gulf. It then meets Highway 25, which loops around the bay and runs on to the west shore of the wild and beautiful Coromandel Peninsula.

The views are stunning, and range from rainforest and jagged mountains to lush fields and exotic beaches. The road follows the coast all the way to the old harbour town of Coromandel, on the far side of the gulf, where the Bohemian atmosphere attracts artists, tourists, and local fishermen.

LUPINS ON THE EDGE OF THE LINDIS PASS

Index

Page numbers in **bold** refer to main entries; page numbers in *italics* refer to illustrations

Acknowledgments

Dorling Kindersley would like to thank the following: US consultant: Lawrence Ulrich. Proofreader: Alexandra Beeden. Photography: Gary Ombler. Indexer: Vanessa Bird. Chapter openers: Phil Gamble. Design assistance: Renata Latipova. Steve Crozier at BCS Ltd. Alan Chandler, Petroliana.co.uk and Rob Arnold, Automobilia.co.uk

The publisher would like to thank the following for their kind permission to reproduce their photographs:

Key: a-above; b-below/bottom; c-centre; f-far; l-left; r-right; t-top

1 Dorling Kindersley: Petroliana.co.uk / Gary Ombler. **2-3 BMW Group**. **4 Alamy Stock Photo:** Chronicle (tr). **5 AF Fotografie:** (tr). **akg-images:** mauritius images / Karl Heinrich Lämmel (br). **Bridgeman Images:** Private Collection / Avant-Demain (tl). **Getty Images:** Art Media / Print Collector (bl). **6 akg-images:** (tl). **Getty Images:** Car Culture, Inc. (tr); Hulton Archive (bl); Tom Kelley Archive (br). **7 Alamy Stock Photo:** Peter Lopeman (tl); David Wall (br). **Rex Shutterstock** Airbus / Italdesign / Handout / EPA (tr); Sipa Press (bl). **8-9 Rex Shutterstock:** Magic Car Pics. **10-11 Alamy Stock Photo:** Chronicle. **12 akg-images:** Heritage-Images / Art Media (bl). **13 akg-images. Getty Images:** SSPL (bl). **14-15 akg-images. 14 akg-images:** Universal Images Group / Universal History Archive (tr). **15 akg-images:** De Agostini Picture Lib. / G. Dagli Orti (br); Heritage-Images / English Heritage / Historic England (tc). **16 Bridgeman Images:** SZ Photo / Scherl. **17 akg-images:** Imagno (clb). **Bridgeman Images:** Look and Learn (cr). **Daimler AG:** Mercedes-Benz Classic (br). **18 akg-images:** Heritage-Images / Oxford Science Archive (cl). **Louwman Museum-The Hague:** (bl). **18-19 akg-images:** Heritage-Images / Art Media. **19 akg-images:** Heritage-Images / National Motor Museum (tl). **Musée des arts et métiers-Cnam, Paris:** photo M. Favareille (br). **20-21 Alamy Stock Photo:** Shawshots. **22 Alamy Stock Photo:** Chronicle (tc). **22-23 Getty Images:** Hulton Archive (b). **23 Bridgeman Images:** Look and Learn. **Getty Images:** Hulton Archive (tc). **24-25 akg-images:** Heritage Images (b). **25 akg-images. 26-27 Getty Images:** Science and Society Picture Library. **28-29 Getty Images:** Kirn Vintage Stock / Corbis. **29 akg-images. 30-31 Getty Images:** Art Media / Print Collector. **32 akg-images:** Heritage-Images / Art Media (tr). **Getty Images:** Science and Society Picture Library (bc). **33 Getty Images:** Science and Society Picture Library (tc, b). **34 akg-images:** G. Dagli Orti (cl). **38-39 Getty Images:** ullstein bild Dtl. (b). **39 akg-images:** Heritage-Images / National Motor Museum (br). **Getty Images:** Science and Society Picture Library (t). **40 Alamy Stock Photo:** Art Directors & TRIP (tc). **Dorling Kindersley:** Gary Ombler / R. Florio (fbr). **National Motor Museum, Beaulieu:** (tr). **40-41 National Motor Museum, Beaulieu. 41 National Motor Museum, Beaulieu. 42 akg-images. 42-43 Getty Images:** Science and Society Picture Library (b). **43 Getty Images:** Heritage Images (tc). **44-45 akg-images:** Interfoto. **46 akg-images. Getty Images:** Bob Thomas / Popperfoto (bc). **47 Getty Images:** Schenectady Museum; Hall of Electrical History Foundation / CORBIS. **48-49 Alamy Stock Photo:** ClassicStock. **50 Getty Images:** Culture Club (br); Stefano Bianchetti / Corbis (bl). **51 Alamy Stock Photo:** Interfoto (br); Universal Art Archive (bl). **52 Alamy Stock Photo:** Motoring Picture Library (cl). **Dorling Kindersley:** Gary Ombler / R. Florio (bl). **Getty Images:** Stefano Bianchetti / Corbis (br). **53 Bridgeman Images:** Private Collection / Avant-Demain. **54 Getty Images:** Culture Club. **55 Alamy Stock Photo:** i car (tc). **Getty Images:** A. R. Coster / Topical Press Agency (bl); Topical Press Agency / Stringer (br). **56 Alamy Stock Photo:** My Childhood Memories (tr); JHPhoto (br); Dinky Art (cl); My Childhood Memories (cra). **Buddy L Toy Museum (buddylmuseum.com):** (crb). **Rex Shutterstock:** Associated Newspapers (bl). **57 Alamy Stock Photo:** My Childhood Memories (tl); Chris Willson (tr); My Childhood Memories (cl); Mike Rex (cr); Paul Cox (crb); JHPhoto (bl). **Mattel, Inc.:** Hot Wheels ® (clb). **Rex Shutterstock:** Jonathan Hordle (br). **58 Bridgeman Images:** (crb); DHM (cl); Granger (bc). **59 Alamy Stock Photo:** Granger Historical Picture Archive. **60-61 Citroën UK**. **62-63 Getty Images:** Topical Press Agency / Stringer (b). **62 akg-images:** WHA / World History Archive (tr). **63 Getty Images:** Chris Graythen (br); Sports Studio Photos (tr). **64 akg-images:** Heritage-Images / National Motor Museum (cl). **64-65 State Library of South Australia:** (t). **65 Getty Images:** Car Culture ® Collection (br). **66-67 Getty Images:** H. Armstrong Roberts / ClassicStock. **68 Alamy Stock Photo:** Heritage Image Partnership Ltd (cl); Lordprice Collection (br). **69 Getty Images:** David Paul Morris / Bloomberg (tc); Topical Press Agency (b). **70 Alamy Stock Photo:** Universal Art Archive (cl). **Dorling Kindersley:** Imperial War Museum, London / Andy Crawford / Imperial War Museum (tr). **70-71 Getty Images:** Science and Society Picture Library (b). **71 Dorling Kindersley:** The Tank Museum, Bovington / Gary Ombler (tr). **72 Alamy Stock Photo:** Daniel Valla FRPS (bl); PjrTransport (tr); dpa picture alliance archive (cl); Falkensteinfoto (crb). **73 Alamy Stock Photo:** Derek Gale (c); Phil Talbot (cra); Phil Talbot (bc). **Getty Images:** Chesnot (br). **radiatoremblems.com:** (bl). **74-75 Getty Images:** Austrian Archives / Imagno. **76 akg-images. Bridgeman Images:** Musee de l'Ile de France, Sceaux, France (tr). **77 Bibliothèque nationale de France, Paris:** Département Estampes et photographie, EST EI-13 (248) (t). **Louwman Museum-The Hague. 78 Alamy Stock Photo:** National Motor Museum / Heritage Image Partnership Ltd (crb). **Getty Images:** Popperfoto (bl). **Mary Evans Picture Library:** Retrograph Collection (cl). **79 Alamy Stock Photo:** Interfoto. **80-81 Giles Chapman Library. 82 Alamy Stock Photo:** Heritage Image Partnership Ltd (b). **Dorling Kindersley:** R. Florio (tr). **83 Reprinted courtesy of the Amherst News:** Reprinted courtesy of the Amherst News (tc). **Getty Images:** Keystone-France (crb). **84-85 TopFoto.co.uk:** Roger-Viollet. **86-87 Bridgeman Images:** Underwood Archives / UIG. **88 Getty Images:** Fay Sturtevant Lincoln / Underwood Archives (br). **Mary Evans Picture Library:** Onslow Auctions Limited (bl). **89 Getty Images:** General Photographic Agency / Hulton Archive (bl). **Mary Evans Picture Library:** Illustrated London News Ltd (br). **90 Getty Images:** General Photographic Agency (b). **Mary Evans Picture Library:** Illustrated London News Ltd (tr). **91 Alamy Stock Photo:** National Motor Museum / Heritage Image Partnership Ltd (t). **Mary Evans Picture Library:** Sueddeutsche Zeitung Photo (br). **92-93 Getty Images:** ISC Images & Archives (b). **92 Alamy Stock Photo:** chrisstockphotography (tl); National Motor Museum / Motoring Picture Library (tr). **93 Alamy Stock Photo:** National Motor Museum / Motoring Picture Library (t). **Getty Images:** National Motor Museum / Heritage Images (cr). **94 Dorling Kindersley:** Automobilia.co.uk (ftl); Petroliana.co.uk / Gary Ombler (tl, tc, ca, tr, ftr, fbr, bc, br). **Getty Images:** Austrian Archives / Imagno (fbl). **95 Dorling Kindersley:** Petroliana.co.uk / Gary Ombler (tl); Petroliana.co.uk / Gary Ombler (tc); Petroliana.co.uk / Gary Ombler (l, tr). **96-97 Getty Images:** ullstein bild. **98 AF Fotografie:** (tr). **Alamy Stock Photo:** Bob Masters Classic Car Images (tl). **Getty Images:** The LIFE Picture Collection / Bernard Hoffman (bc). **99 Getty Images:** Fay Sturtevant Lincoln / Underwood Archives. **100 Giles Chapman Library. Louwman Museum-The Hague. TopFoto.co.uk:** John Topham (tl). **101 akg-images:** mauritius images / Karl Heinrich Lämmel (b). **Art-Tech Picture Agency:** (tr). **102-103 Getty Images:** Keystone. **104 Getty Images:** National Motor Museum / Heritage Images (tr); The Print Collector (cl). **Steve Sexton:** (bc). **105 Alamy Stock Photo:** National Motor

Museum / Motoring Picture Library (tr). **Getty Images:** General Photographic Agency / Hulton Archive (b). **106-107 Giles Chapman Library. 108 Alamy Stock Photo:** Alexander Perepelitsyn (tr). **108-109 Getty Images:** Popperfoto (b). **109 akg-images. Getty Images:** Imagno (crb). **110 Alamy Stock Photo:** John James (bc). **Getty Images:** Fox Photos (br). **111 Bridgeman Images:** SZ Photo / Scherl. **112-113 Mary Evans Picture Library:** Illustrated London News Ltd. **114 Getty Images:** The Asahi Shimbun (tl); Bettmann (bl); Car Culture ® Collection (br). **115 Toyota (GB) PLC. 116-117 Getty Images:** National Motor Museum / Heritage Images. **116 Giles Chapman Library. Motoring Picture Library / National Motor Museum:** (br). **118-119 Bridgeman Images:** Underwood Archives / UIG. **120 Getty Images:** Bettmann (bl); Hulton Archive / Central Press (br). **121 Alamy Stock Photo:** National Motor Museum / Heritage Image Partnership Ltd (br). **Getty Images:** Archive Photos / Transcendental Graphics (bl). **122-123 Alamy Stock Photo:** Charles Phelps Cushing / ClassicStock (b). **122 Getty Images:** Hulton Archive / Fox Photos (tr). **123 Getty Images:** H. Armstrong Roberts / Stringer / Retrofile (tr). **124 Alamy Stock Photo:** NZ Collection (tr); Shawshots (bl). **125 akg-images. Giles Chapman Library:** (tr). **126 Dorling Kindersley:** Automobilia.co.uk (ftr, cl); Petroliana.co.uk / Gary Ombler (ftl, tl, fcl, cr, fcr, br, fbr). **Getty Images:** Bob Harmeyer / Archive Photos (bl). **Museo Fisogni:** (tr). **127 Dorling Kindersley:** Petroliana.co.uk / Gary Ombler (tl, fcl, bl, br). **Museo Fisogni. 128-129 Getty Images:** John Kobal Foundation. **130-131 Alamy Stock Photo:** Interfoto. **132 Getty Images:** Bettmann (bl). **133 Getty Images:** Hulton Archive (tr); Clarence Sinclair Bull / John Kobal Foundation (br). **134-135 Alamy Stock Photo:** Goddard Automotive (b). **134 Dorling Kindersley:** Gary Ombler / The Tank Museum (tl). **Louwman Museum-The Hague. 135 Getty Images:** Keystone-France / Gamma-Keystone (br). **Wikimedia:** Sergey Korovkin / GAZ-61.JPG / CC Attr. 4.0 (tr). **136 Bridgeman Images:** Peter Newark Military Pictures (tr). **Getty Images:** Hulton Archive / Central Press (b). **137 Alamy Stock Photo:** American Photo Archive (t). **Getty Images:** Bettmann (br). **138-139 akg-images:** ullstein bild. **140 Alamy Stock Photo:** National Motor Museum / Heritage Image Partnership Ltd (tr). **140-141 Getty Images:** J. A. Hampton / Topical Press Agency (b). **141 Getty Images:** Underwood Archives (t). **Rex Shutterstock:** Magic Car Pics (br). **142-143 Alamy Stock Photo:** National Motor Museum / Heritage Image Partnership Ltd. **144 akg-images:** Interfoto / TV-yesterday (cl). **Alamy Stock Photo:** Interfoto (cr). **Giles Chapman Library. 145 Bridgeman Images:** SZ Photo / Scherl (t). **Getty Images:** Hulton Archive (br). **146-147 ŠKODA AUTO Corporate Historical Archives / GKM-ŠKODA Museum. 148 Getty Images:** Transcendental Graphics (tr). **148-149 Getty Images:** Underwood Archives (b). **149 Giles Chapman Library. 150-151 Giles Chapman Library. 152 Getty Images:** Fox Photos (tr); Alexander Sorokopud (b). **153 Getty Images:** General Photographic Agency. **154-155 Getty Images:** H. Armstrong Roberts / Retrofile. **156 Getty Images:** Keystone-France / Gamma-Keystone (bl); ullstein bild / Otfried Schmidt (br). **157 Alamy Stock Photo:** Mark Summerfield (br). **Rex Shutterstock:** Magic Car Pics (bl). **158 Alamy Stock Photo:** Steve Sant (cla). **Getty Images:** Ullstein bild Dtl. (bl). **159 The Advertising Archives:** (bc). **Getty Images:** Yale Joel (t). **160 Dreamstime.com:** Rob Hill (br). **Getty Images:** National Motor Museum / Heritage Images (tl). **161 Alamy Stock Photo:** Interfoto. **162-163 Getty Images:** Manuel Litran / Paris Match (b). **162 Dorling Kindersley:** Matthew Ward (tr). **163 Dorling Kindersley:** Matthew Ward (tr). **Getty Images:** Keystone-France / Gamma-Keystone (cr). **Giles Chapman Library. 164 akg-images:** Bernhard Wübbel (c). **Bridgeman Images. 165 akg-images. Getty Images:** Fotosearch (bl). **166-167 Getty Images:** Hulton-Deutsch Collection / Corbis. **168 Neil Pogson, Holden Retirees Club:** (l). **169 Neil Pogson, Holden Retirees Club. 170 Alamy Stock Photo:** Frank Chmura (tc); Lightworks Media (tl); Eugene Sergeev (cla); grzegorz knec (ca); S. Forster (c); YAY Media AS (cra); Peter Horree (cl); Arndt Sven-Erik / Arterra Picture Library (br). **Rex Shutterstock:** imageBROKER / Shutterstock (crb); Global Warming Images / Shutterstock (tr). **171 Alamy Stock Photo:** Carol Dembinsky / Dembinsky Photo Associates (bc); Tom Grundy (tl); Phil Crean A (bl); Nils Kramer / imageBROKER (tc); Ken Gillespie Photography (cb). **Getty Images:** The Washington Post (br). **Rex Shutterstock:** Image Source (tr); Moritz Wolf / imageBROKER (cla). **172-173 Rex Shutterstock:** Magic Car Pics. **172 Getty Images:** National Motor Museum / Heritage Images (bl). **Courtesy Mercedes-Benz Cars, Daimler AG:** (tr). **Rex Shutterstock:** Magic Car Pics (tl). **173 Alamy Stock Photo:** kpzfoto (tr). **Dorling Kindersley:** Deepak Aggarwal / Titus & Co. Museum for Vintage & Classic Cars (tl). **174 Getty Images:** Archivio Cameraphoto Epoche (bc); Ralph Crane / The LIFE Premium Collection (cl). **174-175 Alamy Stock Photo:** Charles Phelps Cushing / ClassicStock (b). **175 Getty Images:** Hulton-Deutsch Collection / Corbis (tr). **176-177 Getty Images:** Car Culture, Inc.. **178 Rex Shutterstock:** Magic Car Pics (cl). **178-179 magiccarpics.co.uk. 179 Alamy Stock Photo:** Pictorial Press Ltd (tl). **Getty Images:** George Torrie / NY Daily News Archive (cr). **180-181 Getty Images:** ullstein bild / Otfried Schmidt. **182-183 Rex Shutterstock:** Magic Car Pics (b). **182 Bridgeman Images:** Private Collection / DaTo Images (tr). **183 Alamy Stock Photo:** Interfoto (br). **Bridgeman Images. 184-185 Getty Images:** J. R. Eyerman. **186 Getty Images:** Illustration by Jim Heimann Collection (tr); SSPL (bl). **187 Mary Evans Picture Library:** David Lewis Hodgson. **188-189 Alamy Stock Photo:** Allan Cash Picture Library. **190 Dorling Kindersley:** Petroliana.co.uk / Gary Ombler (tr). **Getty Images:** National Motor Museum / Heritage Images (bl). **Giles Chapman Library. 191 Dorling Kindersley:** Matthew Ward (br). **Getty Images:** Hulton Archive / Fred Mott (t). **192-193 Alamy Stock Photo:** Mark Summerfield (b). **193 Image created by Simon GP Geoghegan:** (tl). **Giles Chapman Library. © British Motor Industry Heritage Trust:** (br). **194 Alamy Stock Photo:** Interfoto (b). **Serrvill Txemari:** (cra). **195 Autohistorian Alden Jewell:** flickr.com (cra). **Giles Chapman Library. 196 Collection de l'Automobile Club de France:** (cl). **Getty Images:** Octane / Action Plus (bc). **The Royal Automobile Club:** (tr). **197 Alamy Stock Photo:** Courtesy Everett Collection / Everett Collection Inc (bc); Jonny White (br). **Getty Images:** Franck Fife / AFP (tr); Nick Laham (l); FIA / Handout (tc); David J. Griffin / Icon Sportswire (fbr). **198-199 Getty Images:** Harry Kerr. **200 Getty Images:** Hulton Archive (cr); Peter Stackpole / The LIFE Picture Collection (b). **201 Getty Images:** Lessmann / ullstein bild (br); Popperfoto. **202-203 Getty Images:** Henry Groskinsky / The LIFE Picture Collection. **204 Alamy Stock Photo:** Goddard Automotive (bl). **Getty Images:** Yale Joel / Time Life Pictures (br). **205 Getty Images:** John Pratt / Keystone Features (bl); Tom Kelley Archive (br). **206 Getty Images:** Bettmann (t); Bettmann (bl); Rolls Press / Popperfoto (br). **207 Getty Images:** The Asahi Shimbun. **208-209 Getty Images:** National Geographic / Bruce Dale. **210-211 Alamy Stock Photo:** Goddard Automotive (b). **210 Dorling Kindersley:** James Mann / Eagle E Types (tl). **211 Rex Shutterstock:** Associated Newspapers (br). **212 Getty Images:** Geography Photos / Universal Images Group (tr). **Rex Shutterstock:** Daily Mail (b). **213 Alamy Stock Photo:** Matthew Richardson (bc). **Press Association Images:** PA Archive (tr). **214-215 Getty Images:** Gene Laurents / Condé Nast. **216-217 Getty Images:** Yale Joel / Time Life Pictures. **216 Getty Images:** Bloomberg / Andrew Harrer (tr). **217 Alamy Stock Photo:** Keystone Pictures USA (br). **Getty Images:** Henry Groskinsky / The LIFE Picture Collection (tr). **218 Dorling Kindersley:** Heritage Motoring Club of India / Deepak Aggarwal (cr); James Mann / Colin Laybourn / P&A Wood (tl); National Motor Museum Beaulieu / James Mann (cl); James Mann / Peter Harris (br). **Louwman Museum-The Hague. 219 Dorling Kindersley:** Jaguar Heritage / Gary Ombler (cl); James Mann / Ivan Dutton (t). **Courtesy Mercedes-Benz Cars, Daimler AG. Rex Shutterstock:** Camera 5 (br). **220 Alamy Stock Photo:** Phil Talbot (cr). **Getty Images:** Rolls Press / Popperfoto (b); Science & Society Picture Library (cla). **221 Giles Chapman Library. 222-223 Rex Shutterstock:** United Artists / Kobal. **224 Louwman Museum-The Hague. magiccarpics.co.uk:** (tl). **Rex Shutterstock:** Magic Car Pics (bl). **224-225 Rex Shutterstock:** Magic Car Pics. **225 Giles Chapman Library. 226-227 The Scientific Exploration Society:** (b). **226 Alamy Stock Photo:** John Henderson (tr). **227 Richard Emblin:** (bc). **Kelvin Kent:** (tr). **228-229 Getty Images:** Tom Kelley Archive (bc). **228 TopFoto.co.uk:** (tl). **229 Alamy Stock Photo:** Michael Wheatley (br). **Getty Images:** Loomis Dean / The LIFE Picture Collection (tr). **230-231 Getty Images:** Spence Murray / The Enthusiast Network. **232-233 magiccarpics.co.uk. 232 Art-Tech Picture Agency. magiccarpics.co.uk. Rex Shutterstock:** Underwood Archives (cr). **233 magiccarpics.co.uk. 234 Getty Images:** Teenie Harris Archive / Carnegie Museum of Art (cra); Rust / ullstein bild (bl); Time Life Pictures / Pictures Inc. / The LIFE Picture Collection (br). **235 Getty Images:** John Pratt / Keystone Features. **236-237 Giles Chapman Library. 238 Rex Shutterstock:** Tim Graham / robertharding (tc). **TopFoto.co.uk:** Sputnik (b). **239**

Getty Images: Francois Lochon / Gamma-Rapho (br); Behrouz Mehri / AFP (tr). **240 The Advertising Archives. Alamy Stock Photo:** Interfoto (cr); The Print Collector (tl); The Print Collector (ftr); Jeff Morgan 04 (fcr). **Giles Chapman Library. Mary Evans Picture Library:** John Maclellan (tr). **Rex Shutterstock:** Snap (fcl). **241 The Advertising Archives. Alamy Stock Photo:** (cr); Lordprice Collection (tl); Shawshots (tr); Interfoto (ftr); DWImages (br). **Giles Chapman Library. Rex Shutterstock:** Magic Car Pics (bl). **242-243 Fordimages.com**. **244 Dorling Kindersley:** Matthew Ward (tr). **Getty Images:** Rogge / ullstein bild (b). **245 Getty Images:** Ron Eisenbeg / Michael Ochs Archives (tc). **246-247 Getty Images:** Bettmann. **246 BIGFOOT 4x4, Inc:** (bl). **248 Alamy Stock Photo:** Everett Collection Inc (crb); Photo 12 (t); Photo 12 (bl). **249 Alamy Stock Photo:** PvE. **250-251 Hyundai Motor Company**. **252 Giles Chapman Library. 253 akg-images:** Heritage-Images / National Motor Museum. **254-255 Getty Images:** AWL Images. **256 Getty Images:** (br). **Rex Shutterstock:** Sipa Press (bl). **257 Alamy Stock Photo:** Design Pics Inc (bl); dpa picture alliance (br). **258 Alamy Stock Photo:** Dmitrii Bachtub (tl). **magiccarpics.co.uk:** John Colley (bl). **259 Alamy Stock Photo:** YAY Media AS (tr). **260-261 Rex Shutterstock:** Magic Car Pics (b). **260 Rex Shutterstock:** imageBROKER / Martin Siepmann (tr). **261 Getty Images:** FPG / Hulton Archive (br). **Renault (UK):** (tr). **262-263 Giles Chapman Library:** Mitsubishi Motors Corporation. **262 Honda (UK). Peter Nunn:** Mitsubishi Motors Corporation (tl). **263 Giles Chapman Library:** Suzuki Motor Corporation (tl); Suzuki Motor Corporation (tr); Suzuki Motor Corporation (c). **264-265 Used with permission, GM Media Archives**. **265 Rex Shutterstock:** Sipa Press (bl). **266 Getty Images:** David Madison (tr). **magiccarpics.co.uk:** John Colley (bl). **266-267 Getty Images**. **267 Giles Chapman Library:** Lancia (tr). **268 Alamy Stock Photo:** Dave Cameron (bl). **Getty Images:** National Motor Museum / Heritage Images (crb); Chris Niedenthal / The LIFE Images Collection (cl). **269 akg-images. 270-271 Rex Shutterstock:** Sipa Press. **272 Getty Images:** Hulton Archive (cla); Frederic Pitchal / Sygma / Sygma (bc). **273 akg-images:** Sputnik. **274-275 Alamy Stock Photo:** Bhandol (b). **274 Giles Chapman Library. 275 Getty Images:** Ulrich Baumgarten (br). **Giles Chapman Library. 276-277 Getty Images:** Dong Wenjie. **278 Alamy Stock Photo:** Carnundrum (tr); Rod Williams (tc); Adrian Muttitt (c); National Motor Museum / Motoring Picture Library (cr); Tim Gainey (fbl); supermut (bl); pbpvision (br). **Mary Evans Picture Library:** Onslow Auctions Limited (cra). **279 Alamy Stock Photo:** Carnundrum (cr); Mark Scheuern (br); imageBROKER (tc); Goddard New Era (fcl); Mim Friday (fcr). **Bridgeman Images:** Christie's Images (cl). **Mary Evans Picture Library:** David Cohen Fine Art (tr). **280-281 Rex Shutterstock:** Magic Car Pics. **280 Alamy Stock Photo:** Performance Image (tr). **Dorling Kindersley:** James Mann / John Mould (tl). **281 Alamy Stock Photo:** Evox Productions / Drive Images (tr); Performance Image (tl). **Getty Images:** Eric Rickman / The Enthusiast Network (br). **282-283 Alamy Stock Photo:** Design Pics Inc. **284 Getty Images:** Heritage Images (cra); Ullstein Bild (bl). **285 Rex Shutterstock:** Carlos Osorio (br); Sipa Press (t). **286 Alamy Stock Photo:** The Image Barrel (tl). **Getty Images:** The Image Bank / Eric Van Den Brulle (tr). **286-287 Alamy Stock Photo:** Peter Lopeman (b). **287 123RF.com:** tupungato (tr). **Getty Images:** Dorling Kindersley / Dave King (tl). **288 Alamy Stock Photo:** dpa picture alliance (b). **289 Dreamstime.com:** Anizza (br). **Getty Images:** Boston Globe (ca). **290-291 Rimac Automobili**. **292 Getty Images:** John Macdougall (br). **Nissan Motor Company:** (bl). **293 Morgan Motor Company Ltd:** (br). **Courtesy of Volkswagen:** (bl). **294 Getty Images:** Bettmann (cra). **Giles Chapman Library. 295 Alamy Stock Photo:** Everett Collection, Inc. (clb). **Giles Chapman Library. Rex Shutterstock:** Electric / Sony / Kobal (tr). **296-297 Nissan Motor Company:** (b). **296 Alamy Stock Photo:** eVox / Drive Images (tl); Joseph Heroun (tr). **297 Alamy Stock Photo:** Marco Destefanis (tr). **Getty Images:** AFP Photo / Pierre Andrieu (tl); Bloomberg / Jin Lee (br). **298 Alamy Stock Photo:** Jiraroj Praditcharoenkul (b). **299 Alamy Stock Photo:** dpa picture alliance (crb). **Getty Images:** Bloomberg (t). **300 Getty Images:** STR (tl). **300-301 Getty Images:** Bloomberg (b). **301 Alamy Stock Photo:** Dave Ellison (cra); Renaud Rebardy (ca). **302-302 Bugatti Automobiles S.A.S.. 302 Ferrari:** (tr). **Courtesy Mercedes-Benz Cars, Daimler AG. 303 Alamy Stock Photo:** eVox / Drive Images (tl); WENN Ltd (tr). **304-305 Honda (UK)**. **306 Alamy Stock Photo:** Kropp (tr). **306-307 BMW Group UK:** (b). **307 Alamy Stock Photo:** Jeffrey Blackler (br). **Getty Images:** Car Culture (tr). **308-309 Morgan Motor Company Ltd**. **308 Alamy Stock Photo:** Tom Wood (tr). **309 Alamy Stock Photo:** Newscom (cra); Matthew Richardson (tl). **Bristol Cars:** (tr). **310 Alamy Stock Photo:** Steve Lagreca (bl). **Dorling Kindersley:** Tuckett Brothers (cra). **311 Bugatti Automobiles S.A.S.:** (cr). **Dorling Kindersley:** David Ingram, Audi UK (tc); Paul Self / Porsche Cars Great Britain (cla); Paul Self / Honda Institute (bl). **Rex Shutterstock:** Magic Car Pics (crb). **312-313 McLaren Automotive Limited**. **314 Getty Images:** milehightraveler (b). **315 Alamy Stock Photo:** Jim West (b). **Getty Images:** Scott J. Ferrell (cr); John Macdougall (tl). **316 Alamy Stock Photo:** ZUMA Press, Inc. / DARPA (bc). **Giles Chapman Library. 317 Alamy Stock Photo:** Tesla Motors / Dpa (t). **iStockphoto.com:** chombosan (br). **318 Jaguar Cars Limited:** (tl). **Courtesy Mercedes-Benz Cars, Daimler AG. Rex Shutterstock:** Airbus / Italdesign / Handout / EPA (bl). **319 Honda Motor Europe Ltd:** (tr). **Toyota (GB) PLC. Courtesy of Volkswagen:** (b). **320 4Corners:** Hans Peter Huber. **322 Gregory Melle:** (bl). **323 Alamy Stock Photo:** Stanislav Moroz (t). **324 Alamy Stock Photo:** Steve Bly (b). **325 Alamy Stock Photo:** Andre Jenny (tl). **Getty Images:** J.Castro (b). **326 Getty Images:** Doug Steakley (t). **327 Depositphotos Inc:** Maks_Ershov (t). **Getty Images:** Onfokus (bl). **328 123RF.com:** Mirko Vitali (t). **329 Alamy Stock Photo:** Stephen Saks Photography (b). **330 BORIS G:** (br). **331 Amanda & Andrew Prenty:** (b). **Rex Shutterstock:** Eye Ubiquitous (tl). **332 Alamy Stock Photo:** John Michaels (t). **333 Alamy Stock Photo:** Andre Seale (b). **Getty Images:** MyLoupe / UIG (tr). **334 Alamy Stock Photo:** Neil McAllister (tr). **335 Alamy Stock Photo:** allan wright (br). **336 Alamy Stock Photo:** Panther Media GmbH (b). **337 Alamy Stock Photo:** Vito Arcomano (br). **Getty Images:** Sandro Bisaro (t). **338 Alamy Stock Photo:** Eye Ubiquitous (bl). **339 olino.org:** (bl). **Daniel Tengs:** (t). **340 Imagelibrary India Pvt Ltd:** Benjamin gs (t). **341 123RF.com:** alizadastudios (tr). **iStockphoto.com:** Cenkertekin (bl). **342 Alamy Stock Photo:** Maxim Toporskiy (b). **343 Getty Images:** Westend61 (t). **344 Imagelibrary India Pvt Ltd:** Andrey Armyagov (t). **345 Alamy Stock Photo:** Panther Media GmbH (bl). **Tarun Goel:** (tr). **346 Getty Images:** Sean Caffrey (bl). **346-347 Imagelibrary India Pvt Ltd:** Steve Phan (t). **348 Alamy Stock Photo:** Adwo (bl). **Getty Images:** annamir@putera.com / Moment Open (tr). **349 Jez O'Hare**. **350 Alamy Stock Photo:** Stephanie Jackson (bc). **350-351 Alamy Stock Photo:** David Wall (t). **352 Getty Images:** Photolibrary (tl); Puripat Wiriyapipat / Moment (br)

Endpaper images: Front: **Alamy Stock Photo:** ClassicStock ; Back: **Alamy Stock Photo:** ClassicStock

All other images © Dorling Kindersley
For further information see: **www.dkimages.com**

LITS